DIGITAL
TEST
ENGINEERING

DIGITAL
TEST
ENGINEERING

DIGITAL
TEST
ENGINEERING

J. Max Cortner

A WILEY-INTERSCIENCE PUBLICATION
JOHN WILEY & SONS
NEW YORK · CHICHESTER · BRISBANE · TORONTO · SINGAPORE

Library of Congress Cataloging in Publication Data:

Cortner, J. Max.
  Digital test engineering.

  "A Wiley-Interscience publication."
  Bibliography: p.
  1. Digital electronics. 2. Electronic
apparatus and appliances—Testing. I. Title.

TK7868.D5C67 1987   621.3815'48   87-8319
ISBN 0-471-85135-3

Printed in the United States of America

10 9 8 7 6 5 4 3 2 1

To My Father, James C. Cortner
and My Wife, Gwyn K. Cortner

PREFACE

Test engineering has emerged as a subprofession of electrical engineering in the past 20 years. The complexity and integration level of digital electronics have driven up the cost of testing a product to where it exceeds the cost of manufacture in high-reliability markets such as defense and aerospace. Technology has also driven up the complexity of tools used to test the product. Automatic test equipment suites cost millions of dollars to acquire, install, and operate during the life of a product. In response to these economic factors, the demand for engineers skilled in test development, design for testability, and related specialties has risen. This volume is intended to help the senior student or practicing engineer acquire the knowledge to fill that demand as a full-time profession, as a part of a related profession such as design, or as a manager of a team responsible for applying those skills. After reading this text, an engineer should be capable of selecting a test strategy to match the product and the tools to carry out the strategy in accordance with performance requirements.

Often underrated, test engineering requires a considerable volume of specialized knowledge beyond the sound electronics background provided in a bachelor's degree program for electrical engineering. The specific knowledge is ordinarily obtained through experience and the training programs associated with automatic test equipment. The few university courses which exist in the digital test area suffer from the difficulty of staying current in a particularly fast-moving area of electrical engineering and from the lack of compiled sources for the wide variety of information needed in test engineering. Much of the knowledge required is informal and either unwritten or scattered throughout manuals and magazine articles. It is difficult for anyone but a practitioner to recognize the importance of some of the material.

Test engineering is truly an eclectic profession, using skills from diverse areas such as computer science, robotics, optics, mechanics, and economics as well as the foundation of electrical engineering. Although it is ordinarily viewed as an analytical pursuit, opportunities for design occur frequently. Recent innovations include the use of artificial intelligence techniques and the application of physical electronics in sensing technology. Arranging a text for such a diverse subject is challenging. The initial limitation of subject matter to the digital side of testing does not overly restrict the text since most of the automation in testing applies to digital testing. Further selections were made to focus on information which will be of immediate use to a practicing engineer faced with developing tests for digital circuits. The attempted balance is to provide a sound theoretical understanding and practical knowledge gained through experience. In a few instances, a brief discussion of a basic concept is repeated to place it conveniently in context with a broader chapter subject. The intention is to provide concise reference material for the reader who is interested in a specific topic. Details about individual manufacturers' automatic test equipment have been avoided. Such information becomes out of date quickly and is provided in profusion by the manufacturers in any case.

The core chapters deal with the sequence of events encountered in most digital test development efforts where the test subjects are printed circuit boards or integrated circuits. Stimulus Generation, Expected Response Determination, and Circuit and Fault Modeling (Chapters 3–5) explore the alternatives for deriving a test program, while Automatic Test Equipment, Device-Under-Test Interface, ATE Languages, and Diagnostics and Troubleshooting Aids (Chapters 6–9) cover the options in applying the test to the product. The first two chapters provide a background and perspective for the reader and should prove valuable for the manager who wishes a lighter coverage of the subject. The last chapter (Test Planning) is also tuned to the managerial perspective, but should not be dismissed by an engineer who expects to exercise the full range of test responsibilities. Chapters 10 and 11, which cover the special subjects of Memory Testing and Design for Testability, complete the breadth of coverage to enable the reader to deal with all phases of a test program.

For those who want a deeper view of any particular subject matter, the references at the end of each chapter provide specific theoretical support. Details of the IEEE standards on instrument communication (IEEE Std. 488) and on the test language ATLAS can be obtained from that organization. They are too voluminous for meaningful abstraction here. There is no intention to provide exact answers, such as test methods for a 68000 microprocessor. The ingenuity of design engineers and the variety of application environments encountered negates the value of precise test information for a particular circuit. In general, my philosophy agrees with the old proverb "Give someone a fish and you feed him for a day. Teach him to fish and you feed him for a lifetime."

The challenge of test engineering is one of continual learning, and I hope that I have contributed to meeting that challenge by authoring this text. I am indebted to Richard B. Fosmer, who provided encouragement and guidance throughout the project, and to Gary C. Humann, who through his technical reviewing and patient tutoring lent me the aid of his considerable expertise in test generation and fault simulation. My further thanks to my colleagues at Unisys St. Paul Test Engineering for many hints and corrections.

J. Max Cortner

Inver Grove Heights, Minnesota
July 1987

CONTENTS

CHAPTER 3

STIMULUS GENERATION

53

CHAPTER 12
TEST PLANNING   293

GLOSSARY   326

INDEX   333

2   FAULTS AND FAILURE MECHANISMS

SINGLE STUCK-AT FAULTS

CHAPTER ONE

FAULTS AND FAILURE MECHANISMS

The performance of a test implies doubt as to the outcome. There may be great confidence in the circuit or device which is the subject of the test, but if there were no chance of failure, there would be no need to test. Conversely, the nature of suspected faults influences the design of tests. Therefore, it is appropriate to discuss the variety of faults which commonly occur in digital circuits before attempting to describe the methods of test design.

A failure could ordinarily be described as a lack of expected performance. This definition fits digital electronics inasmuch as the expected performance is an output condition (or sequence) in response to an input condition (or sequence). A fault is a physical or intellectual imperfection. More precisely, in electronics a fault is a defect such as an open circuit, a short circuit, or a ground in a circuit, component, or line [1]. Whether faults are mistakenly built into a circuit at the factory or introduced in the field by environmental stress, they usually fall into one of the aforementioned broad categories. A broken bond in an integrated circuit (IC), an overly etched printed circuit (pc) path, or a contaminated connector pin creates open circuits. Solder splashes, missing insulation, or metal scraps create short circuits. More exotic faults will be discussed later in the chapter.

It is common practice in digital electronics to categorize faults by their effect upon the logic values of a circuit rather than to address causes directly. This organization technique will be used in this chapter. An exploration of causes will accompany each category after the fault type is sufficiently defined. The symptoms of each fault category and the propagation of its effects will complete each discussion. Faults will be discussed beginning with the simplest, most common types and proceeding toward more complex categories. Beginning with the single stuck-at faults, the progression to multiple, intermittent, and complex faults evolves by relaxing one or more of the category restrictions which define the single stuck-at fault.

SINGLE STUCK-AT FAULTS

As the name implies, a single stuck-at fault causes one and only one node of a digital circuit to be held at a particular logic value. The stuck-at-1 (s-a-1) fault inhibits the node from switching to a 0, while the stuck-at-0 (s-a-0) fault inhibits switching to a 1. In 3-state logic a stuck-at-off (s-a-Z) fault can be defined which prohibits a logic signal from being switched onto a bus. Figure 1.1 shows the equivalent schematic effect of a typical s-a-1 fault caused by a cut pc board path.

Several alternate failures could occur which would result in the same logical fault. A broken bond between the IC metallization pad and the wire lead to the package input or output could result in an open at either the output of gate A or the input of gate B. Either case would cause an s-a-1 fault for transistor–transistor logic (TTL). For a more detailed examination of the failure–fault relationship and for a look at possible internal failures, the 2-input TTL NAND gate shown in Figure 1.2 will be used as an example.

If either input is disconnected, the respective input node formed by an emitter of Q_1 and a clamp diode charges toward V_{cc} at a rate determined by the value of R_1 (4 kΩ for the 7400 gate) and the combined capacitance of the reversed clamp diode and any attached circuit board path. The output of the gate then depends only upon the other input. If the intact input is a 1 also, the base collector junction of Q_1 will conduct, thus turning on transistor Q_2 resulting in a 0 output. The open input voltage will settle at about 1.2 V in this case since the base voltage of the input transistor is fixed by the V_{BE} drops of Q_2 and Q_4 and V_{BC} of Q_1.

Internal shorts caused by metallization splatters, contamination, or crystal structure abnormalities such as "piping" [3] in Q_3 could cause the output of gate A to be s-a-1. Piping occurs when doping ions penetrate the epitaxial crystal layers in excess along the edges of a stacking fault. Impurities from the preparation of the crystal substrate begin the process by causing the epitaxial growth to occur along different planes. This results in a path for ion penetration and the characteristic triangular shape of the stacking fault region. As one region shorts to another, Q_3 is disabled, and the entire net it drives appears s-a-1. A similar short in the 1.6 kΩ bias resistor would also cause an output s-a-1.

Open circuits caused by step faults in Q_4 or Q_1 result in s-a-1 logic faults. A

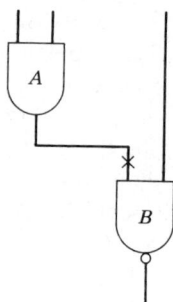

Figure 1.1. The cut line in this schematic results in an s-a-1 fault at the input of gate *B*.

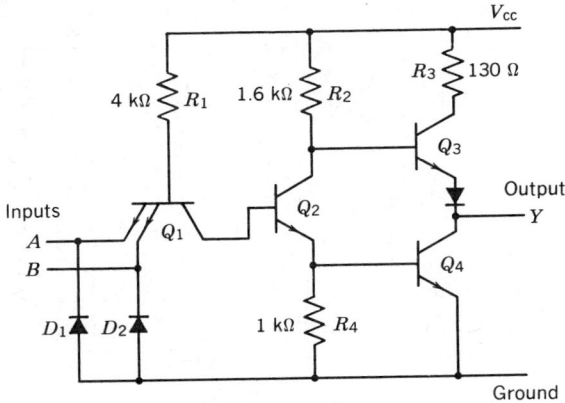

Figure 1.2. A transistor-level diagram of a 7400 NAND gate relates physical defects to logic faults. (Courtesy of Texas Instruments Incorporated [2].)

step fault occurs when metallization does not follow the vertical step from one etched layer of silicon to another. A similar line of analysis leads to a list of failures which cause s-a-0 faults.

The correspondence between failures and logic faults is not a clear cut one. An open in the 4 kΩ bias resistor of the input circuit will cause the output to be s-a-1, but a short in the same resistor will not result in a definite stuck-at fault. Under this condition the effective input threshold will be raised, and the driving gate may or may not be able to meet the new logic high-input condition. Whether the gate appears to have an s-a-0 input or inputs or whether the gate operates normally except for reduced noise immunity cannot be determined without analyzing the characteristics of the driver(s).

Perhaps the most indeterminate defect in common logic is the defective bus enable. The 3-state output of a bus driver gate can be locked into the off state (high impedance) by a shorted enable path. Even this case reduces to a single stuck-at fault (s-a-1) if the bus is pulled up by a load resistor or if sufficient time is allowed for the circuit to drift to a 1 as a result of the leakage currents (I_{IH}) of the driven gates. The case of an open enable path is more complex since the driver will be kept on when it should relinquish the bus. The resulting "bus crash" becomes an interaction of two drivers and is admittedly not a single stuck-at fault. Such complex faults will be evaluated later in the chapter.

The correlation between simple defects such as an open circuit and an s-a-1 fault is dependent upon the logic circuit type. In the emitter-coupled logic (ECL) case shown in Figure 1.3 an open circuit would leave the disconnected input transistor base pulled down toward -5.2 V (a logic 0 in positive logic), turning the transistor off. As in the TTL case for a shorted input the gate reacts as though its input were connected to a driver, forcing a permanent 0. A short to ground in an ECL circuit causes an s-a-1 fault since the ground in the typical negative power supply configuration is a logic 1.

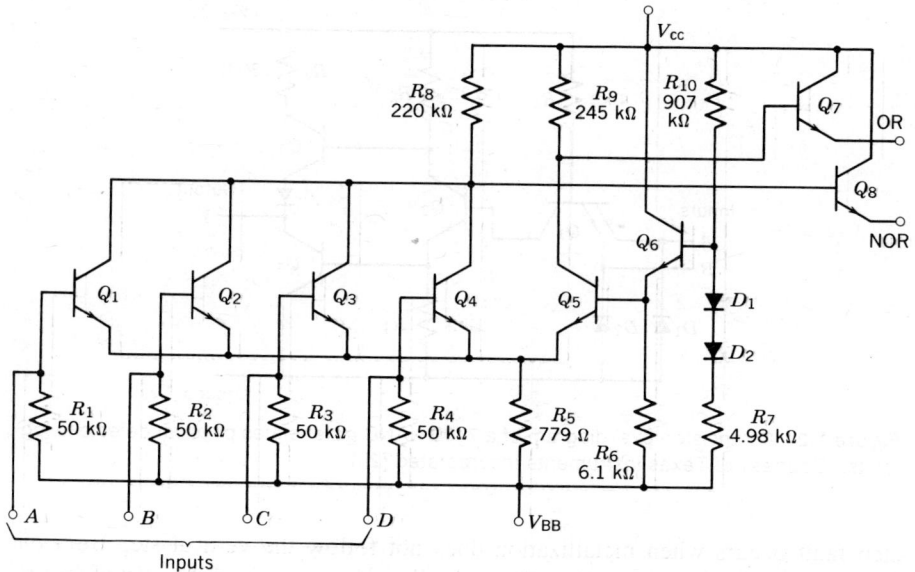

Figure 1.3. Implementing logic functions in differing technologies (ECL above) produces a new relationship of physical causes and logic faults. (Courtesy of Motorola, Inc. [4].)

Many more defects can be conceived which would result in single stuck-at faults. The precise effects of a failure depend upon the logic implementation technology and in some cases the surrounding circuitry. The set of stuck-at faults for ordinary binary circuitry contains only two faults—s-a-0 and s-a-1. The s-a-Z 3-state case is a meaningful concept only at the periphery of the circuit where the current and voltage characteristics can be measured. Internally, the s-a-Z condition translates into a complex fault dependent upon the circuit loading, alternate drivers, and circuit timing. Such faults are covered later in the chapter. The simplification of single stuck-at faults to include only two makes the problem of describing faults and their effects a tractable one.

Fault Propagation

Having relegated the effects of various physical defects to logical faults, it becomes helpful to develop a method of expressing the effects of such faults in a symbolic manner. For the restricted case of single stuck-at faults each fault can only exist in a circuit which is identical to the good circuit schematically, except for the single feature (such as a short or an open) which represents the fault. All other faults must exist in their own separate schematics in a like manner. Since only single faults are to be considered, the usual approach to fault analysis is to compare the good circuit values to those of the circuit containing the fault to be evaluated. Each schematic containing a single fault is termed a fault universe.

Good Circuit

	0	1
0	0	D
1	D*	1

Faulty
Circuit

Figure 1.4. D definition table shows the D algorithm notation for comparing the logic value of a good circuit and a faulty circuit at a single node.

Analyzing the propagation of logic states in a single schematic is a complex task, and maintaining many nearly identical analyses is mind boggling. A simple bit of conventional notation will at least allow comparison of the good circuit state and a single faulty circuit's state on a single schematic. The table of Figure 1.4 enumerates the possible outcomes of a single node comparison at a given instant in time (assuming simple binary logic).

It is important to note that the "D" is a notation for comparative states and not for marking the fault location. The D appears to propagate through the circuit in the same fashion as the binary signals which make it up. In order to predict the propagation of fault effects a set of extended state tables is required for each logic gate in the circuit. These tables include the D and $D*$ (negated D) inputs and enumerate the outputs for good circuit and faulty circuit simultaneously by again employing the D symbology [5]. Predicting the propagation of D is a matter of examining each of the cases represented. In Figure 1.5 the D propagation table for the 2-input logic AND function is shown. The upper left-hand corner of the table is the familiar binary propagation table for the AND. Examine the second column, third row, for the output resulting from an input pair of $(1,D)$.

If the $(1,D)$ input pair is broken into the $(1,1)$ normal circuit inputs and the $(1,0)$ faulty circuit inputs, the outcome is more easily visualized. The normal circuit will produce a 1, while the faulty circuit will output a 0. A 1 in the good circuit and a 0 in the faulty circuit is once again represented by a D. It should be made clear that the AND gate in both circuits was not faulted, but only propagated the differing inputs resulting from a fault which occurred earlier in the faulty circuit's logic.

Conditions may occur which block the propagation of a D or $D*$. A simple situation which extinguishes fault effects (Ds and $D*$s) is shown by row 1, column 3, of Figure 1.5. Since the gates in both the good and faulty circuits have at least one input of 0, the results are identical and independent of the other inputs. The fault effects were blocked by what is termed an insensitive gate. Fault effects will only propagate along paths that are sensitive, that is, paths in which the outcome

	0	1	D	D*
0	0	0	0	0
1	0	1	D	D*
D	0	D	D	0
D*	0	D*	0	D*

Figure 1.5. D notation propagation through an extended AND truth table tracks good and faulty circuits simultaneously.

of the next gate depends upon the state of the node presently exhibiting a fault effect. Naturally, when the boundary of the circuit is reached, the fault effect propagation process is finished and the effects are observable on primary outputs of the circuit.

If observation of the circuit states is limited to primary outputs (which is very often the case) the fault cannot be detected until one of its effects is propagated to such a circuit boundary.

Along the way, fault effects may cancel one another. For example, if the inputs to the AND gate of Figure 1.5 are given a D and a D^*, the situation of row 4, column 3, is invoked. The good circuit gate would see a $(1,0)$ input pair, while the faulty circuit gate would see a $(0,1)$ input pair. The results are identical. This phenomenon is caused by reconvergent fan-out. Since the faulty circuit is assumed to have only one fault, the D and D^* must have originated from the same fault site. Fan-out from the fault site carried the effects along different paths, but for the gate in question the paths converged.

The "functional" test method for digital circuits consists of choosing an input vector or a string of vectors that excite the fault, thus creating a D or D^*, and then sensitizing at least one path to an observable point. The methods of accomplishing these ends will be discussed in later chapters. It is important to realize, however, that this sequence of events constitutes a test for a fault.

The analysis of fault effects assumes that the logic state of the good and faulty circuits is known. Any circuit containing storage elements such as flip-flops or memories can have unknown nodes, particularly when the circuit is powered up and before an initializing sequence has been performed. Unknown logic values may be propagated through any sensitive path, and no fault effects can be associ-

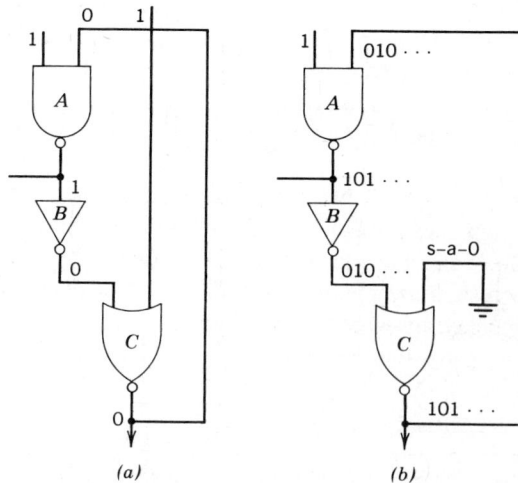

Figure 1.6. The good circuit results in stable logic states. An s-a-0 fault causes oscillation.

ated with any node of unknown value. The comparison of an s-a-1 and an unknown good circuit is unknown since the good circuit may be at a 1 also. If an unknown is propagated to an insensitive gate input, the output will be known. Thus, unknowns can be blocked and eventually eliminated from the circuit.

Unknowns may occur in the faulty circuits as a result of the fault. Usually, this occurs when an initializing signal is blocked by a stuck-at node. A peculiar case which also occurs is the oscillatory fault effect. The circuit depicted in Figure 1.6 displays this confusing fault effect.

For the input conditions shown, the normal circuit has a stable output of 0. The s-a-0 fault on gate C, however, causes the faulted circuit to oscillate. Whether or not the faulted circuit has the correct output depends upon when that output is sampled. A very tricky situation indeed! Usually, the output of the faulty circuit is considered to be unknown, and no fault detection is possible.

Undetectable faults occur in the cases just mentioned and in the case of redundant logic. Redundant logic consists of at least two circuits which perform an identical function. Two gates whose inputs and outputs are connected in parallel to increase available drive constitute the simplest redundant circuit. From a logical standpoint it is impossible to discover if one of the circuits is malfunctioning. Of course, the net being driven may require both circuits to successfully reach a logic threshold, but this test uses parametric information, not simply logic. A more common occurrence of redundant circuits is in error correction. A circuit which calculates a result in three identical branches and then votes to determine the correct answer has undetectable faults unless the correction circuitry can be disabled, allowing each branch to be examined separately. Thus far, the examples have been drawn from combinational circuits.

Sequential circuitry is treated similarly, except that the D and D^* values must be stored occasionally, as are normal binary results. It is quite possible for a fault effect to be clocked into a flip-flop and to finish its propagation to a primary output at a later time when the original fault-activating input conditions are no longer present. The interpretation is unchanged. The good circuit has stored a 1 and the faulty circuit has stored a 0 (or vice versa). Therefore, the state of the flip-flop is a D (or D^*).

The effect of faults on sequential circuits cannot be treated altogether so lightly. Maintaining a record of the stored Ds and D^*s becomes a major task. Subtle effects also arise. A fault effect entering the clock input of the simple data flip-flop shown in Figure 1.7 has an indirect effect on the output Q since the outcome depends upon the present and past states of the input data.

At some arbitrary beginning of the sequence (t_0) the data input is a 1, but since the past inputs are unknown, the output Q cannot be determined. In the good circuit the clock arrives at t_1, and the input 1 is latched and propagated to the output Q. However, in the faulted circuit the clock line does not change states, and the output remains unknown. The tendency for faulted circuits to become unknown due to fault effects on control lines is a common occurrence.

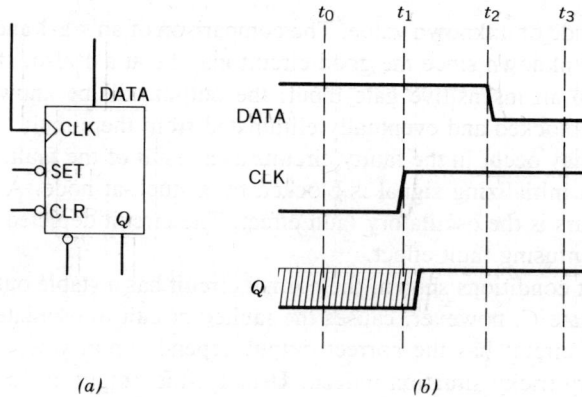

(a)   (b)

Figure 1.7. Proper operation of a flip-flop whose past data are unknown until a clock edge is sensed.

Fault Equivalence

Before examining further fault classes a simplification to the listing of single stuck-at faults will be covered. The concept of fault equivalence is introduced to remove repetition from the fault enumeration and test generation process. In a straightforward fashion the single stuck-at faults of a given circuit can be listed by applying an s-a-0 and an s-a-1 to each logic gate input and output. While this list is sufficient, it is considerably more than necessary. Many faults in the list produce effects upon the circuit which are indistinguishable from other faults. Such faults are termed equivalent.

The NAND gate of Figure 1.8 is labeled with six single stuck-at faults. Those input s-a-0 faults which are crossed out are equivalent to the output s-a-1 fault since, from the truth table, either input s-a-0 will result in an output of 1 without regard to the other input. Any test which would detect an s-a-1 on the output would also detect either input s-a-0. There is no need, therefore, to list more than four faults for this gate. This 33% savings becomes significant for large circuits.

Faults which are equivalent because of their relationship through a gate or device are always equivalent to one another without regard to the location of the gate in circuitry. These faults are often called the "prime" faults of the gate. A nonprime but equivalent fault example equates the faults at the output of a gate and the single gate input it drives. If more than one gate input were driven by the output in

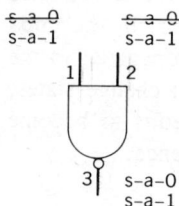

Figure 1.8. Input s-a-0 faults of the NAND gate are covered by the output s-a-1 fault.

question, the input faults would not be equivalent to the output fault. An open circuit at one of the inputs would not disable the other driven gates, for example.

MULTIPLE STUCK-AT FAULTS

The single stuck-at fault model is admittedly a simplification of reality, although it is a simplification commonly made. In actual circuits, particularly those newly manufactured, multiple faults are frequent. Consider a few of the common manufacturing mistakes and this becomes evident. An incorrect IC usually creates faults on more than one circuit node. Overetching a pc layer will probably cause more than one open circuit.

If the faults are noninteractive, the situation remains straightforward. Two open circuits on different bits of a communication bus may not interact. In this case the faults can be considered singly since a test for one will not be affected by the other. Even if the same set of inputs is used to detect both faults simultaneously, the faults will be distinguishable since their outputs will give faulty conditions on two different bits of the bus. The only confusion may occur when the diagnostic technician notices the two faulty outputs and attempts to find a single cause.

Interactive Multiple Faults

Interactive multiple faults are quite another matter and can lead to inaccuracies in test coverage and diagnostic messages. If the communication bus described above were fed into a parity checker whose output was the only observable output of the circuit, the conditions which revealed both faults previously could fail to uncover either fault. Consider the case in which the test conditions caused 32 zeroes to appear on the bus. Even parity would be indicated by the good circuit. If two of the lines of the bus were s-a-1, the parity would still be even. Of course, if the actual bus lines were observable, both faults would be detected.

To observe the propagation of multiple faults in detail, examine the state table in Figure 1.9, which shows the outputs for an EXCLUSIVE-OR circuit when all cases (both good and faulty) are considered for both inputs. This case is peculiar in that, like the parity circuit, any time there are two faulty values input to the circuit its output is correct. In effect, a single error will be transmitted, while a double error will be corrected. The detection of the fault through this path is therefore blocked. Perhaps another test condition will block only one of the fault effects before it reaches the EXCLUSIVE-OR, and the other fault will be transmitted to eventual detection.

	0	1	D	D^*
0	0	1	D	D^*
1	1	0	D^*	D
D	D	D^*	0	1
D^*	D^*	D	1	0

Figure 1.9. The extended truth table for an EXCLUSIVE-OR shows interaction of multiple faults.

Any test which purports to be unconditionally thorough must detect all combinations of multiple faults. The techniques for predicting fault behavior and determining test conditions or sequences apply to multiple faults straightforwardly. The *D* method of comparing a good and a faulty circuit can be easily applied if the faulty circuit has more than one fault, although the propagation process entails the resolution of more complex cases, such as those in Figure 1.9. The number of distinct multiple faults is given by $n^2 - 1$ where n = number of single faults. This number obviously becomes very large for realistic circuits (which commonly contain thousands of faults). Addressing the effect of each of these faulted conditions becomes a lifetime occupation even at computer speeds.

Due to the unmanageable number of multiple faults, a compromise is usually sought. Circuits can be designed for which a 100% single-fault coverage test also is guaranteed to detect all multiple faults [6]. However, the design restrictions for such circuitry are severe, and in general a 100% single-fault coverage test of any circuit will detect most of the multiple faults. The exact correspondence depends upon the circuit implementation and is therefore not very predictable. The more sequential a circuit becomes, the more unlikely it is that a set of single fault tests will cover all multiple faults.

Bridging Faults

All the varieties of faults discussed thus far can be thought of as variations in number and duration of simple "single stuck-at" faults. Several fault classes which are of interest cannot be so described. In general logic circuits, intercircuit shorts are the most frequently occurring complex faults. Memory circuits display faults such as disturb failures and read only memory (ROM) link healing. With the prevalent use of programmable logic in the form of gate arrays, the improbable-sounding fault of erroneous gate function has become an occasional occurrence. Finally, the metal oxide semiconductor (MOS) technology contains some peculiar soft value faults due to the transistor implementations of logic gates.

The bridging fault may be caused by a misplaced component lead which touches another circuit path or excess copper on a pc board which was not removed during etching. Solder bridges or splashes (Figure 1.10) are the most common cause of intercircuit shorts. As time, temperature, and electric fields work on the metals in ICs, ion migration can create strange stalagmites which grow out of the surface and contact other metal or silicon regions of the circuit, as in Figure 1.11.

The effect on the circuits involved depends upon the logic implementation technology. For TTL a wired-AND gate is created in that if either circuit driver is a 0 the entire common net will be a logic 0 (Figure 1.12). In the case of ECL the wired-OR gate is created, with the effect that any one will drive the combined net to a 1. In any case the result is a function of both circuits, and an effective test cannot be devised without considering both nets simultaneously.

Perhaps the most difficult part of deriving tests for intercircuit shorts is deciding which circuits are likely to become shorted together. Since the number of possible

Figure 1.10. A solder splash has bridged several circuits on this pc board.

Figure 1.11. Metal migration due to electric fields at elevated temperatures creates bizzare tentacles which can short ICs. (Photo courtesy of Sperry Corporation).

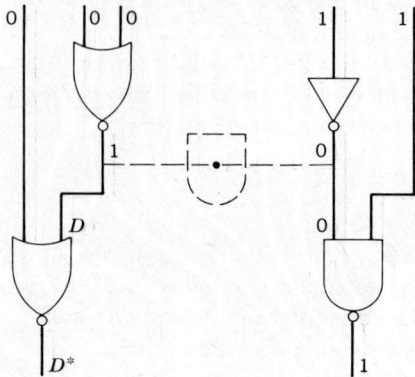

Figure 1.12. Intercircuit shorts create a wired-and in TTL technology circuits.

intercircuit shorts is again proportional to n^2, checking all of them is impractical. A common way to decide is to check those circuits which connect to adjacent package leads or input/output (I/O) connector pins.

Even when the list of intercircuit shorts to be included in the fault list has been determined, the procedure for deriving test conditions is akin to the multiple fault problem. A set of conditions must be found which will cause differing states on the two interconnected nets. Then the effect of the differing states must be evaluated using the wired-AND or other appropriate algorithm. Finally, if one of the nets is determined to have a value of D or $D*$ the fault value must be propagated to a primary output of the circuit. Since the fault under consideration is a short circuit rather than an active device, it is sufficient to detect its action upon either net. The effects are bidirectional, and the faults created by either circuit operating through the short upon the other, although quite different in effect, are equivalent.

The preceding discussion was tacitly limited to bridging between nodes of the same logic rank, or between two circuits which thereafter propagated to different outputs. When bridging connects the output of a circuit to its inputs, a feedback bridge has been established. A simple example is shown in Figure 1.13.

In fact, the opportunity for feedback bridging is readily available to as great an extent as parallel bridging. An examination of the pin configuration of the 7408 AND gate DIP package shows outputs adjacent to inputs.

The equivalent circuit reveals that the simplest of combinational circuits can be converted to a sequential latch by feedback bridging. If input A is driven to a logic 0 by the previous rank of logic, the output of the AND gate will become 0 regardless of input B. In the presence of the feedback bridge the output will in turn force input pin 1 to remain 0 after the stimulus at A has been returned to a logic 1. Not only has the previous value of A been latch, but it cannot be cleared. A test for this fault must first establish a condition for forcing A to a 0, and then return it to a 1 before sensitizing a path to check its value of C. Of course, much more complex feedback situations can occur, and the difficulty of testing for their presence becomes complex also.

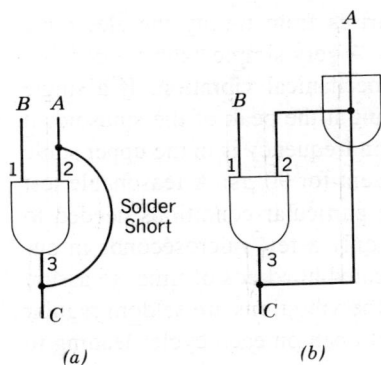

Figure 1.13. Shorts between inputs and outputs create a wired-and gate with feedback effects.

INTERMITTENT FAULTS

Electronic circuitry, no matter how fast, takes times to perform its function. The testing process also requires time and frequently is performed at a slower rate than actual operation. If a fault occurs or disappears within the test length, it is termed intermittent. Generally, any fault which occurs and then disappears again without overt repair action is intermittent, but for test purposes, only the time of test condition applications is pertinent.

The mechanisms which cause intermittent faults are environmental in nature. It is difficult to conceive of an intermittent fault by merely examining the schematic for a logic circuit. Only by considering the operation of a physical implementation of the circuit do the circumstances become reasonable. Although it sounds unlikely that a shorted output transistor creating an s-a-0 would heal itself, if the short were the result of a loose metal particle inside the package, a vibration might remove the short temporarily.

Vibration

In fact, vibration is a prime cause of intermittent faults. Conductors sway and touch, pc boards flex and crack, and solder connections fatigue and give way. In operation the vibration may come from nearby machinery or the vessel, aircraft, or vehicle to which the circuitry is mounted. Low-frequency vibrations of a very regular nature are experienced in ships due to the propeller motions and the action of waves on the hull. Subsonic and supersonic airflows over an aircraft fuselage create a random vibration with a broad spectrum of energy. The resonant effects of the pc board, its mounting, and the components themselves accentuate some frequencies. Even equipment built for the home may encounter 10g shocks (10 times the acceleration of gravity) from being thrown into a closet.

Tests for electronic assemblies expected to perform under high-vibration conditions can be carried out under simulated environmental conditions using mechanical or electromechanical shakers. Sophisticated equipment is available to recreate nearly any vibration pattern in the audio or near supersonic frequency range.

The difficulty of testing for intermittent faults arises from timing the electronic tests to coincide with the mechanical occurrences. A very simple example consists of a pc board which flexes under a sinusoidal mechanical vibration. If a single circuit path is cracked and becomes an open circuit at the peak of the sinusoid, it may be open for 10% of each cycle. If the vibration frequency is in the upper sonic region, then the fault is present for 10 μs and absent for 90 μs. A reasonable test sequence might last for several seconds, but the particular conditions needed to detect the particular open circuit occurrence are only a few microseconds in duration. The entire sequence would have to be repeated hundreds of times to assure a reasonable chance of detecting the fault. Since the vibrations are seldom regular (due to resonance distortions), the circuit may not open on each cycle, leading to the need for even more repetitions.

The repetition of a lengthy test sequence may not at first seem to be a problem. Modern automatic test equipment (ATE) can easily be configured to repeat a sequence ad infinitum. Performing the test does take time, however, and the total number of circuit assemblies which can be tested per hour with a given amount of test equipment is severely limited by such practices. As will be seen in later chapters the economical use of expensive equipment is a prime concern. Therefore, the repetitious tests are usually performed at a higher level of assembly, such as in a complete computer where the auxiliary equipment investment required is small. At higher levels the tests may be performed by cycling a program, for example, which would require only an I/O device to start the procedure.

Thermal Stress

Thermal stress causes intermittent faults also, but the time period of intermittence is much longer. Most thermal problems are caused by the differences in expansion rates among the materials used to manufacture electronic devices. The copper path of a pc board expands at a rate of $17 \times 10^{-6}/°C$ [7] while the glass used in the fiberglass and epoxy substrate expands at a rate of $9 \times 10^{-6}/°C$. In the process of warming up from ambient (25°C) to an operating temperature of 100°C a 1-cm section of copper would grow to 1.001275 cm, while the glass to which it is affixed would grow to only 1.00675 cm. Actually, each material creates stress on the other, which results in a compromise growth and strain. Such strain can create cracks and other effects which result in electrical faults. If the temperature of the assembly is reduced, the cracks may close, temporarily correcting the fault.

In this case a single pass through the test sequence has a good probability of detecting the faults if the temperature of occurrence has been reached. The catch is the length of time required to stabilize the assembly at a given temperature. Even though the assembly may be immersed in a preheated (or precooled) medium, the thermal inertia of a circuit board and the limited heat conductivity of its outer layers combine to create a temperature lag in its interior that may require minutes to hours to overcome completely. Thus, the assembly must be "soaked" at each test temperature for minutes to assure a reasonable opportunity for occurrence and detection of intermittent faults. A simple test could well require hours if three temperatures were used (cold, ambient, hot), again occupying expensive test equipment.

Since the test sequence need only be applied at three points rather than continuously during temperature cycling, soaking chambers and temperature-controlled handling equipment can be attached to the test in an assembly line fashion. When the circuit assemblies have stabilized at the appointed temperature, they can be moved into position, connected, and tested without leaving the temperature chamber. This is often referred to as "hot hands" testing.

At the far extreme of intermittent behavior are the so-called soft failures. Causes such as electromagnetic interference, infrequent power glitches, and cosmic radiation can create spurious faults of short duration (nanoseconds). These faults are nearly impossible to isolate. Fortunately, most of these faults occur very infrequently and are temporary in effect. In these cases the effort is to recover from the fault rather than to isolate and repair it. Fault tolerance addresses these techniques of detection and correction. Further discussion of soft faults follows, with particular attention to their occurrence in semiconductor memories.

MEMORY FAULTS

Probably the most sought-after complex fault group consists of memory faults. The pressure for extreme miniaturization of memory circuits has led to susceptibility to various interdependence effects. These effects cause what are commonly termed "neighborhood faults," and the tendency toward any particular failure depends upon the physical arrangement of the paths and storage elements within the IC. In addition to the thermal and power supply variance environmental factors, time becomes a factor in dynamic and nitride-insulated "write seldom" memories. Magnetic memories such as core and wire may have limits to the number of times they can be read without being rewritten. As an example, the typical test sequence for a dynamic IC random access memory (RAM) will be explored.

Placement of 64K, 256K, or in the future even more memory cells on a single die requires a cell design that leaves minimal space between all elements. In fact, that spacing is a feature of the etching, crystal growth, and impurity deposition processes which produce the ICs. If a process with higher resolution is available, the dimensions will be reduced to take advantage of it for new designs. With such small (5×10^{-7}) spacing, effects such as charge migration and field effects become critical to operation.

The storage mechanism is a capacitor (as shown in Figure 1.14) consisting of the substrate material (the grounded plate), a silicon dioxide insulating layer, and a second layer of silicon deposited on top in the spade pattern to form the upper plate. The silicon also forms a path for charging or sensing individual capacitors. The neck which connects the capacitor to the charging line is turned into a (MOS)

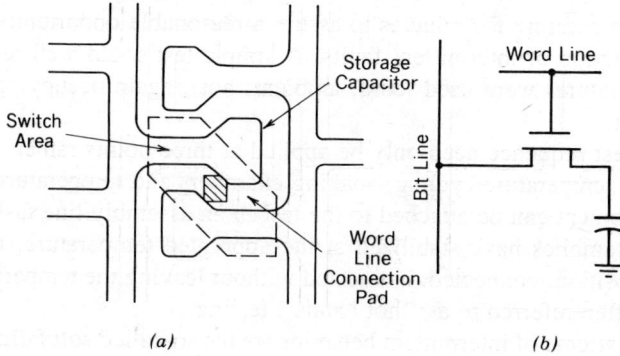

Figure 1.14. The mask outline of a dynamic memory cell and its corresponding circuit diagram.

transistor by covering it with a thin insulating layer of silicon dioxide followed by a deposited metal gate. All the gates in a horizontal line are connected to form a word line for accessing a string of capacitors simultaneously, while each silicon bit line connects one bit from each word to a charging amplifier and a sensing amplifier.

Unfortunately, the internal implementation of a memory chip is usually at the discretion of the manufacturer, so a test which depends upon the structure of the IC may not be effective upon the next shipment received. For example, if vendor A were to sell a 64-bit memory implemented as 2 bit lines and 32 word lines, it would appear as in Figure 1.15a. Vendor B might use 8 bit lines and 8 word lines, but since both vendors would have multiplexed the bit lines into one data line "in" and one data line "out," the chips would be interchangeable.

A test designed to detect the sensitivity of a bit to the fields generated by the

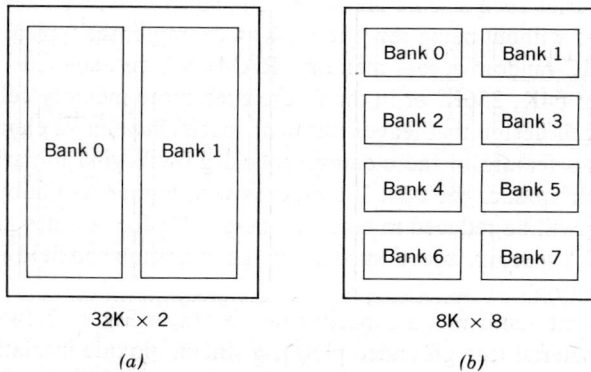

Figure 1.15. Alternative internal structures for a 64K RAM chip. Structure-dependent faults from one do not apply to the other.

charging and discharging of adjacent bits in the same column would not be effective on all the bits of an alternate layout, as in Figure 1.15, since the intended bits may not be adjacent. Exhaustive tests which check for interference between any two bits in the memory array eliminate the layout dependence, but the cost is a lengthy test sequence.

Rather than immediately commit to a lengthy test for such an elusive fault set, the test sequence usually consists of carefully structured sets of input patterns which exercise fundamental faults first. A simple address pattern consisting of writing each row's address into its data and then checking by a series of reads was a popular test for shorted or open address lines when most memories were constructed with broad data paths and short arrays. A 256 × 8 memory could be tested this way. Now that memories are usually very vertical in form (16K × 1 bit, 64K × 1 bit, etc.), a different technique must be applied. Such details will be covered in a later chapter.

The sequence of structured tests probably would move on to a checkerboard pattern if the memory had more than a data line. Such a test would place each data line in a 0 condition while its adjacent lines are 1s, and vice versa. Thus, not only would open or shorted data lines be detected, so would bridging faults between the data lines. The test sequence would then move to appropriate patterns to check each cell for operation and perhaps for intercell disturbance.

The essence of this procedure is that each test pattern selected is intended to detect a different fault among a large group of fault types. The inclusion and position of a test pattern in the sequence depends upon the characteristics of the device, its faults, and the test patterns themselves. Less time-consuming, and therefore less expensive, tests will be performed first if they are likely to detect frequently occurring fault types.

Hard memory faults such as stuck address lines, stuck data lines, or interline shorts are treated as input, output, or bridging faults in ordinary logic. Pattern-sensitive faults such as adjacent cell disturbance are time consuming to detect, but they will appear repeatably under the proper sequence of inputs. The most elusive memory faults are soft faults. Soft faults cause a read error, but cannot be repeated after rewriting the cell.

As cell sizes shrink, soft faults caused by cosmic radiation become statistically noticeable. Figure 1.16 illustrates this unusual mechanism. When a cell is written to a 1, the metal plate of the capacitor is connected to a current source and a charge is built up between the plate and the substrate material of the IC. The gating transistor is cut off before the current source is removed such that the charge is trapped in the cell capacitor. The maximum amount of charge deposited depends upon the voltage capability of the current source and the value of the capacitor.

The charge in a semiconductor region is somewhat complex, but it can be approximated by $q = Cv$ where v is the voltage across the plates of the capacitor. In this case, assume that sufficient time has been allowed for the voltage of the capacitor to reach the voltage of the source. Later memories have been designed, for the convenience of the user, to operate on the same voltage as the surrounding

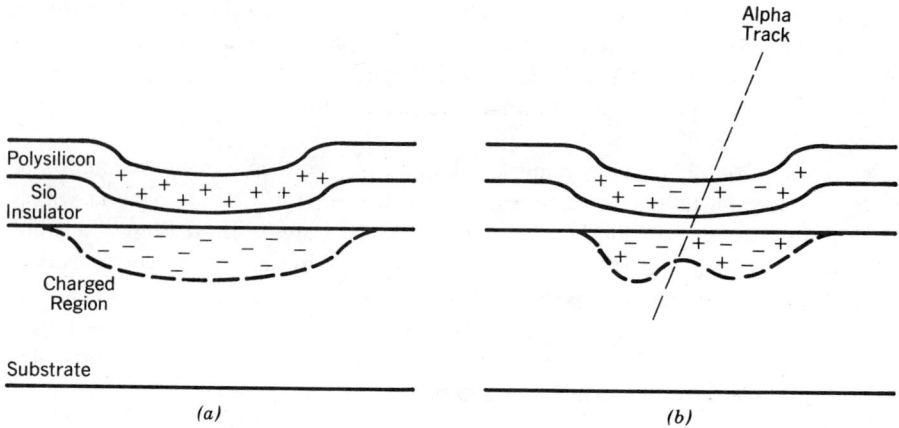

Figure 1.16. Cross section view of a dynamic memory cell before and after being struck by a cosmic alpha particle. (a) Written cell. (b) Alpha particle discharge.

logic (usually 5 V). The capacitance for parallel plates is: $C = eA/d$ where A is the area of the plates and d is the separation. The constant e has a value of about 3.4×10^{-11} $C^2/N\text{-}m^2$. In state-of-the-art memory cells the thickness of the insulating layer of oxide may be 0.02 μm and the plates may be 5 μm on a side. Solution of the equation results in a capacitance of 4.3×10^{-14} F or 0.043 pF. The charge on a capacitor of this size at 5 V is 2.15×10^{-13} C, or only 1,341,600 electrons.

The ionizing effect of an energetic heavy cosmic particle such as an alpha particle is within the range of the charge on a storage cell of such memories. As a cosmic alpha particle passes through the cell, it dislodges electrons from atoms in the oxide and substrate. These electrons are temporarily free and will migrate to neutralize the capacitor's charge due to its potential field. Although the cell probably will not be totally discharged, its charge may fall below that necessary to trigger the sensing amplifier during the read operation. Information will be lost, and a fault will have occurred. Very little, if any, permanent damage is done by cosmic alpha particles, however, and once the cell charge is restored during a write operation, it will function normally. Soft faults such as alpha particle disturbs are statistically scarce occurrences and impossible to test.

FIRMWARE FAULTS

Stuck-at faults and simple faults in general do not assume that a new input can be added to the function of a circuit by a fault. Instead, one of the function variables is replaced by a 1 or a 0. Intercircuit shorts result in the addition of a variable, as wire ANDing of two signals alters them both. More complex and much less visible function-changing faults occur within programmable devices such as programmable ROMs (PROMs) and programmable logic arrays (PLAs).

PROMs can be represented as an array of OR gates driven by an address decoder. Each address combination results in the activation of a single word line (W_3 in Figure 1.17). In the unprogrammed chip, all the OR gates have a connection to all the word lines, and all outputs will be activated. During programming the connections between selected word lines and OR gates are.broken to create the characteristic logic function desired. In effect, the address decoder is a row of AND functions with each word line as an output, while the programmable array of OR functions completes a classical canonical Boolean function realization between inputs and outputs. Each distinct address is a term made up of address bit variables. Each output may include or exclude each term from its equation.

Stuck-at faults apply to the address inputs and the outputs of the PROM in the classical fashion, but the internal fault representation is not so clear. It can be looked upon as a failure to connect or disconnect a term of the equation. An adequate test would consist of checking each output for activation when its connected terms have been invoked and then assuring that its output is not activated when other terms are invoked. Since PROMs usually have large address fields and small data fields, the addresses are usually sequenced through all combinations while the data field is checked for correctness at each step.

PLAs also display a fault type which is characterized by the connection or lack of connection of terms to a Boolean equation. A popular PLA is arranged internally as eight OR gates each driven by seven AND gates. Each AND gate has true and complement inputs from 16 chip inputs. The programming is accomplished by disconnecting the opposite input to that which is desired to influence the output equation. Since the AND of a variable and its complement is equal to "0," all unprogrammed terms will remain "0." In bipolar IC technology the connections are made as fusible links similar to those in PROMs. In MOS technologies the connections are transistors. The matrix location which may be optionally occupied by a connecting device is called a crosspoint, and the missing or extra devices manifest themselves as crosspoint faults [8].

The architecture of the PLA differs from the PROM in that there are only 56 potential terms to its Boolean equation, rather than the 64K addresses suggested by its inputs. A test which considers the logic in terms of this equation will be shorter by far than an exhaustive combinational manipulation of the inputs. Consider the equation: $F5 = (A3^* \cdot A6 \cdot A7^*) + (A1 \cdot A2 \cdot A3 \cdot A6^*)$ where $F5$ is an output function, $A\#$s are input variables, the * indicates negation of the preceding variable, the $\cdot$ is used for the logical AND function, and the + designates the logical OR function. If the function were constructed of simple gates, the test sequence might appear as in Table 1.1.

The first two input patterns test the OR gate output for s-a-0 and each of its inputs for s-a-0. The next three patterns test AND1 (the AND gate represented by the first term) for inputs s-a-1 and output s-a-1. The inputs of the OR gate are tested for s-a-1 also. The final four patterns similarly test AND2. Some crosspoint faults will be detected by this test, notably the crosspoint faults consisting of missing crosspoint devices since a missing device will be a stuck fault for one of the nodes exercised.

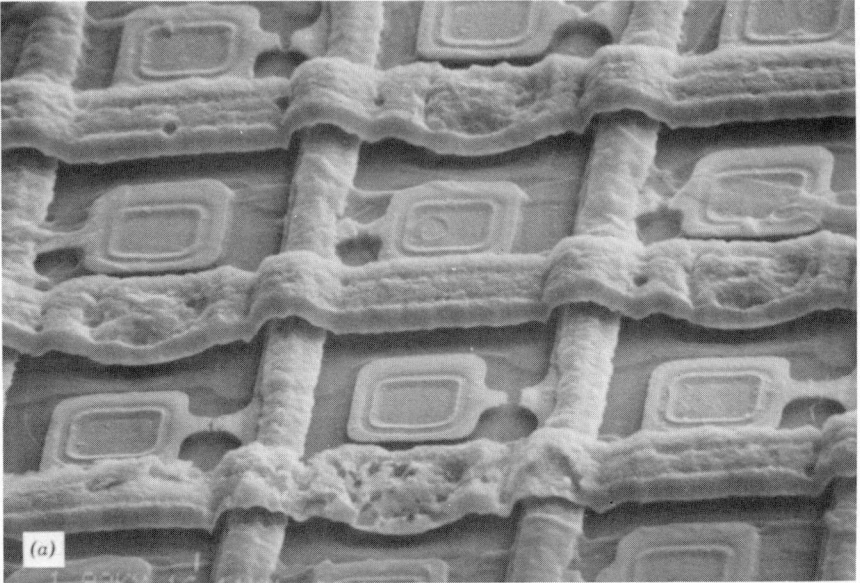

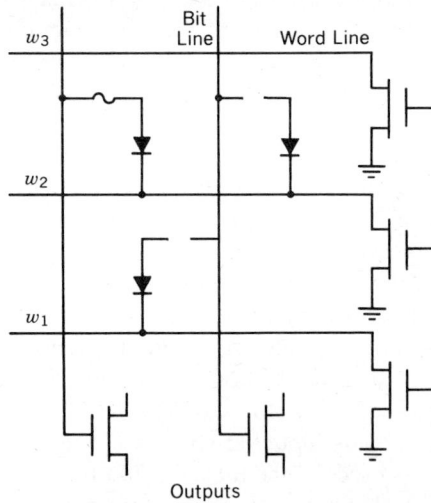

Figure 1.17. The microphotograph of a programmed PROM (Courtesy of Sperry Corporation) reveals opportunities for shorts which are not as clear in the equivalent logic diagram. (a) PROM. (b) Logic equivalent.

TABLE 1.1   Test Patterns for Simple Gate Realization of $F5 = (A3^* \cdot A6 \cdot A7^*) + (A1 \cdot A2 \cdot A3 \cdot A6^*)$

A1	A2	A3	A6	A7	AND 1	AND 2	F5
0	0	0	1	0	1	0	1
1	1	1	0	1	0	1	1
0	0	1	1	0	0	0	0
0	0	0	0	0	0	0	0
0	0	0	1	1	0	0	0
0	1	1	0	1	0	0	0
1	0	1	0	1	0	0	0
1	1	0	0	1	0	0	0
1	1	1	1	1	0	0	0

The class of crosspoint faults also includes the possibility of extra crosspoint devices being present. No tests were generated which included input $A5$, but a crosspoint fault is possible which would add $A5$ or $A5^*$ to an equation. Therefore, a complete test for crosspoint faults must include patterns of inputs which hold a term inactive with a sensitized path to the circuit output, while uninvolved inputs to the package are toggled one at a time. Then each term is made active, and again the uninvolved inputs are toggled. This procedure detects single crosspoint faults only. More detail on functional test generation will be provided later in the text. This example is discussed to show the distinction between classic stuck-at faults and crosspoint faults.

TIME DEPENDENT FAULTS

Intermittent faults are time dependent in a sense, but their dependence is sometimes erratic and usually linked to a physical rather than electronic phenomenon. There are several types of logic faults which have a predictable dependence upon time. High impedance logic implementation technologies such as MOS exhibit faults of this type when gates become stuck off or open. Dynamic memories, which depend upon charge storage, suffer from refresh faults, which are dependent upon the cycle time of the refresh operation. Some fast logic structures depend upon techniques such as precharge between accesses and are therefore subject to recovery faults if inadequate time is allowed between cycles. Finally, all logic elements and the conductors between them exhibit delay between input and output, and these delays can cause transients when signals from different paths interact.

High-Impedance Faults

Bus logic, which consists of more than one driving element available to a conductor (not simultaneously), created the need for a third state of logic in which no active transistors provide drive to the output. This ''off'' or ''Z'' state disconnects a driver from the bus to allow another to supply signals without interaction. In the TTL logic technology the technique is termed 3-state. In MOS technology a similar effect is accomplished by the transmission gate or T-gate. The new fault class created is stuck-at-off (s-a-Z, to avoid confusion with s-a-0).

Consider the following situation. Two drivers capable of 3-state drive a single input as shown in Figure 1.18. Suppose the inputs begin as InA = 0, EnA = 1 (this signal turns on driver A), InB = 1, EnB = 0 (driver B is off). Node N is held at logic 0 due to driver A. If the inputs are changed to (0, 0, 1, 1), then the node N will be driven to a logical 1 by driver B. The waveform will appear similar to that in the top of Figure 1.18b. The driver reverses the current coming from the input of gate C (TTL logic assumed) and supplies enough current to overcome leakage such that the voltage rises well above the threshold for a logic 1. If, however, the EnB line were s-a-0, externally or internally, neither driver would be active and the node N would float, as shown in the bottom of Figure 1.18b. I_{OL} current from the input of C would eventually charge the line capacitance and raise the voltage to the logic 1 threshold, but much more slowly than in the intended case. In addition, the voltage would not exceed the threshold since at that point the input junction of C is reverse biased and current falls off sharply. If there are no pull-up resistors attached to the node, the voltage will eventually settle at about 1.7 V.

The slow rise time of the signal at node N aggravates the time dependence of the circuit result. It is assumed that in any case a finite amount of time must be allowed before the output can be sampled via gate C, but with the presence of the fault gate B s-a-Z the time is an order of magnitude longer. If the sample is taken microseconds after the input change, the output will probably be correct, but if it is taken in hundreds of nanoseconds after the input change, it is likely to be wrong. Modeling such faults requires knowledge of the test method as well as the circuit and component characteristics.

MOS technology logic with its inherent high off impedance aggravates the s-a-Z fault, resulting in node storage by capacitance of the old value for millise-

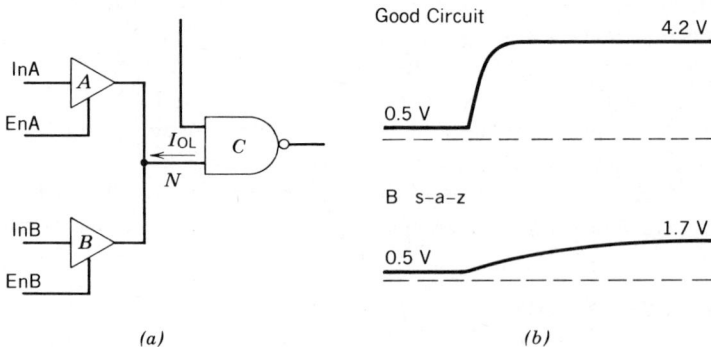

Figure 1.18. A simple two-driver bus can exhibit slow rise time under the influence of an s-a-2 fault. (a) Simple 3-state bus. (b) Waveforms at node N.

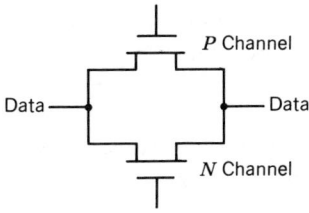

Data ——— [P Channel] ——— Data

N Channel

Figure 1.19. In MOS technology two complementary transistors are arranged in parallel to form a transmission gate.

conds. If the test sequence is arranged to set a node condition, reverse the condition, and immediately check its value in close sequential steps, s-a-Z faults can be detected. However, the surrounding logic through which inputs and outputs must be propagated can easily slow the process and obscure the results. Further complications arise when the MOS T-gate is involved. The T-gate consists of a bidirectional switch made of two complementary MOS field effect transistors (MOSFETs) in parallel, as shown in Figure 1.19.

Both transistors are switched in unison to create a low impedance path which is insensitive to the data line potential or a high impedance [9].

If one of the transistors is shorted, the path will not be disconnected and signal mixing will create complex results dependent upon the data and drivers on either side. If one of the transistors becomes open, the path impedance in the on condition becomes sensitive to the voltage potential of the data path. For the case of enhancement devices with the N-channel device inoperative, a logic 0 will be passed unhindered, but a logic 1 will reduce the gate to drain potential, increasing the channel resistance. A weak 1 will result. Whether the weak 1 is detectable or not depends upon the nature of the driven circuit and the timing of the test. The voltage of the weak 1 will approximate a normal 1, but its current capabilities are much lower. If a weak signal is wire ANDed with a normal signal, the results may be different than if normal signals were wire ANDed in the same way.

Refresh Failure

Dynamic memories and charge coupled devices are unique structures whose function depends upon the capacitive storage of charge similar to that encountered in the s-a-Z conditions just discussed. The storage cells are chip regions arranged to provide a defined capacitor accessed by gating transistors. The transistors create a low-impedance connection to a digit line through which a current is passed to charge the capacitor. Then the transistor is turned off to isolate the capacitor until the cell is read. Under ideal conditions the information would remain indefinitely, and some metal nitride oxide semiconductor (MNOS) devices store charge for years. In most fast access memory arrays, however, storage time is in the range of milliseconds, as leakage current to the surrounding substrate carries away the charge.

In order to retain information each cell is periodically read and rewritten in a cycle called "refresh." Allowable time between refresh cycles (which usually refresh one row of the array at a time) depends upon the structure and composition of the device. Excess leakage in a given region of the device may cause cells to bleed off charge faster than the specified interval, resulting in a refresh fault. Since leakage and the amount of charge originally deposited in a cell during a write are dependent upon power supply voltages and operating temperature, a true refresh test must be performed under the worst combination of these parameters to assure operation over the entire allowable range. (See "Schmoo Plotting" in Chapter 10.)

Memory refresh faults are not the only occurrence of refresh faults. Some MOS circuits have single storage cells or strings of cells incorporated as latches, shift registers, or FIFOs. Communication ICs such as universal synchronous receiver–transmitters (USRTs) often contain similar circuits. The user is warned of a minimum clock rate below which operation is uncertain. Essentially, the chip experiences refresh failure below this rate.

Delay Faults

All logic elements exhibit delay. A finite amount of time must pass after an input change before the output changes. The delay consists of the capacitive and inductive charge times and the electron movement time. The technology used to implement the logic element is of paramount importance, but the conductors which interconnect the elements and the topology of the circuit itself affect the delays significantly. In most instances the inputs to the circuit in question do not arrive simultaneously, resulting in skew. Therefore, the possibility for failure in a circuit due to the inherent delay of its elements exists. Worst case design practices synchronize inputs to a circuit with a carefully controlled clock signal and arrange the logic to assure consistent results in spite of various circuit delays. There are component failures which could result in a change in these delays, although the occurrence of such failures is comparatively rare.

One such example consists of an open collector TTL driver which is provided with a discrete pull-up resistor on the circuit board. If the pull-up resistor is omitted during manufacture or becomes open circuit due to environmental stress, the rise time of the output will be lengthened greatly. In effect, the propagation time for a signal which causes the output to transition from logic 0 to logic 1 will be much greater then normal, while the 1-to-0 propagation time will be unaffected. Other conditions, such as incorrect IC type (S instead of LS), in the driven gates of a net or improper termination of a line, could overload a driver and affect its propagation delay.

Figure 1.20 is a simple example of a logic glitch or transient caused by the difference between fall-time and rise-time waveforms of otherwise synchronous signals. In ordinary TTL circuits such transients are only a nanosecond or so in duration and seldom have time to reach logic 1 height. A defect such as the examples given could produce a pulse of tens to hundreds of nanoseconds with sufficient energy to clock a flip-flop erroneously. Similar glitches are caused by the differing arrival times in signals from ATE when lengthy cables are involved.

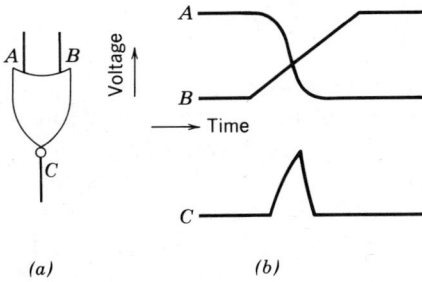

Figure 1.20. Rise- and fall-time differences contribute to glitch generation in even simple combinational gates.

(a)   *(b)*

Within a circuit, reconvergent fan-out brings signals to logic elements with differing delays. Clock skew is the most frequent occurrence.

Unavoidable transients caused by designed-in logic hazards or input skew from the ATE can sometimes be blocked before the glitch reaches an edge-triggered device or a primary output. If a number of inputs to a gate change simultaneously such that a transient is generated, desensitizing the gate by holding one of the inputs until the others have settled inhibits the glitch. For example, if one of the inputs of a 4-input NAND gate is held low until all the other inputs have settled and is then raised to 1, no glitch can be generated. This is the principle of clock gating and is often used during tests to inhibit false transients. The clock signal should not be so broad as to disguise slow rise-time transients, or these faults cannot be detected.

Completely asynchronous inputs to a circuit (such as communication inputs) are usually synchronized by a flip-flop clocked by a sampling clock from the system. This does not eliminate entirely the possibility of coincidence of the data and clock changes. Flip-flops have setup and hold-time requirements, and although one or the other may be practically zero there is always a small period of vulnerability as the clock changes. A small but finite possibility exists that this coincidence of signals will occur, and in most flip-flops the result is an oscillation that may not settle. The behavior has been termed metastable [10].

Although such failures may occur, they cannot be classified as faults since they are the result of a statistical occurrence which is not abnormal and does not arise from a defect in the circuit.

PARAMETRIC FAULTS

Digital circuits are composed of components which are essentially analog amplifiers constrained to operate in a switching fashion over a well-defined range. Within the circuit the parameters of operation are seldom given a second thought. The design rules for fan-out loading and circuit compatibility are simply adhered to, and the voltage, current, and power levels are predicted by the logic family guidelines. At the circuit boundaries, however, the parameters of input and output are less well defined, and specifications are applied to each particular case. This gives rise to the opportunity for a malfunction which causes failure to meet these specifications.

Strictly speaking, when the entire system is operated as a whole under the extremes of the environment for which it was designed, the parametric fault must appear as a functional fault at least once for it cannot be considered a fault. It may manifest itself as a defect or imperfection, but that which does not fail is not a functional fault. Subassembly testing must deal with the circuit boundary in some way, either implicitly or overtly. If the environment in which the subassembly is to operate and the circuitry to which it is to be connected is well defined, carefully chosen loads for outputs and controlled signals for inputs satisfy the need for parametric testing implicitly. Details of load and signal selection will be discussed later in Chapter 6.

If the circuit's surroundings in operation are not well defined, a set of specifications for its inputs and outputs must be drawn up. The ICs in a logic family are excellent examples. Each parameter is defined and given minimum and maximum values, and the conditions under which these values hold are spelled out. Just as in the case of ICs a subassembly may be so specified when it is a building block to be used in many different systems as designs evolve. The application of conditions and measurement of parameters are also subjects of a later chapter.

SUMMARY

Fault classification and modeling are the origin of digital testing. A very large number of fault classes can be defined, but only a few are presently considered worthy of attention. The single stuck-at fault is classically the beginning point for test coverage, but certain multiple stuck-at faults (in particular, bridging faults) have frequently been included in the fault list for years. As new technologies evolve, peculiar faults such as the s-a-Z MOS faults are added to the list.

Static faults give way to time dependent faults in the s-a-Z faults and in dynamic memory applications. Although elusive, parametric and delay faults can be defined and tested. Most of the exotic fault classes are restricted by economics to application only when the environment of the circuit necessitates it. The average manufacturing test for a logic circuit such as a pc board consists of a thorough coverage of single stuck-at faults, with some attention given to bridging faults on buses and other straightforward parallel structures. A thorough understanding of the classical stuck-at faults is an excellent beginning for test engineering.

REFERENCES

1. J. Markus, *Electronics Dictionary*, McGraw-Hill, New York, 1978.
2. *The TTL Data Book, Vol. 2*, Texas Instruments, Dallas, TX, 1985, pp. 3–4.
3. F. Barson, "Emitter Collector Shorts in Bipolar Devices," *IEEE J. Solid-State Circuits*, **SC-11**(4), 505–510 (Aug. 1976).

4. *The Semiconductor Library, Vol. IV, MECL Integrated Circuits*, Motorola, Phoenix, AZ, 1974.

5. J. P. Roth, "Diagnosis of Automated Failures: A Calculus and a Method," *IBM J. Res. Dev.*, **10,** 278–291 (July 1966).

6. M. Breuer and A. D. Friedman, *Diagnosis and Design of Digital Systems*, Computer Science, Woodland Hills, CA, 1976.

7. R. Resnick and D. Halliday, *Physics*, Wiley, New York, 1966.

8. C. W. Cha, "A Testing Strategy for PLAs," in *Proc. 15th Des. Automat. Conf.*, IEEE, 1978, pp. 326–334.

9. R. L. Wadsack, "Technology Dependent Logic Faults," in *Proc. COMPCON '78*, IEEE, 1978, pp. 124–129.

10. Y. K. Malaiya and R. Narayanaswamy, "Modelling and Testing for Timing Faults in Synchronous Sequential Circuits," *IEEE Des. Test*, **1**(4), 62–74 (Nov. 1984).

CHAPTER TWO

DIGITAL TESTING OVERVIEW

The essential procedures of digital testing, and a few of the difficulties that may be encountered, can be seen in an ordinary incident. Consider a customer about to purchase the most pervasive digital electronic device in the modern world, the calculator. The customer selects one of the models from the showcase and asks to inspect it. The calculator is turned on. The clear buttom is pushed (usually several times just to be sure). Several of the functions are exercised using simplistic data so that the answers will be obvious. As long as the display is suitable and the feel of the keys is comfortable the only issue remaining is the price.

Although the thoroughness is woefully inadequate, the customer feels compelled to ''test'' the product before buying. The steps of the test were power up, initialize (clear), input test sequences, and observe expected responses. The last two steps were interleaved and continued until some level of confidence (in this case minimal) was established. The same procedure is used in industry and generally termed a functional test. The planning of each step and the determination of thoroughness are much more rigorous. This chapter will develop the procedure and strategies employed while leaving detail of complex methods for each step to later chapters.

The intention is to provide a background by describing the relationship of digital testing to its environment. The next section places testing in the more general manufacturing process, and the section titled ''Test Structure'' relates the various approaches to testing with the logic to be tested. Thus, we begin at the front door of the factory and refocus attention on continually smaller realms of the test engineer until the circuitry level is reached.

TEST PROCESS

The calculator example can be readily generalized as a sequential circuit and more generically as a finite state machine. In actual test engineering applications the circuit being tested may well be a finite state machine, but it will seldom be so well bounded. The circuit to be tested often constitutes only a portion of the final machine or system. Its boundaries are well defined in terms of physical limits, but not in functional terms. Other parts of the machine provide inputs or respond to outputs in order to complete its function. The circuits discussed will not be assumed to be complete in the sense of performing an obvious function.

The manipulation of circuits which may only be part of a working machine requires interfacing at the voltage and current level rather than at the button and display readout level used in the calculator shopping example. In addition, the number of test vectors applied to each circuit to be tested multiplied by the number of units tested in a typical manufacturing day precludes manual application. The test process utilizes ATE in one form or another in most instances. Even small operations use semiautomated programmable instruments for "bench testing." The process shown pictorially in Figure 2.1 represents an electronic subassembly manufacturing and test operation.

The components used in the subassembly process are assumed to be pretested, either in another department or by the vendor. The finished subassemblies are held in inventory until needed to assemble the complete machine, which will be tested again before it is delivered to the customer. The tests applied at each level in the manufacturing process do not test for the same defect set, although there may be some overlap. As discussed in Chapter 1 the fault sets to be tested are those which may have been created in the last manufacturing operation. The tradeoffs concerning placement of test stations and selection of testing criteria within the manufacturing flow are primarily economic decisions and will be discussed thoroughly in Chapter 12. Several aspects of the test environment will be examined here, and the manufacturing flow helps to focus on the reasons for these details.

Testing is typically employed on a high volume of product in the manufacturing environment. This text is written under the manufacturing environment assumption. Although the field repair or product qualification test environments have many of the same elements, manual test methods may be viable in the latter arenas due to lower volumes and broader schedules. ATE has dominated the manufacturing

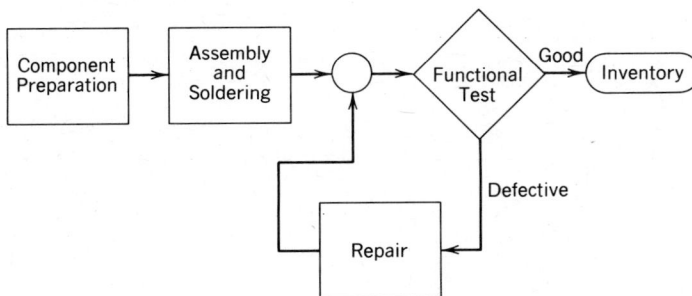

Figure 2.1. Typical manufacturing/test flow. Functional testing directs defective circuits to repair.

environment for several years, and it is the preparation of tests for automated application that will receive the majority of attention herein.

ATE is often complex and expensive, costing from $100,000 to $5 million for a single installation. The investment must be spread across a large volume of product if it is to yield an acceptable return, so the test times are kept short. Also, if possible, a number of different subassemblies should be tested on a single test station unless the volume of a single line is sufficient to keep the station busy. The result is an emphasis on speed and flexibility in ATE design. In turn, the flexibility places a burden on the test engineer to provide the specific hardware and software needed to adapt general-purpose equipment to specific circuits to be tested.

The process of creating such a test capability is shown in Figure 2.2. The process uses automated test generation (ATG) tools related to the computer-aided design (CAD) tools used to create the circuit which is the product and the test engineer's problem. In fact, the process inputs are ideally drawn directly from the information files which are the results of the CAD process, and often, the two processes are merged. The CAD circuit description file shown in the upper left-hand side of Figure 2.2 details the interconnection of devices in the subassembly and identifies the device according to some standard mnemonics. The test engineer accesses this information through the graphic representations of the documentation, while the test engineer's tools access the same information directly through computer channels.

The most common tool used for digital test generation is a program or system

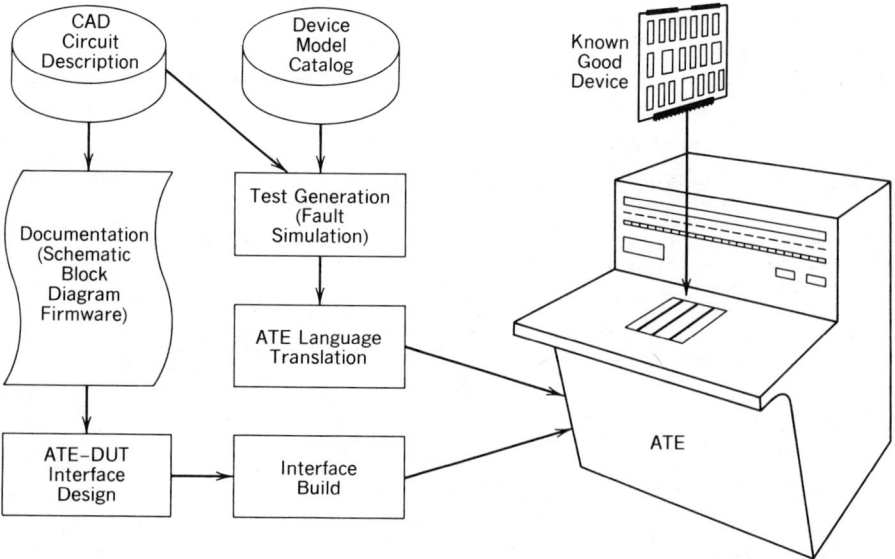

Figure 2.2. A flow diagram of the test development process. Both hardware and software resources contribute.

of programs which performs fault simulation. Fault simulation will be covered in detail in a separate chapter, but to enable a better understanding of the process, some of its theoretical background will be described in this chapter. For now, let it suffice to say that fault simulation enables the test engineer to grade the test sequence and provides guidance as to which areas of the circuit remain untested.

Modeling

The fault simulation and ATG programs used to generate tests need individual descriptions of each circuit component which describe the function of the components under all possible input conditions. For a NAND gate a simple truth table is adequate, but the description of complex components such as controllers is formidable. Complex components are modeled in one of three ways, depending on the requirements of the fault simulator and the economics of creating a model. In the simplest approach the circuit of the component is reverse engineered to yield a gate-level representation of an equivalent circuit to be used as a model. The equivalence is a matter of question since it is difficult to derive the exact circuit used in the component, and different vendors may use differing circuits in any case. Accuracy is usually limited to functional accuracy with only a reasonable representation of timing characteristics.

An intermediate level of modeling uses other models of MSI or LSI devices already in the library to build the equivalent circuit. This technique is sometimes called "macro" modeling. The amount of labor involved in creating the model is greatly reduced since the engineer is removed to the point of thinking in familiar functions such as count, shift, and so on. The accuracy of the model can suffer, however, since the intermediate devices usually implement extra functions which may not be required. Postprocessing of the model by a computer program can reduce this problem by removing circuitry which is not used in a given instantiation.

The highest level of modeling is a direct representation of the functions of the component in the language of the fault simulation program. Register transfer language (RTL) representations are often employed as an intermediate form to improve the user friendliness of the system to the test engineer. TML™ (Teradyne Modeling Language) is an example of such a modeling tool. It is sold in concert with the LASAR™ test generation system.

Test Generation

Information from the model catalog and the CAD circuit description files is merged to form a complete model of the circuit for fault simulation. As input vectors are provided, the model is used to predict the correct outputs and to grade the vectors by reporting the faults which are detected by the input sequence. Additional information to be utilized during test operation, such as the expected state of all circuit nodes during each test step and the output differences which result when a specific fault is first detected, is recorded as test generation progresses. At this point the method of deriving input vectors for test and the details of test generation cannot be covered in detail. An individual chapter has been devoted to each of these important areas.

Actually, the test generation block is a nest of several blocks similar to that in Figure 2.3. The fault simulation path through the outer block is usually the most time consuming, but the other operations may be required for some circuits. The partitioning step directs the test approach to utilize the most appropriate method on each functional feature of the circuit and may divide the circuit into subcircuits if the circuit size is too large for efficient fault simulation. The test program is then compiled from the results of each section and a shell or preamble which sets up the correct power conditions and defines the interface with the ATE.

No matter how the input vectors and resulting output vectors are derived, the format of their representation will probably not match the language of the test equipment. Test generation tools are as general purpose as possible, and so are ATE installations. A translator is therefore required to format the test generation output to be properly interpreted by the ATE input. In addition, aspects of the test which may not be provided by the strictly digital test generation system must be added to complete the test. These include power supply voltages and currents, timing, and parameters of the logic interface such as which voltage ranges represent a logic 1 and which a logic 0.

Device-Under-Test Interface

The physical aspects of adapting the device under test (DUT) to the test system (ATE) complete the preparation of test capability. The DUT interface adapter serves several roles. It physically supports or restrains the circuit board, IC, or

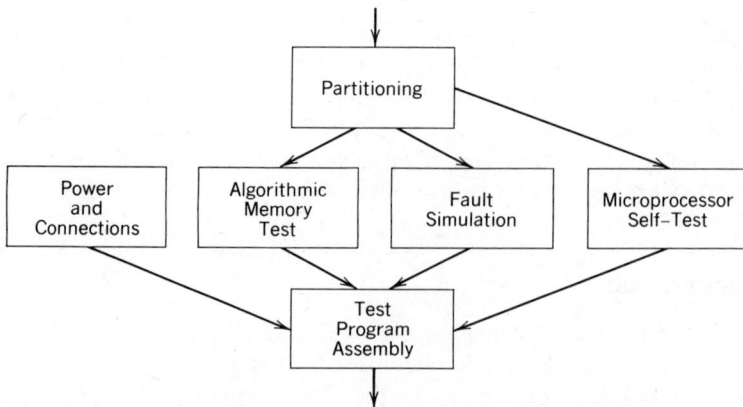

Figure 2.3. Details of the test generation core process. Several alternatives are available in creating a test.

module that is to be tested. It connects the plugs, jacks, and test points of the DUT to the ATE power supplies, logic drivers, and logic receivers. If necessary, the interface adapter provides cooling or physical stimulus to the DUT.

Signal preconditioning and line termination are, however, the greatest concern to the test engineer when the interface adapter is designed. CAD database information about the physical characteristics of the circuit interfaces must be matched to the ATE capability and properly conditioned or loaded to accurately and repeatably register the correct logic state when the corresponding test step is applied. Unreliable results in the test application due to noise, crosstalk, or distortion in the interface negate the value of a carefully prepared test sequence. Specific problems and solutions will be discussed in Chapter 6.

Proofing

Throughout the lengthy procedure of test generation there are numerous opportunities for error: model inaccuracies, interface wiring errors, programming mistakes, and inaccuracies in the tools used. In order to verify the correctness of the test a known good board (KGB) is tested. The chances of an error which would cause a good circuit to fail are greater than the chances of missing a fault, so the KGB testing approach to test verification is usually sufficient. In addition, the KGB test verifies the physical aspects such as connector polarity in the interface and voltage margins for the ATE programming.

In cases where the test generation method is in question or in cases where the reliability and accuracy of the test and its associated diagnostic messages are important, fault insertion may be performed as part of proofing. A sample of the faults which are to be detected is randomly selected and simulated by modifying the KGB circuitry. Some faults can be easily inserted by opening a circuit path, while others, such as shorting an output, must be applied with care to avoid damaging too much of the KGB circuitry. For most circuits the total number of possible faults is so great that it is difficult to insert a statistically significant sample. The fault insertion procedure is therefore giving way to more theoretical studies of the test generation method. Confidence in the correctness of the method must be established for acceptance of tests since meaningful verification of the results is beyond human ability.

Long-term statistics of the field failures which occur after subassemblies have been tested give the final measure of success. The results must be viewed as the outcome of a total test program and not just the last test performed on a product before shipping. Although component, subassembly, and final product tests may have much in common, they are seldom identical. Fault classes are introduced during assembly, leading to a need for differing tests. In contrast, some test procedures are impossible or impractical on larger circuits, but may be useful on components. Chapter 12 will cover specific issues of the distribution of test capability.

TEST STRUCTURE

The problem of deriving a test for digital circuits is actually a set of problems which depends upon the elements of circuit structure present in the circuit under consideration. Present test technology has not resulted in a cure-all procedure for generating tests for all circuits. Test strategy which results in an effective test for memory arrays has little meaning when applied to sequential logic circuits. VLSI functional blocks must be treated functionally, while combinational gate arrays can be tested pseudorandomly. Before beginning test generation, a process of classification and, if possible, partitioning of the circuitry should be performed to divide the problem into subcircuits for specific test technique application.

Partitioning and Strategy

The division should be along functional lines whenever possible and along lines of physical constraint whenever necessary. For example, the first cut at organizing a test for a single-board computer is to separate the test generation for the on-board RAM and ROM from the processor test. The second order of priority may be to divide the RAM test into segments based on the addressing capability of the particular ATE to be used. The techniques for partitioning a circuit for test are independent of the reason for the effort. The scheme that will be described will begin at the highest level of structure for the typical PC board circuit and proceed toward the gate level. Only one or two of the methods for partitioning a circuit shown will be applied in the typical case, but the range of techniques shown is intended to exemplify most cases.

The test approach can best be planned by examining the block diagram of the circuit to be tested. The functional description and flowchart are extremely helpful for heavily sequential circuits. The structure of circuits constructed of standard commercially available ICs may be obvious from the circuit layout and parts list, but the increasing popularity of application-specific ICs (ASICs) is making that case the exception. The single-board computer block diagram in Figure 2.4 is an example of an easily partitioned circuit.

Major blocks can be immediately identified (microprocessor, memory manager, RAM (3), ROM, I/O [Serial, Parallel, and direct memory access (DMA)], and clock generator). A table of test approaches versus blocks (Table 2.1) regroups the blocks into partitions for test.

The success in applying the chosen test approach to each block depends on the degree of independent access afforded by the circuit and the availability of the test technique in the test environment. A secondary list of test methods should be kept in mind. The self-test chosen for microprocessor testing is not available for most examples popular in the early 1980s, but may become available as ASICs rise in popularity and production costs necessitate the test at the IC manufacturing level. A backup plan would place the microprocessor and probably the memory manager in the functional logic test category. Each test method will be discussed in detail

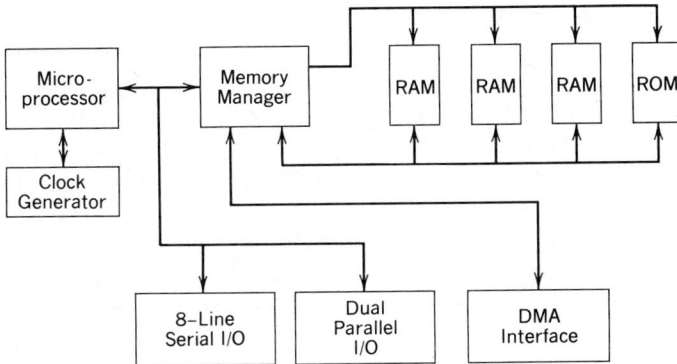

Figure 2.4. A single-board microcomputer diagram can be divided into functional blocks to guide the test approach.

in later chapters. The process of interest here is the separation and control of specific circuit partitions.

Access to the inputs and outputs of a given partition is critical to testing it separately from the remaining circuitry. In the single-board computer example the buses which connect the partitions form natural access channels as well as demarcation lines between blocks. If the buses are available to the test equipment through interface connectors or test points, the only remaining concern is gaining control of the blocks to assure that there is no contention for bus use during the test. The microprocessor will be placed in an interrupted state, which turns off its interface drivers, and other ICs on the buses will be treated likewise. Each partition can be turned on while it is tested and returned to the off state while testing the remaining partitions. Under this assumption, the memory address and data buses could be connected to the test equipment and a set of write and read commands could be applied to exercise the RAM arrays with checkerboard data, address patterns, and so on. ROM addresses could be applied sequentially while the outputs are compared to a known good firmware listing.

TABLE 2.1. Alternative Test Strategies

Primary Test Approach	Block
Algorithmic memory tests	RAMS
Functional logic test	Serial I/O Parallel I/O DMA I/O Clock generation
Self-test	Microprocessor Memory manager
Learned pattern	ROM

When direct access is not provided, the circuits between the interface connectors and the partition under test must be rendered transparent to allow effective partitioning. Transparency can be either combinational or sequential. If the circuit can be manipulated to a mode in which its data outputs are directly affected in a predictable state for state relationship with its data inputs, the circuit is combinationally transparent. A multiplexer or an adder between the interface and the partition can be made combinationally transparent if their control signals can be manipulated. If a sequence of input events is required to create a given output from an intervening circuit block, but all necessary outputs can be created, the block is sequentially transparent. A counter or shift register is an example of a sequentially transparent circuit blocks.

Test generation under these circumstances is a matter of developing the test sequence for the partition to be tested and then transforming the test input and output vectors to reflect the effect of the transparent circuits through which they must pass. In the case of combinational transparency the transformation is a direct one, such as the inversion of the signals by a buffer. For sequential transparency the transformation is sometimes a repetitive application of a control loop with the original test sequence imbedded within it. The procedure may, however, be non-linear and quite complex. Accessing a circuit through a protocol I/O port is likely to be a nonlinear process, as control modes and data modes are traversed to achieve data transfer.

In the example of Figure 2.4 the DMA interface could most probably be made combinationally transparent or at least linear sequentially transparent with respect to the memory partition. Input vectors could be communicated to the address lines with perhaps a single clock pulse on an input latch, and the data lines could be accessed by enabling a bidirectional buffer. Accessing the memory through a serial I/O port would constitute sequential transparency since each input or output vector would need to be encased in the string of events which selects port direction and clocks in (or out) data. Obviously, the latter would not be preferred over the former since the serial port access would result in many times the test sequence length and would require considerable mental acumen to accomplish.

The degree of ease with which sensible partitions can be identified, isolated, and controlled in a circuit is the first measure of its testability. Microscopic measures of testability such as the controllability or observability of each circuit node are similar concepts on a more intricate level. Chapter 11 will discuss both levels in more detail. For the present, consider that in a practical sense testability of a circuit depends on the access afforded the test engineer both in the hardware as circuit paths and in the information accompanying the circuit. If the information about the circuit's functions is complete and easily understandable, the partitioning and test scheme selection will be straightforward and as effective as the circuitry allows.

Register-Level Analysis

During partitioning (or block diagram analysis if only one test approach applies), elements of the circuit below functional block level were examined. The buffers, shift registers, and other similar elements which had to be considered while determining if partitions could be accessed are elements at the register-level. Before proceeding to allocate test techniques which operate at the basic gate level, the register level structure should be examined for insight into possible test strategies.

Fault set definition is often fixed at the register level when only the "chip pin" faults are selected. These faults include the stuck-at faults on IC package interface connections and sometimes the intercircuit shorts between adjacent package pins. These faults are a more realistic set than the gate-level faults if the test is intended to detect manufacturing defects on pc boards. Chapter 1 covered the alternatives from a physical standpoint.

The register level is the level where data paths and control lines can be defined separately and seen to interact. RTLs are methods of describing structure and function at this level. Tests can be generated directly at the register level in algorithmic fashion similarly to the way in which software testing is performed. If a register has several modes of operation (shift left, clear, load, and decrement), a test which exercises these modes with appropriate data on the data path will detect the faults associated with the circuit. The problem arises in determining the confidence of the test sequence since RTL compilers and simulators available at present do not analyze potential faults. Nevertheless, the thinking which accompanies manual test generation graded by gate level simulation or other methods is usually carried out at the register level.

Register-level test generation can be carried out manually for regular logic devices such as memories, but most logic requires analysis at the gate level or at least a mixed treatment. Any logic circuit can be treated at the gate level, but regular structures are more economically treated as registers or block functional devices. Gate level analysis has a rich history since the simpler circuits of a dozen years ago were based on nearly the same logic gates used today. The theoretical test approaches of initialization, control, and observation apply to gate-level circuitry in general. Even though the details are often hidden in fault simulation software, the concepts in the next few sections are the basis of most test generation systems.

Gate Level Analysis

At the gate level the reason for partitioning is seldom related to the function of the various parts of the circuit. Instead, the desire is to allow the sheer number of gates which must be considered at each test generation step to be reduced. If simulation or other computer-aided test generation is to be employed, the storage capacity of the computer or other resources may be too small to manipulate the entire circuit at once. If a circuit contains 10,000 gates, the fault simulation software will need several times that much memory to describe the connectivity and state of the circuit as well as the possible faults associated. 256K words of memory used for a 10,000-gate circuit would not be unusual. Partitioning the circuit makes the problem fit in the allotted space, although multiple passes through the program will be needed. After the results have been obtained, they must be fit together to represent the entire circuit for the test.

The partitioning and reassembly of circuits at the gate level is often automated, using a procedure as follows. Each primary output of the circuit is examined in turn, and a file of gates which are connected directly or indirectly to the output in question is constructed. The subcircuit made of these gates is a cone, and although each output has a cone, the cones often overlap. That is, a given input may affect more than one output. Figure 2.5 shows a cloud of logic gates whose outputs appear at the bottom and whose inputs are shown at the top.

The two shaded cones are independent in that they share no inputs. The cone between them shares inputs with both. Ideally, independent cones can be found, allowing manipulation of a cone for test generation without concern for the surrounding logic. In most circuits there is a high degree of interaction, and signals such as clocks or clear lines are shared by almost every cone. Additional rules are applied which artificially identify cones by restricting the state of surrounding logic during test generation for a given cone. For example, if a multiplexer is held in a single select state, only one of its input sets need be included in the cone. The others cannot influence the output. The necessary state of the multiplexer controls is a boundary condition for the cone and must be attainable independently of the other cone inputs.

After cones have been sorted from the logic, they can be combined to make partitions of a target size optimized for test generation. Efficiency of the partitioning and of the test generation is improved if partitions are chosen using cones with maximum overlap. Boundary conditions are a prime consideration in combining cones since cones with conflicting boundary conditions cannot be placed in the same partition. Furthermore, cones with similar boundary conditions should be combined as often as possible to reduce the number of boundary conditions which apply at the partition level. The ideal result leaves evenly sized partitions of logic which have distinct boundary conditions and minimal overlap.

If all partitions are independent of one another, the test sequences developed for each partition can be applied to the circuit simultaneously. When partitions overlap, and particularly when the boundary conditions in the overlapping area conflict, the test sequences must be applied to one partition and then to the other.

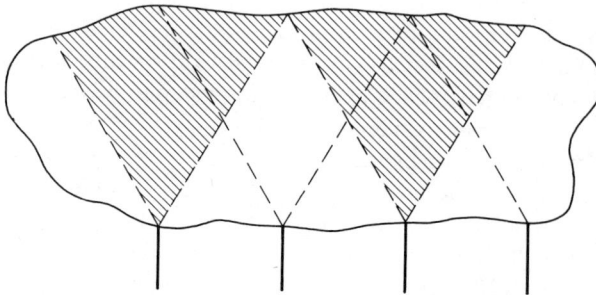

Figure 2.5. An abstract representation of gate-level partitioning with cones of influence for several primary outputs.

In the worst case each set of test vectors would be applied while the rest of the circuitry not in the partition for which the vectors were developed is held in an appropriate state to meet the boundary conditions. The overall test sequence generated using partitioning is nearly always longer than the test set developed for the entire circuit taken as one.

Now that the circuit has been analyzed from block level through register level and into gate level, the test generation solutions most efficient in each case can be brought to bear. Before detailing the method used to generate tests in practice, the theoretical basis for test generation in gate-level circuits will be discussed. The concepts carry through to automated methods particularly well.

Although the concepts of initialization, control, and observation which will be covered next apply to any circuit, it is helpful to consider circuits made of gates or MSI devices such as counters and latches. The examples are circuits which might be found in a larger circuit implemented as a single large scale IC or as a pc board. These basic circuit analysis and test techniques apply most clearly to such circuits, and extensions are usually attempted on larger circuits only with the aid of computer programs.

INITIALIZATION

Although it is certainly possible to perform useful functions with purely combinational logic circuits, many circuits of interest contain sequential elements. Flip-flops, memories, latches, counters, and registers are commonly included in ICs and pc boards. Unless the designer makes special provision for a ''power up master clear circuit'' to preset the value of each storage element in the circuit, the state of some parts of the circuit will be unknown when the power is first turned on. A power up master clear is infeasible in memory arrays, and is often neglected in other areas if the system outside the circuit boundaries is intended to provide values to the storage elements before using their outputs. The test engineer must usually consider the circuit as an entity separate from its supporting system.

The task is to assign a known value to each storage element in the circuit. There are two ways to accomplish this end. The first is to force a given value or set of values into each storage element, and the second is to determine what value each storage element has assumed by chance after power up. Determining the circuit values is seldom sufficient in itself. Beginning circuit values are critical to the test sequences which will follow initialization, therefore, a particular circuit state, rather than merely a known one, is the real goal. The determined state may be used to manipulate the circuit to the correct starting state, as will be demonstrated later.

Synchronizing Sequences

If the circuit is simple and the designer is thoughtful, all storage elements will be asynchronously resettable either automatically on power up or through a single pattern of input signals (direct initialization). Otherwise, it will be necessary to apply a sequence of input vectors (a set of input values applied in a single step is called a vector). Such a sequence which assures that a circuit is left in a specific state no matter where it started is called a synchronizing sequence. The circuit in Figure 2.6 contains both directly and indirectly initializable elements.

Because the 4-bit counter's clear line is accessible at the circuit boundary, the internal storage and the corresponding output values associated with the counter become known a short propagation time after the inputs for step 1 are applied. Since the lower two bits of the counter (Q_0 and Q_1) serve as address bits for the addressable latch and the write enable (WE) is active in the first vector, the output of the latch also becomes known. The circuit is not initialized at this time, however. Only one of the four storage elements within the latch has been written. When the counter is advanced by cycling the clock as shown in steps 2 and 3, the latch output becomes unknown, reflecting the unknown value of cell 1. Repeated toggling of the WE alternated by clocking the counter writes all four cells, and the circuit becomes known. The sequence could be shortened by writing the same data in all four cells. This would eliminate the need for returning the WE to the inactive state during counter clocking since the address hazards created by the counter transition would not result in incorrect data being written into adjacent cells. The sequence shown is safer and more general in nature.

After the circuit in Figure 2.6 has been powered up, but before the first vector is applied, there are 16 possible states for the counter and 16 possible states for the addressable latch for a total of 256 possible starting states. After the first vector the number of possible circuit states has been reduced to 7 since the counter and one of the latch bits are known. At the end of the synchronization sequence, a properly constructed version of the circuit with no defective parts will be in only one possible state, and the test can begin. In fact, the test may already by partially complete if the outputs of those parts of the circuit which became known early in the sequence were observed and compared to the expected values. It is not nec-

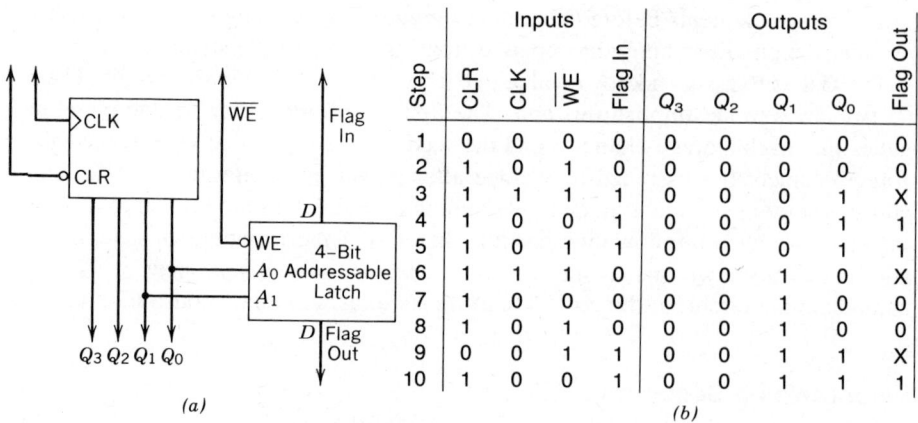

	Inputs				Outputs				
Step	CLR	CLK	$\overline{WE}$	Flag In	Q_3	Q_2	Q_1	Q_0	Flag Out
1	0	0	0	0	0	0	0	0	0
2	1	0	1	0	0	0	0	0	0
3	1	1	1	1	0	0	0	1	X
4	1	0	0	1	0	0	0	1	1
5	1	0	1	1	0	0	0	1	1
6	1	1	1	0	0	0	1	0	X
7	1	0	0	0	0	0	1	0	0
8	1	0	1	0	0	0	1	0	0
9	0	0	1	1	0	0	1	1	X
10	1	0	0	1	0	0	1	1	1

(a)   (b)

Figure 2.6. A simple sequential circuit which requires a relatively complex synchronization sequence. (a) Example circuit. (b) Initialization sequence.

essary for the entire state of a circuit to be known before valid tests can be performed, but valid tests cannot be performed on a portion of the circuit which is unknown.

Homing Sequences

A homing sequence differs from a synchronizing sequence in that utilizes the observed outputs of the circuit rather than blindly applying a set of input vectors. A simple example of a homing sequence acting on the circuit of Figure 2.6 would clock the counter repeatedly without the clear or latch WE being active. As the counter is clocked its outputs and the output of the latch are recorded, producing a table of the circuit states. After four count cycles have been completed, the state of the circuit is known. This is an example of a fixed homing sequence. The adjective "fixed" refers to the fact that the sequence of inputs is independent of the outputs of the previous steps. Only the endpoint of the sequence is decided by the output comparison. As mentioned earlier the knowledge of which of the 256 possible states corresponds to the circuit at hand is of limited use unless a battery of 256 individual test sequences, each starting at a different state, is available to proceed.

The "adaptive" homing sequence is more often used to determine the state of a circuit, as it can also manipulate the circuit into a desired state. In the adaptive homing sequence any or all of the input vectors may depend on the output values observed previously. The simplest example is a flip-flop connected as a frequency divider (Figure 2.7).

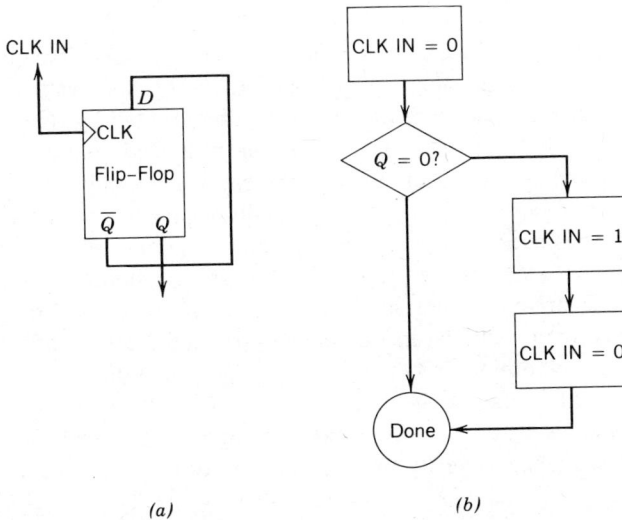

(a)   *(b)*

Figure 2.7. A single flip-flop connected as a frequency divider must be initialized by an adaptive homing sequence. (*a*) Frequency divider. (*b*) Flowchart.

The first step of the sequence is to set the clock to a logic 0 and observe the output. If the output is not a 0, then the clock is toggled. If the output is a 0, then nothing need be done. Although this example is trivial, the principle can be readily adapted to any circuit which cycles through all possible states (such as a counter).

When the behavior of the circuit is not cyclic or contains more than one cycle, a heuristic adaptive homing pattern is required. Information from the behavior of outputs when specific input vector sequences are applied is used to decide which of a number of predefined input sequences to apply next. The process is continued until the circuit converges on the desired state. Such schemes are seldom used in reality since they require real-time decision making during the test operation. In addition, the number of test vectors needed to cover all possible contingencies is much greater than the number required to use a fixed homing sequence or a synchronizing sequence.

It has become obvious by now that initialization of a circuit requires control of the circuit. The control or stimulus of the circuit is necessary throughout the test, and often the process of initialization may merge into testing as parts of the circuit take on known states. Initialization is treated separately because it deals with fundamentally different strategies than the remainder of the test sequence. Beginning with unknown values and attempting to establish a desired state is usually much more difficult than moving the circuit from one known state to another. It is theoretically possible to begin each control sequence as though the circuit were unknown and work toward each desired test state separately. The waste of information creates extremely long test sequences and generally impractical tests.

CONTROL (STIMULUS)

There are two phases to each test sequence for a circuit. First, the circuit must be driven to a particular state; then, it must be proven by observable evidence that the state was reached. The state of a circuit is determined by its outputs and its internal storage cells. During the discussion on initialization various methods for determining the final state of a circuit were given. The assumption involved is that the circuit is correctly implemented and thus will behave as predicted. The switching theory which is used to develop these approaches does not make provision for observing the final state of the circuit, it only proves that a correctly operating circuit would be driven to that state. The other half of testing the circuit is to manipulate the circuit such that its outputs will provide evidence of the state it was in at the end of the homing or synchronizing sequence. Whatever the strategy for selecting the circuit states to be tested or for physically executing the test, all digital functional tests are made up of sets of a sequence to establish a circuit state, followed by a sequence to verify the achievement of the state in question.

Random Patterns

Manipulation of the internal states of a circuit through the timely application of input vectors requires similar strategy whether the intent is to achieve an internal state or check that it was achieved. Combinational circuits (particularly small combinational circuits) can be manipulated successfully with casual approaches such as random application of inputs. Large numbers of input vectors will be required for even mediocre-sized circuits, and even then there is no precise way of knowing the thoroughness of the test. The 74LS181 used by Chin [1], as an example, was estimated to require 502 random test patterns to achieve 95% confidence of detecting all faults, while the manually derived 135 patterns shown to detect all faults were rated at only 54% confidence. If all possible input patterns were presented, the result would be 2^n test steps where n is the number of inputs to the circuit. An n-bit counter is often used for a test generator in such simple cases.

For very shallow sequential circuits, a similar scheme may be effective. Random inputs will generally give a better result in sequential cases than the application of all possible input vectors because such exhaustive testing does not repeat vectors, and repetition may be needed to activate sequential elements. In either case the number of vectors required to have a high probability of activating all circuit states is exponentially related to the sequential depth of the circuit. Even when all states are activated, the probability of having primary circuit outputs reflect the state achievement is also inversely proportional to the sequential depth in an exponential fashion.

The two-chip circuit of Figure 2.8 points out the effect. Due to the nature of the counter, the circuit actually has a sequential depth of 5. Arranging a test would be simple, but for illustrative purposes we first attempt to use the exhaustive technique. Eight vectors are applied in binary order (although any order could be chosen without appreciable improvement). Judicious selection of the input order has assigned the least significant bit to the clock, the middle bit to the latch gate, and the most significant bit to the clear. The first observation is that of the first four vectors only one results in a unique test. Assuming a logic 0 opens the latch gate, the first vector (000) will permit verification of the cleared state of the counter. The next three (001,010,011) are wasted since the counter is still held clear and does not respond to clocks. During the last four vectors, the clock will be toggled twice, but the latch will only be open to pass data during one of those counts. The entire 8 vectors have only verified 2 states of the counter's possible 16.

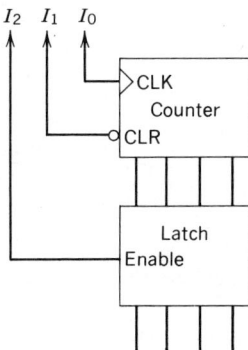

Figure 2.8. The counter in this two-chip circuit results in a surprisingly large sequential depth of logic.

In the example just given the solution is quite simple, and it is tempting to rearrange the input vectors, repeating a vector sequence when required. First, initialize the counter using the clear; then, return it to the inactive state. Now repeat the sequence of toggling the counter clock and opening and closing the latch 16 times, and the test is complete. Heuristic control sequences can be originated by humans for some circuits, but the general case of a circuit with hundreds of counters, latches, adders, and other functional blocks is mind boggling. A systematic approach which can be carried out by computers is necessary for complex circuit control. Any such technique which applies a strategy to achieve specific circuit states is termed deterministic.

Deterministic Patterns

Deterministic strategies involve changing the state of a circuit from its present state to some desired step. If the change cannot be accomplished in an input vector, an appropriate intermediate state is chosen, and the circuit is driven to that state before the process is applied again. This recursive methodology is ideal for computer problem solving since the task is thereby divided into simple repetitive subtasks. Choosing each intermediate state and maintaining the overall objective are the heuristic components of the approach. The simple repetitive subtasks are usually logic gate evaluations performed by truth table lookup or by functional algorithm solution. A commonly used strategy for connecting the subtasks together to achieve the state is called "backtrace".

The circuit illustrated in Figure 2.9 might be part of a greater circuit for which a test is being derived. Suppose that the circuit node number 13 is to be driven to the state of 1. The circuit connectivity description reveals that node 13 is the output of gate C, which is a NOR gate. The truth table corresponding to a NOR gate shows only one set of conditions which result in a 1 output; that is, both inputs must be logic 0. The circuit has been traced back one level.

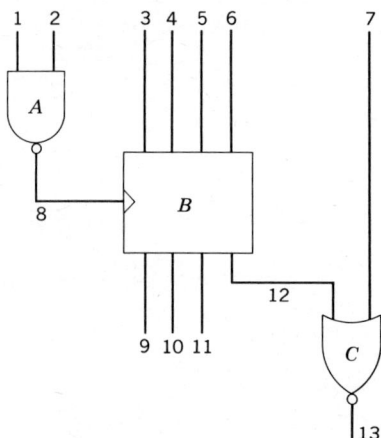

Figure 2.9. Sequential circuit diagram. Deterministic pattern generation backtraces from an output to inputs.

Assume that the node marked 7 is directly accessible. The value 0 will be placed in the input vector for that test step and circuit input. But the vector is not complete unless a check of the present state of the node 12 also reveals a 0. If not, the functional description or circuit model of the device labeled *B* will be consulted to determine what input sequence is necessary to achieve the subgoal of a 0 state for node 12. In this subtask the node 6 must be driven to a 0, and the node 8 must be driven to a 0 and then to a 1. The truth table of gate *A* is consulted, and the first vector is completed with a 0 placed in the node 6 input position and 1s placed in the positions for nodes 1 and 2. Then the next vector is formed by selecting node 1 or node 2 to be driven to a logic 0.

During the backtrace procedure, the present state of the circuit would be consulted at each node, and the procedure could be shortened if the present state were correct for a portion of the circuit. Conflicts may arise when reconvergent fan-out causes two branches of the circuit to be affected by the same node. Alternative choices from the truth tables which have already been examined may be necessary to resolve the conflict. Such a startover of the backtrace process does not constitute a forward trace unless a circuit node value is selected and the truth tables of a driven circuit are consulted to propagate the effect of the node.

Path Sensitization

Forward trace is usually used in combination with backtrace for the path sensitization procedure. Controlling the state of a circuit node is a process of finding a path from a directly controllable node such as a primary circuit input to the node in question and determining along the way which associated gate inputs are needed to enable a change in the primary input to affect the node. A simple sensitive path example is a single NAND gate. If a given input I is to affect the output, then all other inputs must be a 1. Under these conditions, if I has a value of 0, the output of the NAND will be a 1, and if I has a value of 1, the output will be a 0. When at least one of the other inputs is a 0, the input I will have no effect on the output, which will be a 1.

The process of path sensitization for a complex circuit calls for repeated application of the backtrace procedure. At each gate the inputs which do not lie on the path must be specified in order to sensitize the path. Each of these constraints must be achieved by backtracing from the gate input in question to primary inputs, specifying the circuit values along the way.

The forward trace portion of path sensitization is susceptible to reconvergent fan-out also. While a given path from a circuit node to another node is being sensitized, a second path between the same two nodes may also be sensitized if reconvergent fan-out exists. If the two paths have equal parity of inversions, the paths merely duplicate one another. In this case a change in the originating node value will propagate along both paths and result in the same value at the receiving node. When the number of inversions on one path is odd while the parity of inversions on the other is even, the effect of the originating node will cancel itself when it arrives by the two paths. The result is to render the overall path set insensitive.

Distinguishing Sequences

After the desired state has been achieved in the circuit under test, (CUT), the control strategy is to derive a sequence of inputs which will allow evidence of the circuit's state to be observed externally. When such a sequence is successful in conclusively identifying the state of the machine when the sequence was started, it is termed a distinguishing sequence. The necessary components of a simple test are now in place. A screening or "go–nogo" test for a digital circuit can be approached on a microscopic level at this point as a set of homing sequences which activate each node in the circuit to both its possible states interlaced with distinguishing sequences which allow detection of the states at primary circuit outputs.

Distinguishing sequences propagate the correct response from a correct rendition of the circuit, but the same wrong response may be gotten from several possible erroneous versions of the circuit. Troubleshooting circuitry is much easier if some diagnostic information is provided by the test. If the distinguishing sequence is also a machine identification sequence, the differing faulty versions of the circuit will yield distinct outputs for the test sequence. A machine identification sequence for some of the possible faulty circuits may not identify others, although their outputs will be distinguished from the good circuit. This results in a measure of test quality called diagnostic resolution. The ratio of the number of faults for which a distinguishing sequence exists in a test vector set to the total number of faults defined for the circuit is termed the comprehensiveness of the test.

Hazards

The preceding discussion was somewhat antiseptic in that the test environment was not brought into play. The application of signals was conveniently assumed to happen at the appropriate moment for the cause in mind. ATE combined with the interface adapter can add considerable delay and skew to the signals. In fact, some ATE architectures are arranged to change only one input at a time to avoid uncertainty about the order of application. When input vectors are applied "broadside" or simultaneously, the possibility of hazards, or transients, exists. Even properly designed circuitry may be susceptible to hazards under test conditions since the test sequence may not resemble any intended operating sequence.

The simplest example of a hazard is created by changing the inputs of a 2-input NAND from a $(0,1)$ state to a $(1,0)$ state in one step as shown in Figure 2.10. Since the likelihood of both changes being exactly simultaneous is small, the gate will momentarily pass through either the $(0,0)$ state or the $(1,1)$ state. In the former case the output will not change, but in the latter the output will try to achieve a 0 value for the duration of the transition overlap. A "glitch" will be produced if the transition is sufficiently skewed. The cause of the problem is termed a hazard, and the result is a glitch or transient.

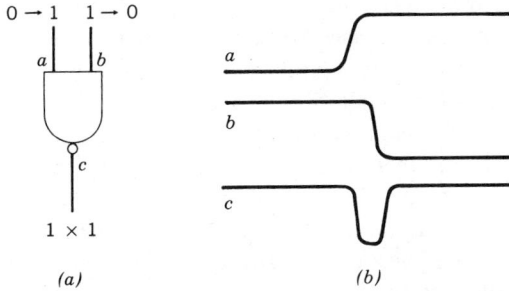

Figure 2.10. Input changes to even a simple gate create hazards when intermediate states are traversed. (*a*) NAND gate with hazard sequence. (*b*) Resulting transient.

If the line bearing the transient fans out, the transient will be amplified and spread through the net. As long as the net only feeds combinational circuits, the transient will die out and the circuit will resume the correct value. But if a clock input is encountered, the transient will cause a permanent side effect, and the circuit will be left in the wrong state after the transient has settled.

Changing only one input at a time reduces hazards, but the test sequence may become extremely long for complex circuits. Tuning the test equipment for simultaneity is impractical. When possible, the test engineer should follow intended operating sequences and observe the relative timing provided to avoid triggering hazards. Such careful procedure may be unnecessary if the test generation tools such as fault simulation are constructed to detect transients and warn the test engineer or reject the input sequence.

OBSERVATION (RESPONSE)

In previous discussions the results of the test were loosely referred to as evidence of the circuit's internal state. The procedure used to distinguish the good circuit from possible faulty ones can be modified somewhat if a particular method of observing the circuit behavior is assumed. The distinguishing sequences described above assume that only a few of the circuit's nodes are directly accessible during the test. These nodes were termed primary outputs. The physical restrictions of circuit construction are assumed to prevent easy access of other circuit nodes.

Physical Observation

In actual practice, several methods of observing the circuit are used, and several methods of comparing the expected results to the actual values are employed. For ICs, initial testing may employ microscopic probes which contact a few dozen selected points on the die or wafer. Final testing is even more restricted, normally to the package pins only. The functional test representation given above is fairly representative in both cases since thousands of circuit nodes with thousands of possible faults are examined through dozens of access points.

At the pc-board-level the scale of nodes and faults to be considered in the test is necessarily altered. To maintain the problem at a tenable level, only the nodes between IC packages and their associated faults may be considered. Only practicality and the massive number of circuit elements (gates or transistors) involved in a VLSI populated board prevent testing under gate-faulting assumptions. Theoretically, a very thorough test can be derived that tests each gate in its place, but the complexity of the test generation task and the length of the resulting test sequences usually lead to the IC boundary fault assumption.

If the pc board connectors intended to connect its circuitry to the remainder of the system are the only means of access during a test, the test is called a functional test. If an array of probes is utilized to contact nodes on the board not normally accessible to the remainder of the system in which the board is intended to function, the test is termed an "in-circuit" test. The precise details of physical connection and test vector application will be covered in Chapter 6.

Comparison

No matter how the circuit states are observed, they must be compared to some standard or expected values in order to perform a test. Simply noting that all the observable points assume logic 0 and logic 1 values at some time during the test proves very little about circuits of ordinary size in today's technology. Once the method of deriving a set of input vectors which constitute a test has been chosen, an appropriate method of checking the results can be picked to match. Table 2.2 below shows several techniques, their relative implementation cost, and their relative comprehensiveness. The nature of the circuits to which the technique might best apply is described in the last column.

The first method pairs the application of random values on the circuit's inputs with a simple check for the presence of both logic values at each of the circuit outputs at some time during the test sequence. Confidence in this test method for

TABLE 2.2. Qualitative Comparison of Test Techniques

Input Generation	Output Comparison	Cost	Comprehensiveness	Applicable Circuits
Random	Activity monitor	Low	Low	Small combinational
Pseudo-random	KGB	Medium	Low–medium	Intermediate combinational
Manual	Stored in-circuit	High	Medium–high	Intermediate sequential
Heuristic	Stored simulation	Very high	High	Complex sequential

any but the simplest combinational circuits is lacking since there is no monitoring of test completeness and no diagnostic information is provided. The second method is somewhat better.

Pseudorandom patterns which have been randomly generated are then stored and applied repeatably to the inputs of the CUT and a good circuit physical realization during each test pass. The KGB outputs may be compared in real time with those of the CUT, with differences indicating failure of the CUT. The comprehensiveness of the test sequence can be attained by inserting a statistically significant sample of faults into a circuit board and noting which faults produce differences in output compared to the KGB. Diagnostic data can be gathered by recording the string of values at probed nodes within the circuit of the KGB.

Several weaknesses are evident in this method, however. If the KGB cannot be initialized easily, the outputs will not be reliable for use in comparison. Also, if the comprehensiveness of the test is found to be lower than desired, the only recourse is to lengthen the sequence of pseudorandom patterns and repeat the rather time-consuming fault insertion routine. Although this method can be applied to sequential circuits, its effectiveness is reasonable only for combinational circuits of a medium complexity (less than 1000 gates).

The in-circuit test technique will be explained in a later chapter more thoroughly. It essentially allows the circuit to be treated as a disconnected collection of components or small subcircuits for testing purposes. Manual deterministic input vector sequences can be composed for each subcircuit, and the expected outputs can be manually determined and stored. This divide-and-conquer approach reduces the labor of test generation and improves the comprehensiveness of results, but the equipment necessary to apply and observe the test values is complex and expensive. In the final analysis the parts of the circuit are determined to be present and functioning individually, but the overall circuit has not been proven to operate correctly as a whole.

The most thorough (and expensive) method applied to digital circuits is the deterministic functional test. Input sequences and output results are calculated using computer simulations of the circuitry. The inputs, corresponding outputs, and internal node information helpful in circuit troubleshooting are captured during test generation. Sophisticated ATE applies the sequences of thousands of test vectors and monitors the test results before reporting success or pointing out specific failures. The completeness of the test is limited only to the accuracy of the simulation models and the patience of the test engineer. Sequential circuits may be tested rigorously, but any change in the circuit design usually requires a complete regeneration of the test.

Data Compaction

As the thoroughness of the test sequence or the complexity of the CUT increases, the number of data required to perform the test and troubleshoot the failures increases rapidly. The heuristic method listed last has data requirements that can mushroom rapidly to thousands of pages for a complex circuit. Complete enumeration of all values of all circuit nodes for each test step has given way to other schemes in the interest of portability of tests. If the outputs of the pseudorandom KGB technique are used to drive counters, a record of the number of transitions from 0 to 1 (or vice versa) can be kept. This yields a single number for each output of the circuit rather than a set of outputs for each test step. Intermediate data compaction can be accomplished by recording only the changes in circuit state from one test step to the next. These techniques and similar ones reduce the data storage needs, but at some loss of accuracy or readability of the information.

The information provided during the test generation process usually includes both go–nogo test data and diagnostic information. The ATE storage is not often large enough to waste, so the information is split into files for application at appropriate times. The input vectors and expected response vectors will be assembled into a matrix for rapid application to the connector pins of the CUT. The diagnostic data may be kept in a separate file or merely printed out for manual reference when a failure is encountered and the test mode changes from go–nogo to troubleshooting. Since a typical yield for go–nogo or screening tests is 70–90%, it is not economical to manipulate the diagnostic data for each application of the test. Further discussion of data structures used by ATE and the economics of testing will occur in later chapters.

TEST COMPLETENESS

Measurement of test completeness is a controversial task. The usual measures employed are similar to the percent figures mentioned earlier. Completeness is described as a percentage fraction detected of a specific class of faults such as single, IC-package pin-level, stuck-at faults. Comparing two tests without understanding the fault grading method is fruitless. Fault collapsing may be employed to eliminate equivalent faults on one but not the other. One test grade may only count gate output stuck-at faults, while the other counts both input and output faults. Less standard fault sets such as adjacent line short faults may change the statistics. It is not at all correct to conclude that if a given fault class is not measured during the test it is not covered by the test.

The structure of the circuit may preclude testing some faults, and the characteristics of the simulator may not acknowledge others. Reconvergent fan-out is an instance often encountered in which gates of the circuit may not be testable. A test engineer who has taken the time to understand the circumstances of the test may sometimes suggest a modification of the fault grade based on known occurrences of one or more of these situations. As in many endeavors, the last 10% of the test often requires 90% of the effort, so a thorough understanding of what constitutes an acceptable test and how the measurement is to be made is important.

The ultimate success of a test is determined by the failure rates experienced by the users of the tested circuits. In the case of the calculator the customer would be unhappy if he discovered that 10% of the functions did not work. In the usual subassembly test situation the customer is a unit test department, and the economics of troubleshooting board or chip failures in a unit will determine the threshold of an acceptable incoming failure rate.

TESTABILITY

The complexity of the circuits to be tested and difficult-to-test circuit designs have been alluded to as a major consideration in the test effort. A feedback path to the design department called "design for testability" (DFT) has been shown to be effective in reducing test costs and improving manufacturability of circuit assemblies. A complete chapter is devoted to explanation of the various schemes which have been proposed or implemented. The general idea is to select circuit structures which avoid test weaknesses or to provide circuit structures, for test purposes only, which circumvent tedious control or observation sequences. The addition of auxiliary connections (called test points) to the circuit is the most common testability feature.

Testability is usually subjective since the tools to be used for test generation and performance vary and each has a different set of requirements. Attempts at objective measures of testability center around counting the number of gates which must be passed through to reach a given fault from a primary output (combinational controllability) or to traverse from a fault to a primary output (combinational observability). If sequential circuitry is involved, the number of clock transitions necessary to open either path may be included in the calculations to give sequential controllability and observability. The value of such measurement is relative in that it serves to point out problem areas in the circuit. Attempted use of these measurements in estimating test cost has had only limited success, usually only when the measurement technique and the test generation technique were carefully matched.

Advanced testability design regimens have been used with success when the design could be carefully controlled. Techniques such as LSSD (IBM Corporation's level-sensitive scan design technique) limit the use of sequential elements in a logic design to a specific latch which is converted to a shift register during the test to allow serial access to all internal circuit states. Essentially, the circuit can be partitioned into combinational blocks of logic between ranks of scan latches. Several good automated techniques are known for generating tests in combinational circuits, but sequential techniques are lacking. These and other DFT techniques make use of test generation characteristics to minimize the overall design–manufacturing–test cost.

SUMMARY

Digital testing has been outlined as a process, and the results have been shown to fit into the manufacturing process. Circuit analysis establishing the relationship of blocks or partitions of circuitry to the intended test methods is the first step in designing a test for digital electronic subassemblies. The information gathered during this phase is also needed to generate tests for many of the subcircuit blocks. Schematic and component description data feed the fault simulation programs which model the behavior of the circuit and predict the circuit outputs. These programs act on the basic rules of logic and the principles of path sensitization to determine which faults will be detected by a given test sequence. Using these tools, the test engineer or perhaps an ATG program exercises initialization, control, and observation until a sufficiently thorough test sequence has been attained. The test output is translated and applied through ATE to a KGB to verify its correctness.

REFERENCE

1.   C. K. Chin and E. J. McCluskey, "Test Length for Pseudo Random Testing," in *Proc.* 1985 *Int. Test Conf.*, IEEE Comput. Soc., pp. 94–99.

CHAPTER THREE

STIMULUS GENERATION

As the previous chapters have introduced, the underlying concept of a digital test is the selection of a set of input vectors which reveal a difference between good circuits and faulty ones. The input vectors are sets of logic values applied to the primary inputs of the CUT. Each set is applied approximately simultaneously and is intended to stimulate the circuit toward some goal logic state. If all primary inputs are given the value 0, that is an input or stimulus vector. If three primary inputs are then driven to a 1 in concert, that is another stimulus vector, and so on. The entire sequence of changes makes up a test when conducted in the given order.

The strategy for selecting test vectors is complex and dependent upon many factors, including circuit complexity, sequential elements present, test application method, DFT features, and practical limits of cost and time. A number of algorithms will be explored, followed by a comparison of applications and performance. Understanding the techniques may be of great importance if a choice of tools is available for test generation. Even if the stimulus generation method is imposed by the situation, a knowledge of its strengths and weaknesses will help maximize the efficiency of use.

FAULT ACTIVATION

Whether stimuli are to be generated by an automated algorithm or manually, the goal is to activate faults and then to establish a path of propagation for the fault effect to reach an observable point (primary output). Fault activation is achieved when the logic state of the faulty circuit differs from the logic state of the good circuit at some point. In the case of single stuck-at faults, the activation of an s-a-1 fault is achieved when the node corresponding to the fault location is driven to a 0 state. The good circuit will achieve the 0, while the faulty circuit is s-a-1, resulting in a difference to be propagated as a D^* (not D) in the notation adopted earlier.

Activation of stuck-at faults is usually restricted to single faults, but a multiple fault assumption does not change the procedure significantly. The only additional aspect is the coordination of fault activation for two or more single faults such that they are active simultaneously and their effects can be propagated together. Conflicts in control and observation path sensitization are more prevalent under the multiple fault assumption, but the techniques for resolution are the same.

Bridging faults share some of the activation difficulties of multiple faults and are more often included in the desired fault set. When two distinct circuit paths are bridged (shorted together), the fault is active only when they have differing values. If path A and path B have either a 1 or a 0 together, the faulty circuit paths A' and B' will also have the same values and the fault will not have been activated. Path A must have a value of 0 while path B has a value of 1, or vice versa, to activate the fault. In either case the faulty circuit paths A' and B' will both have the value 0 (for TTL), owing to the action of the "wired-AND" gate present due to the bridge. In the D notation, one of the paths will have a logic value of 0, and the other will have a value of D. D denotes a value of 1 in the good circuit and a value of 0 in the faulted circuit.

Fault activation for non-stuck-at faults can become a procedure involving time as well as logic values. The test for sense amplifier recovery in a memory chip involves writing a column and immediately reading the same column. Other memory faults exemplify the need for a thorough analysis of the fault effect before attempting to create a test. The physical arrangement of memory cells, for example, dictates which pattern of accesses will activate "nearest neighbor disturb faults." Although the concepts of fault activation and test generation are the same, the details of memory tests are sufficiently different from those of logic circuits that a separate chapter will discuss memory testing. The remainder of this chapter will be restricted to general logic (sometimes called random logic) test generation.

Fault Selection

Before activating faults and beginning a test, the fault set is usually enumerated as a list of unique faults. The faults to be tested are selected from the list, and if the test is successful, they are removed to a second list as detected faults. The ratio of faults on the detected list divided by the original total faults list is the fault coverage or comprehensiveness of the test sequence and is usually expressed as a percent.

Selection of the next fault from the list and the ordering of the list are seemingly minor details, but they can affect the credibility of the statistics and the efficiency of the test generation algorithm. Faults on the list are identified by some location in the circuit and the fault condition such as "gate A06 output 3 s-a-0." This refers to the gate which is part of the package located at grid reference A06 and which drives pin 3 of that package. The fault condition is an s-a-0 on that output. A natural method of creating a list of similarly named faults would be to list all the faults associated with the package at A01 and then proceed to the next grid location, and so on. Unfortunately, if the gates of package A01 are involved in a difficult-to-test portion of the circuit, considerable time and effort may be expended trying to test the first several faults in the list which are all associated with A01 before moving on to easier faults. The list can be randomized to reduce this effect, but ordering can also work to benefit in some cases. A better treatment is to modify the selection of the next fault from the list based on the success of the last attempt, jumping ahead by a dozen or so to select new territory. When the end of the list is reached, the selection can start at the top again so that faults are not permanently skipped.

Random selection from the list may also be employed if the test generation effort is only intended to be a sample. Sampling may be used to estimate the testability and expected cost of testing a circuit. Derivation of a quick low-cost test to verify the test approach and equipment to be used could also make efficient use of random fault selection. In very large circuits, only a sample of the faults will be considered, due to computing or time limitations, in the hope that a statistical sample will represent the test coverage of the remaining faults. Some faults may be activated while the conditions are being arranged for others.

Backtracing

Achieving fault activation is a matter of selecting appropriate primary input values to propagate the desired logic condition to the site of the fault. The site and conditions are selected as above, and the necessary primary inputs are derived by working backwards (backtracing) through the circuit. Suppose the selected fault to be activated is the output of gate C in Figure 3.1 s-a-0. (The node is labeled f in the figure.)

The essential parts of the truth tables for the three gates in the circuit are shown in Figure 3.1a. Blanks may be taken as Don't Cares such that line 5, which refers to AND gate B, could be taken as $(0, 0) = 0$ or $(0, 1) = 0$. Either case is true.

	a	b	c	d	e	f	
1	0	0		0			
2	0	1		1			A
3	1	0		1			
4	1	1		0			
5		0			0		—
6			0		0		B
7		1	1		1		—
8				0		0	
9				0	0		C
10				1	1	1	—

(a)

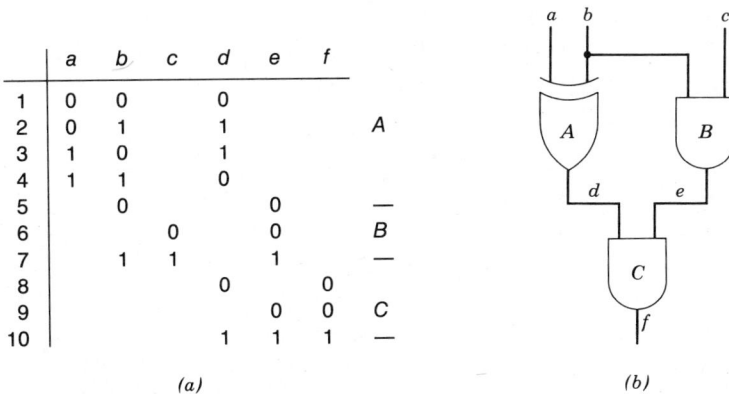

(b)

Figure 3.1. Circuit diagram and its combined truth table, known as a singular cover. (a) Singular cover. (b) Logic diagram.

This compacted table is termed the singular cover of the circuit. Backtracing will be used to select values from this table to guarantee a logic 1 on node f and activate the s-a-0 fault.

Beginning at column f of the table, note that only line 10 results in a value of 1 for node f. Since this is the necessary value for f and f is the output of an AND gate, it is reasonable that both d and e must be given a value of 1. Moving back along the circuit path for d, the gate A is encountered. Under the column for d in the singular cover section for A, two lines (2 and 3) are found to give d the value 1. Choosing line 3 temporarily will assign input a the value 1 and input b the value 0. Backtracing node e to the gate B brings line 7 of the cover into play. No other line will result in a 1 at node e. The value for b specified for the line is a 1, and this conflicts with the previously assigned value. In order to resolve the conflict the path from d must be backtraced with the added stipulation that a 1 must be assignable to input b. The result is the selection of line 2 conditions to satisfy the node $d = 1$ requirement. A backtrace from e will now be successful since line 7 can be selected from the cover. The inputs $a = 0$, $b = 1$, and $c = 1$ give a test for node f s-a-0.

In this simple example the choices to be made while backtracing could have been almost unconscious since they involved only three values and occurred at the primary input level. Even with such simple choices the possibility of a conflict was present. Most circuits involve many choices which set the backtracing off on particular routes which may have to be abandoned later if conflicts arise.

The circuit described by Figure 3.1 contains an example of reconvergent fan-out. Input b affects node f through two paths. One path passes through gate A and node d, while the other goes through gate B and node e. Both paths converge at gate C. If a circuit contains no reconvergent fan-out, there will be no conflicts during backtrace. Unfortunately, reconvergent fan-out is a common occurrence in useful circuits. It is not true, however, that reconvergence implies conflicts, only that it allows them to occur if the logic is so structured and the state selections happen to be made on one of the paths to conflict with later conditions.

The fault activated in the example of Figure 3.1 was on a primary output of the circuit, so a test was incidentally made when the fault was activated. In general, the fault will not be directly observable, and fault activation will not immediately result in a test.

PATH SENSITIZATION

Once the fault is activated, a path of propagation for the fault effect must be opened from the fault site to a primary circuit output. Since the faulted node bears a differing state than its equivalent in the good circuit, it is not sufficient to assure that either a logic 1 or a logic 0 will propagate. Both states must propagate to assure detection of the fault. The D notation used in Chapter 1 is appropriate for this purpose since both cases can be investigated simultaneously.

The adder circuit of Figure 3.2 will be used as a second fault activation example and a path sensitization example. Opening a path for fault propagation is termed "path sensitization."

Activating the fault s-a-0 at node g is the first task. Backtracing will be used to determine which inputs are necessary to assure a logic 1 at node g of a properly operating circuit. Line 22 of the singular cover shows the rather obvious requirement for a 0 at COUT. Line 21 further requires that nodes a, b, and c must be 0s. Examining lines 1, 2, 4, 5, 7, and 8 reveals that two of the three inputs must be 0 for a, b, and c to all be 0. Randomly, lines 1 and 2 are used, and inputs x and y are given the value of 0. In the good circuit, node g is now at logic 1, while the faulty circuit node g is s-a-0.

Only one path from node g to an output exists, and that path passes through gates H and J. Examining lines 25 and 26 in the singular cover table shows that if g is a 1, a 1 will appear at h in any case. On the other hand, if g is a 0, as in the stuck fault case, d must be a 0 for the stuck 0 to propagate. If d is a 1, the output of H is determined without reference to node h. This means that if d is 0, then the 1 from node g in the good circuit will cause a 1 on node h, while the faulty circuit s-a-0 value will cause a 0. The path through gate H is thus sensitized to the value of node g.

A second backtrace procedure is now instituted to find the necessary inputs which assure a 0 on node d. Lines 10, 11, and 12 of the singular cover suggest that a 0 on any of the inputs is sufficient. Since both x and y are already assumed to be 0, no change is required at this time. Path sensitization resumes from node h.

Lines 28 and 29 of the cover describe the output SUM in terms of the node h, and line 27 shows that if node e is a 0, the output will be determined without node h. If node e is a 1 and h is a 1, as in the good circuit, the output SUM will be a 1. If node e is a 1 and h is a 0, as was propagated by the faulty circuit, the output will be a 0. Thus, the path through gate J is sensitized to node h when $e = 1$.

The final backtrace looks for the input conditions necessary to assert a 1 on node e. In the cover table, lines 14, 15, and 16 reveal that at least one of the inputs must be a 1 to achieve the goal. Since both x and y were previously set to 0, the only choice which would not require beginning again is to set CIN to a 1. Since CIN was not previously assigned, no conflicts have been created, and since the activated fault has a sensitive path to the output SUM, the procedure has been successful in finding a test for node g s-a-0.

The manual method for generating test vectors just described was deterministic. The values selected were determined on the basis of rules which lead to well-defined goals. The first goal was fault activation, and the second was path sensitization. For some classes of circuits such a conscious effort is unnecessary. Combinational circuits and some simple sequential circuits can be tested using randomly generated patterns. The advantages include much lower test generation costs, portability of test generation equipment, or even inclusion of test generators in the CUT (see Chapter 11 with reference to built-in self-test). The disadvantages are relatively long test sequences and uncertain fault coverage, the combination of which leads to poor fault coverage in complex circuits.

RANDOM PATTERNS

Before proceeding with deterministic methods for test generation, consider the random and pseudorandom concepts touched upon in Chapter 2 in light of the immediately preceding ideas of fault activation and path sensitization. For combinational circuits, explicit enumeration of input value combinations constitutes a complete valid test. Since there are no internally stored states in a purely combinational circuit, the outputs and the values of all internal nodes depend upon only

	x	y	C I N	a	b	c	d	e	C O U T	g	h	S U M
1	0		0									
2		0	0									A
3	1	1	1									—
4	0					0						
5		0				0						B
6	1	1				1						—
7		0					0					
8					0		0					C
9		1			1		1					—
10	0							0				
11		0						0				D
12					0			0				
13	1	1			1			1				—
14	1								1			
15		1							1			E
16					1				1			
17	0	0		0					0			—
18						1			1			
19							1		1			F
20								1	1			
21				0		0	0		0			—
22									0	1		G
23									1	0		
24							1				1	—
25										1	1	H
26							0			0	0	—
27								0			0	—
28										0	0	J
29								1		1	1	—

(a)

Figure 3.2. An adder circuit diagram and its corresponding backtrace tables. (a) Singular cover. (b) Logic diagram. (c) Cubes of gates H and J.

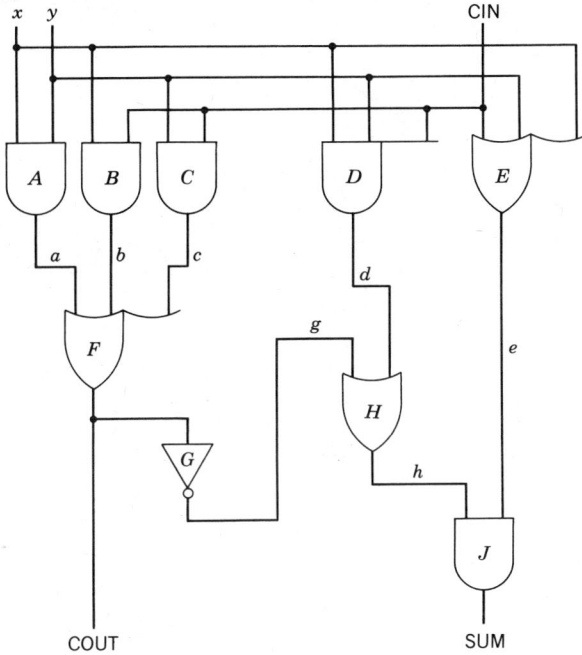

(b)

	g	d	e	h	SUM
1	0	D		D	
2	D	0		D	
3			1	D	D
4			D	1	D

(c)

Figure 3.2. (*Continued*)

the inputs of the circuit. This actually applies to static states of the circuit. Transients depend upon the previous state of the inputs and their present state. Then, for a static test, no faults can be activated or paths sensitized unless the required inputs constitute a single vector. Therefore, since all single vectors are among the explicit enumeration, all tests are generated.

As seen in the previous chapter, however, the number of input combinations rapidly rises for even moderately complex circuits. A simple 32-bit parity generator circuit would require 2^{32} or 4295 million input vectors to achieve a complete enumeration test. Although the test is 100% thorough, high-speed testers seldom apply input vectors at a rate of more than 10 MHz, and a test time of over a minute for a simple circuit would not be profitable in a production environment.

Random pattern testing takes advantage of the characteristic shape of the fault coverage yield curve shown in Figure 3.3. The actual numerical relationship is dependent upon the circuit and the patterns, but the shape is always similar for combinational circuits. The rapid rise in the beginning suggests that a reasonable

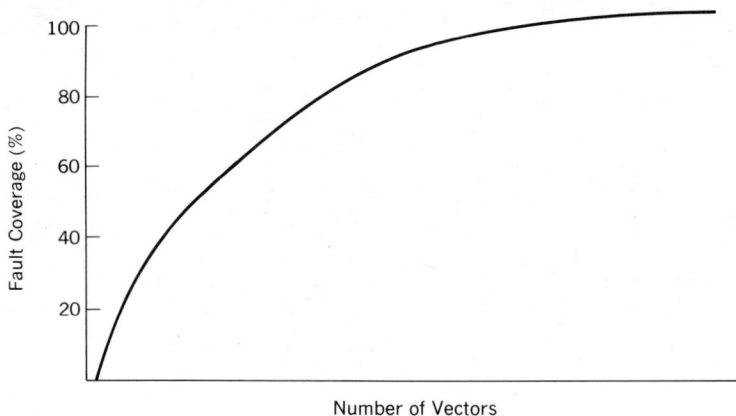

Figure 3.3. Randomly generated vectors achieve partial fault coverage rapidly, but many vectors are needed for thorough coverage.

fault coverage can be achieved with substantially fewer patterns than required for 100%. Even if no less than 90% coverage is acceptable, the test length is considerably shorter than for enumeration. The 74LS181 used as an example by Chin and McCluskey [1] required 388 random test vectors to detect 90% of the stuck-at faults, while 16,384 vectors constitute complete enumeration of the input possibilities. (For later comparison, the minimal deterministic test set contains 135 vectors.)

Purely random selection of input values has the drawback that the output of the circuit is difficult to predict. A repeatable set of test vectors originally selected at random and recorded eases the problem by allowing a number of comparison techniques to determine the correct response. The linear feedback shift register (LFSR) [2] is a convenient device for acquiring such a sequence.

The LFSR (Figure 3.4) is called a polynomial divider in theoretical terms and performs division by a binary polynomial on a series input string. In this application the input string is an infinite string of 0s and the divisor is represented by whether or not the feedback is connected to each stage of the shift register. The divisor can be viewed as a binary polynomial of the form

$$c_r x^r + c_{r-1} T x^{r-1} + \cdots + c_1 T x^1 + c_0$$

where the coefficients c are 1 if a connection exists from the feedback to the exclusive-OR of that stage and a 0 if no connection exists. The power of x is determined by its rank in the line of flip-flops. The output of the first flip-flop on the left is 1, the next is rank 2, and the last is rank r.

In practice, the flip-flops are initially loaded with a binary number, which is the "seed" for the pseudorandom number, and further serial input is left at 0. The string of flip-flops (and thus the rank of the polynomial) is determined by the

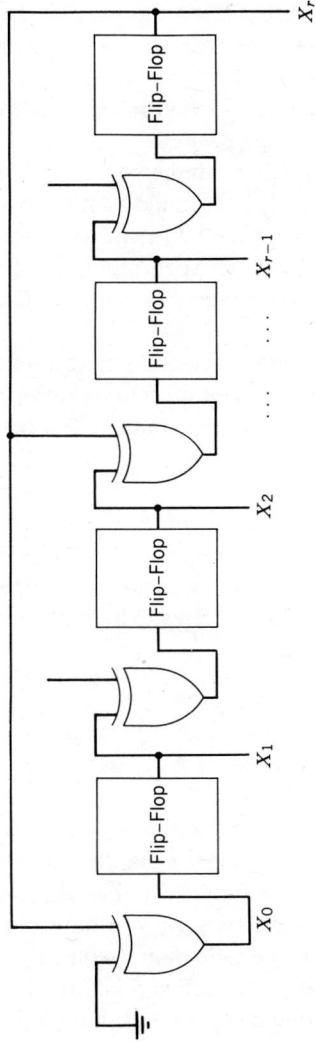

Figure 3.4. Linear feedback shift register. Each time the flip-flops are clocked, the outputs produce a new number in a complex pseudorandom sequence.

number of input values needed in the vectors to be generated. If the circuit to be tested has 34 inputs, then 34 flip-flops will be assembled, and an appropriate interconnection and seed will be provided to generate a sufficiently random number. The length of the sequence of clocks applied to the flip-flops is dependent upon the statistical degree of fault coverage desired.

The shape and severity of the fault coverage versus number of test vectors curve depends upon several factors. Malaiya [3] found a milder dependence on the structure of the pseudorandom generation method than on the testability of the circuit itself. The number of ranks of logic between input and output along paths of a combinational circuit is a rough determiner of testability. Chapter 11 will expound more thoroughly on the subject.

Sequential circuits rapidly become impractical to test with pseudorandom techniques as their complexity increases. For an intuitive understanding consider the testing of the carry output of a 4-bit binary counter. If a load control is available, the carry could be tested with only a few vectors manually selected by a test engineer. The counter is loaded to all 1s and then counted upward one more count. The probability of the carry being tested by a pseudorandom sequence is small even if the circuit consists only of the counter with the clock, clear, load, and four inputs all available. In fact, forcing the clear and load inputs to an inactive state after clearing the counter and clocking it repeatedly will give a shorter, more comprehensive test.

MIXED CLOCK PSEUDORANDOM PATTERNS

An improvement in the performance of pseudorandom input vector generation on sequential circuits can be made by observing that the clock inputs of such circuits are typically changed in value more frequently than the data inputs. Rather than generating random vectors with equal frequency of movement for all bit positions, a degree of order in which some bits toggle more frequently than others is desirable for sequential circuits. Replacing the polynomial divider (LFSR) by a set of counters operating from a base frequency, as in Figure 3.5, can give this freedom of control.

A similar seed value can be preloaded into the counters to alter the absolute value of the counts at any point in the sequence. The sequence of counts is more or less "pseudorandom," depending on the arrangement of counters and their individual moduli. The base frequency is typically applied to the clock line seen as most active in typical circuit use. Higher-frequency outputs drive function control lines such as write/read lines and multiplexer selects, while intermediate-frequency outputs drive data lines and low-frequency outputs drive asynchronous set and clear functions. The best opportunity for counting or shifting a register before the clear resets it is thus afforded.

The need for accurate testing and the reduction of test times becomes critical for complex combinational and even moderate sequential circuits. The pseudo-

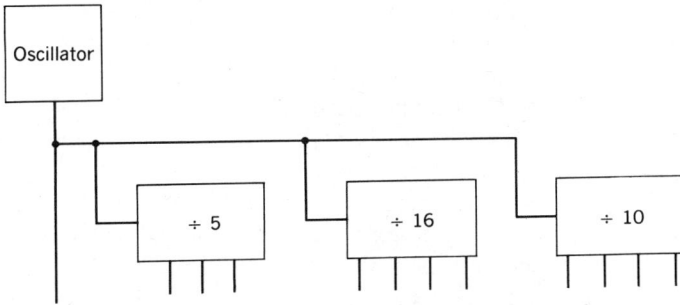

Figure 3.5. A string of counters with a common clock produces structured test vectors.

random techniques have found limited use, but deterministic testing has received most of the attention. Applying computers to the problems of fault activation and path sensitization improved performance and reduced test generation cost.

COMBINATIONAL ALGORITHMS

Fault selection and backtracing are easy algorithms to describe and implement in computer languages, but path selection is not so direct. The correct path to sensitize was obvious in the examples. Ordinary circuits are, however, much more complex, and the fan-out from some nodes may eventually affect thousands of gates. The computer program working at gate level does not see the best path as directly as the engineer may see it. Consider the difference between finding the way out of a mouse maze by viewing it from the top versus finding the exit to the funhouse at the carnival from inside.

Early attempts at automating the path sensitization process resulted in programs which did not guarantee success. They merely tried to find a path until some preset time ran out and then moved to the next fault. The program complexity is proportional to the square of the number of gates in the circuit since for each gate in the path to be sensitized another backtrace must be performed. J. P. Roth provided the first solution to the uncertainty of success with his ''D algorithm'' approach in 1966 [4]. The D algorithm makes use of the D notation introduced in Chapter 1 to investigate all possible paths until a successful one is found, thereby guaranteeing success. The process is not any less arduous than path sensitization, but it is easily described in programming languages, and it always finds a path if one exists.

Roth's D-Algorithm

The method is dependent upon the use of D cubes such as those shown in the table of Figure 3.2c. These are essentially tables of propagation for fault effects through the gates in question. Once the fault is activated, a D or D^* will appear in the representation at the fault site. As a path is sensitized, the D will be propagated along the path as a string of D or D^* values until it reaches an output.

To compare the methods, the example of Figure 3.2 will be used again. It is unnecessary to go through fault activation as a separate step for the D algorithm. The necessary inputs to achieve activation and those needed to propagate the fault effects will be calculated in the same ''consistency'' step at the end of the algorithm. Begin instead with the assertion of fault activation of an s-a-0 fault at node g. In the good circuit, node g is now at logic 1, while the faulty circuit node g is s-a-0 giving the representation of D for the value of that node on the logic diagram. This requirement is recorded in a table of requirements of the logic level containing the faulted node for later use by the consistency-checking operation.

The objective of the D algorithm is to propagate all D or D^* values outward from the fault site until a primary output is encountered. Each step pushes the ''D frontier'' farther, and each step contributes a table of possible sets of requirements for non-D inputs which must be satisfied by backtracing to the primary inputs. Only one fan-out from node g exists, and the drives gate H. The D cubes of only gates H and J were included in the figure, as they are the only ones needed in this example. Line 2 of the D cube table has a D under the node g column and shows that, if node d is a 0, then the D will propagate noninverted to node h. This means that if d is 0, then the 1 from node g in the good circuit will cause a 1 on node h, while the faulty circuit s-a-0 value will cause a 0. All Ds in a given circuit have the same value at the same time, while all D^*s have the opposite value. A temporary table is established to hold the potential input conditions which propagate the D at this level. Because of the simplicity of this example, only one path of propagation exists at this logic level, and only one condition is possible for propagation. The requirement of a 0 on node d is the only entry in this table.

Propagation is now attempted at the next level. Again, there is only one gate in the fan-out, so gate J is examined. Using the D cubes of lines 3 and 4, it is obvious that only line 3 propagates a D from node h to the output SUM. The conditions necessary for this propagation are entered in the requirements table for this level in the circuit. The table will have only the entry ''node $e = 1$'' since no alternative assignments will accomplish D propagation. The ''D phase'' is now complete since the present level of all D paths (there is only one) is a primary output.

The ''consistency phase'' of the algorithm is called to find (if possible) a set of primary input values which achieve at least one of the sets of requirements for each level of the circuit. The D drive can be thought of as progressing through the circuit like a wave. It always moves forward toward primary outputs and investigates the ways in which the fault effect can be propagated. The consistency phase is a reflection of that wave which starts at the logic rank which drives primary outputs and determines which of the possible conditions for propagation at each level is consistent with the other gates in the level and with other levels.

The consistency phase proceeds on a level-by-level basis using a backtrace procedure to test each of the sets of conditions for gates of a given level against the singular cover to find a set of values for the previous level which do not conflict. In this case the first requirement is a 1 on node e. Although gate F is of the same logic rank as J, it is not involved in the D propagation, so its requirement is x (unspecified) and immediately satisfied. Lines 14, 15, and 16 of the singular cover give three possible ways to achieve a 1 on node e, and since none of them conflicts with other assignments, they are recorded in a table of possible assignments for gate E.

A backtrace procedure is now instituted to find the necessary inputs which assure a 0 on node d. Lines 10, 11, and 12 of the singular cover suggest that a 0 on any of the inputs is sufficient. Again, the possible solutions are recorded. On this level the original assertion of fault activation is found, and the requirement of a 1 at node g must be fulfilled to activate the fault g s-a-0. The singular cover for gate G contains only one line which accomplishes this, so a 0 on node COUT is placed in the requirements for this level.

The next level contains only gate F and primary inputs, so the singular cover lines for gate F are used to find possible ways of forcing a 0 on node f. Line 21 is entered as the only possibility. It requires that nodes a, b, and c all be forced to a value of 0.

The level containing gates A through E has now been encountered, and gates D and E have tables of possible conditions to be met in input assignment. The tables are shown in Table 3.1.

The consistency algorithm will select one of the requirements from gate D and attempt to assert it. Since all inputs are primary, the value 0 will be assigned to input x. Next, the requirements for gate E will be examined, and the first requirement will be rejected since it would place a 1 on x and conflict with the previous assignment. The first line in the table will be marked as unsuccessful, and the second is tried. A 1 can be successfully asserted on y since that value has not been previously specified and does not conflict with any earlier selections. The next gate to be driven is gate C, and the first line must be rejected since it conflicts with

TABLE 3.1   Consistency Goals

x	x	CIN	
0			
	0		Requirements for gate D
		0	—
1			
	1		Requirements for gate E
		1	—
	0		Requirements for gate C
		0	—
0			Requirements for gate B
		0	—
0			Requirements for gate A
	0		—

the value just assigned to y. The second line of the requirements for gate C is chosen, and the input CIN is asserted to a 0. The first requirements for gates B and A have already been satisfied due to the 0 asserted on input x. The test has been found, and the process is halted.

In more difficult cases the table of possibilities associated with a gate may be exhausted without resolving a conflict. When this occurs, the assumption made for the gate a level below is rejected, and another selection is made from its table. If all requirements tables at all levels are exhausted without resolving a conflict, then no test exists for the fault.

Formally, the D algorithm proceeds as follows:

1. Select a fault site and assign a D at the site.
2. Propagate the D along all available paths using the D cubes of encountered gates and assigning necessary values to the non-D inputs of those gates. (usually termed the D-drive phase).
3. Backtrace level by level to find the necessary conditions to activate the fault and to find the inputs needed to propagate the Ds, resolving conflicts whenever possible (termed the consistency check phase).

Note that the algorithm is arranged to find all possible paths and then check the consistency of the necessary driving conditions. This may result in more than one sensitive path, but more interestingly, it results in all possible paths being investigated. Paths which require impossible combinations of inputs are inconsistent and will be eliminated.

The example circuit in Figure 3.6 is a case in which D propagation cancels to create a different kind of conflict, which must be resolved also. If the fault site is

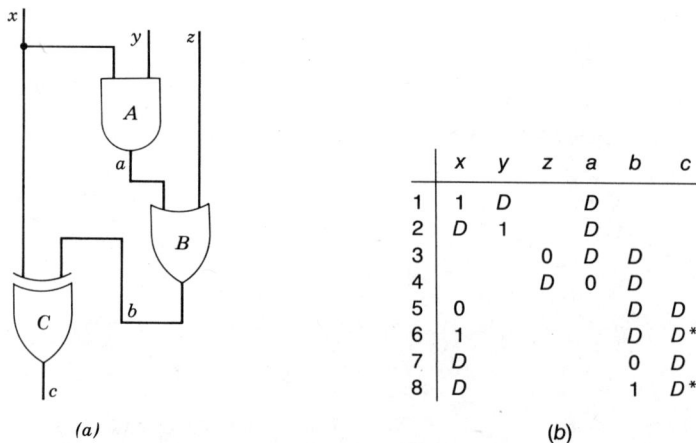

	x	y	z	a	b	c
1	1	D		D		
2	D	1		D		
3			0	D	D	
4				D	0	D
5	0				D	D
6	1				D	D^*
7	D				0	D
8	D				1	D^*

(a)   (b)

Figure 3.6. Reconvergent fan-out in the logic diagram leads to conflicting D propagation. (a) Logic diagram. (b) D cubes.

the input x, then D drive will proceed along both fan-out routes from input x. In the first rank, line 2 of the D cubes will be used to propagate the D through gate A, and line 7 or 8 may be used to propagate the D through gate C, either inverted or normally. But, since consistency checks are not performed until all D frontiers have reached a primary output or have died out, the required inputs of a 1 for y and a 0 for b will be noted only. Line 3 of the cube table will be used to propagate the D through gate B, and finally line 5 or 6 can be chosen to propagate the D on node b to the output. The required assertions are a 0 on z and a 0 on x if line 7 is chosen in the latter case.

The D drive is now complete, and consistency checking begins. Backtracing is trivial in this case since most of the gate inputs asserted are also primary inputs. The first gate encountered, however, is a problem in that there are two requirements for gate C. Since both inputs to the gate have D values propagated to them, neither assertion can be made true. The D cubes of an EXCLUSIVE-OR gate do not provide for propagation of two D inputs. The definition of D in a circuit is that a good circuit shall have 1 at each D location, and the faulty circuit shall have 0. In either case since the Ds of each circuit all share the same value at the same time, the output of the EXCLUSIVE-OR will be a 0 in both the good and faulty circuits. The D propagation has blocked itself unless one of the intervening gate conditions is changed. Choosing line 7 from D cube table and forcing the assertion corrects the problem in this case by setting y to a 0 and blocking the D propagation through gates A and B.

It is precisely the same kind of circuits as aforementioned that are the D algorithm's greatest weakness. Exclusive-OR circuits which contain reconvergent fan-out are unfortunately quite prevalent in computer circuitry. Comparators, parity generators and checkers, and error correction circuits are made from these circuits. The method of propagation D drive and investigating consistency in breadth-first fashion throughout the logic leads to considerable thrashing when these circuits are involved. Consider propagating a fault through a parity tree in which each gate becomes a potential path and each gate's value change reverses another down the tree. The D algorithm may be required to try most of the input combinations before the correct one is found. For many parity trees this could be several million combinations. Even at supercomputer speeds the test generation process becomes uneconomically long.

PODEM Algorithm

In 1979 P. Goel proposed a different algorithm called PODEM (path oriented decision making) [5] aimed at avoiding this problem by discovering which paths were doomed and eliminating them from consideration early in the process. The basis for PODEM is implicit enumeration of all possible input combinations. At first glance this concept sounds even more time consuming than the D algorithm, but several significant improvements to the selection of goals and the detection of dead ends result in high efficiency in general.

The decisions are made and kept track of by a binary tree whose nodes represent primary inputs to the circuit and whose branches provide the alternatives of a 1 or a 0 choice for the above node. The tree is constructed by adding a primary input as it is selected for investigation, and it is pruned by removing primary inputs if neither branch from them results in a test. If the latter is the case, then the value of the input is irrelevant to the test and should not be considered further, regardless of the combinations of values on the remaining primary inputs. Housekeeping mechanisms such as a list of all untried primary inputs and a flag at each node which notes that the alternative value for the node has been previously rejected are required. The basic forward propagation method of using D cubes and the back-tracing method using the singular cover are used in a way similar to that in the D algorithm.

Return to the adder example of Figure 3.2 again. The initial list of primary inputs is x, y, and CIN. For convenience the same fault (node g s-a-0) will be assumed. Ancillary heuristic algorithms aid PODEM in selecting the next fault to be tested and the most appropriate input to select first. These algorithms are helpful, but not necessary to use in the PODEM method. The objectives of the algorithm in general are to force a level on the node to be tested that differs from the stuck-at level being tested and to propagate the fault effect (D or D^*) at least one more level through the circuit toward a primary output. Although the algorithm results only attempt to propagate one path to an output (the test is finished when the first D or D^* reaches a primary output), multiple paths are not prohibited.

The first objective is to assert a 1 at node g. In this simple example the input x is chosen arbitrarily and set arbitrarily to a 0. The objective comes into play only when evaluating the path branches. The node and branch representing the decision are shown in step 1 of Figure 3.7.

The implications of asserting a 0 on input x are then investigated with respect to the question "Is the value of node g forced to the stuck-at value (0)?" All unassigned inputs are assumed unknown. The output of gate A is implied to be a 0 due to line 1 of the singular cover, and line 4 implies that gate B will output a 0. Gate D outputs a 0 due to line 10. The only remaining fan-out of x is gate E. There are no lines in the cover which specify the output of E on the basis of input $x = 0$ alone. The output of E remains unknown. The next rank of gates to be implied consists of gates F and H since only these gates' inputs changed. The output of F is unspecified since the value of gate C is unknown and A and B are both 0. The output of H is also unspecified since no value has been asserted at g at this point. The forward implication stops, but the initial assertion of a 0 on x is allowed to stand since there is no evidence that it conflicts with the goal of a 1 on node g.

Primary input y is next chosen and given the value of 0 as shown in the step 2 decision tree of Figure 3.7. The forward implications are investigated assuming $x = 0$, $y = 0$, and CIN = unknown. The outputs of gates A and B are unchanged and remain 0. The output of gate C can be asserted to a 0 due to line 7 of the cover. Gate D remains unchanged at a 0, and gate E is still unknown due to the

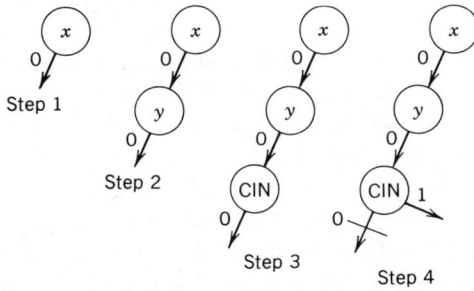

Figure 3.7. A decision tree used by PODEM to find test vectors for the circuit of Figure 3.2.

unknown input from CIN. Progressing to the next level, the output of F is now determined as a 0 due to line 21. The output of G, which is node g, the test node, is implicated in the next level to be a 1 due to line 22. The objective of fault activation has been completed successfully.

The next objective is to create a path of gates from the fault (D at node g) to a primary output such that the fault effect is propagated. Before selecting another input, the possibility of a path already existing as a result of prior selections is examined. The value of node d is a 0, which satisfies line 2 of the D cubes, but the value of node e is unknown, precluding a definite path through gate J.

The next (and in this case last) input is selected, and a value of 0 is arbitrarily chosen. The node and branch are added to the decision tree as shown in step 3 of Figure 3.7. The implications are investigated, and the gates B, C, and D remain unchanged due to previous selections of x and y. Gate E drives node e to a 0. This value intersects with the propagation path, but it forces the output to a 0 rather than the D or D^* desired. Since there are no other implications along other paths, the input assertion is rejected. The flag is set to show the expenditure of a choice for the node, and the other choice is selected. Step 4 of Figure 3.7 shows this marking and the new assertion.

The implications of the 1 assertion on CIN are investigated, and as before, the outputs of gates B, C, and D are unaffected. Gate E drives a 1 on node e due to line 16 in the singular cover. The 1 on node e intersects the sensitive path since the other input to gate J is a D. In this case the output of gate J is a D, as shown in line 4 of the D cube table, and the input assertion is not rejected. The fault effect has reached the primary output SUM, so the test is successful and the sequence is stopped.

It is interesting to note that the D algorithm and the PODEM algorithm arrived at different test vectors for the same fault. Both input vectors constitute a test, so each is valid. The difference in performance between the two algorithms is not meaningful for such a small example. The example is meant only to show the mechanical differences in the algorithms. PODEM is a depth-first investigation of path sensitization requirements, while the D algorithm is a breadth-first search.

9-V Algorithm

The PODEM algorithm implements a depth-first search of the paths affected by a particular change in an input variable. The process is directed by an enumeration of the possible inputs with modifications of the order of selection made by heuristic means. Another approach to higher efficiency involves a modification of the D calculus and an algorithm closely driven by the basic goals of fault activation and path sensitization. The 9-V (9-value) algorithm [6] uses forward implication and backtracing in methods similar to those in the D algorithm, but each gate's progress in the D drive is checked for consistency before proceeding. The check is against partial results for the remainder of the circuit using the 9-V calculus as the tool to make meaningful decisions without overly restricting later options.

Ordinary D calculus as defined by Roth uses a 5-value description of the state of good and faulty machines. A node in the circuit can have a value of 0, 1, D, D^*, or U under Roth's system. In the case of 0 and 1 the good and faulty machines share the same value. A node state of D implies that the good machine has a value of 1 while the faulty machine has a value of 0 at the same node. The symbol D^* means the exact opposite of D (good $= 0$, faulty $= 1$). The unknown value U implies that either or both of the machines is at an unknown value. The unknown or unspecified value is applied to all nodes before the test sequence begins. The 9-V calculus is aimed at resolving the U into more useful subsets.

The nine values expressed are 0, 1, D, D^*, $0/D$, $1/D$, $0/D$, $1/D^*$, and U. The four values added are subsets of the others. They express an uncertainty, but limit the values which can be assumed to the pair given. That is, if a node has a value of $0/D$, then the results of path sensitization so far only require that the node be either 0 or D in the final assignment. Further application of inputs to the circuit may restrict the value and reduce the $0/D$ to a 0 or a D.

Rather than reexamining the complete adder problem, consider a single NAND gate which falls on the proposed propagation path for a fault. The processes for setting up the problem are similar to those of the D algorithm in that a fault is selected and an attempt is made to propagate its effects using forward implication and backtracing. Suppose a 2-input NAND gate is encountered such that one of its inputs is driven by a node with the value D while the other is unknown (U). The 5-value calculus of the D algorithm would implicate an output of U. In the 9-V calculus, however, the analysis would be as follows. The unknown input can be any of the four values (0, 1, D, D^*) at the moment. The truth table for the NAND is then consulted (Table 3.2).

Observe that the row corresponding to an input of D has results of 1 or D^* regardless of which input value heads the column. The output of the NAND gate can be stated as $1/D^*$ rather than U, removing some of the ambiguity from future decisions.

The path sensitization of the 9-V algorithm proceeds similarly to the PODEM technique in that a single path is traced. The forward implication is performed on all gates, however, creating a wave-like frontier as in the D algorithm. The basic procedure is to pick a gate from the present frontier and attempt to select its un-

TABLE 3.2   NAND Truth Table

	0	1	D	D*
0	1	1	1	1
1	1	0	D*	D
D	1	D*	D*	1
D*	1	D	1	D

specified inputs to force the output to a D or D^*. The backward implications are resolved by backtracing one level using the 9-V calculus. The previous inputs are specified as broadly as possible using a $0/D$ or similar specification when sufficient, rather than arbitrarily selecting a 0 or a D. The effects of the selection are propagated forward through the circuit, and conflicts are resolved if possible.

The process might be applied to the NAND gate whose truth table is shown, for example, to investigate the possibilities of forcing the $1/D^*$ output to a D^* for further propagation. Reexamining the table, inputs of 1 or D will both produce a D^* output. The input ambiguity is reduced by requesting the backward implication process to justify a $1/D$ input, but no unnecessary arbitrary choice further restricts the process at this time. The backtracing continues, and if successful, a $1/D$ is placed on the node, the ouput of the NAND gate is set to D^*, and forward implication alters the remaining values of the circuit to reflect the ambiguity reduction. If the backtrace is unsuccessful, then it is useless to pursue this path for sensitization. If a value of $1/D$ cannot be guaranteed at this node for this level of ambiguity, then no further reductions of ambiguity can improve the condition. The path is abandoned, and a new gate from the present frontier is selected.

Combination Techniques Comparison

Several valid techniques to generate stimuli for combinational circuits have been presented. Others are expounded from time to time which address the application of test generation to particular circuit configurations or attempt to improve performance of certain aspects of the process. The overall scheme has remained. A fault site and condition are selected, and the input conditions necessary to activate the fault and/or propagate the fault effect are derived if possible. The activation and propagation steps may be addressed either separately or together, but the inputs required for each must be justified with the other in any case. Since the circuit's addresses are combinational, the fault activation and the propagation of effects must be simultaneous for a test to occur.

Two criteria are used for comparison of techniques in general. Completeness expresses whether or not the algorithm guarantees that a fault will be detected if a test is possible. This is a theoretical criterion since the time involved for derivation is permitted to become very large if necessary. Efficiency evaluates the time needed to arrive at a test vector. The measure of efficiency ordinarily expressed as required computation time for test generation of a given number of vectors (or a given

number of faults detected) on one or more bench-mark circuits. The meaningful-
ness of these comparisons is dependent on the similarity of the bench-mark circuit
to the problem circuits to be encountered in actual test generation. In addition, of
course, the times will vary depending on the architecture and speed of the computer used to carry out the algorithms.

In the paper by Goel referenced earlier a table of bench-mark results revealed
run-time ratios of $34.5:1$ (D algorithm : PODEM) when an error correcting circuit
was the example. The ratio reversed to $1:2.6$ for a programmable logic array,
although the average of 14 circuits including arithmetic logic units, multiplexers,
and decoders was $6.8:1$. The PODEM algorithm was derived to solve the error-
correcting circuit type, and the bench-mark results show its success.

SEQUENTIAL CIRCUIT METHODS

Very few interesting computing machines are constructed of only combinational
circuits. In fact, theoretical limits exist which restrict the number of ranks of logic
which can be used without a holding register to resynchronize the signals. The
delay variations between logic elements due to manufacturing and wiring differences result in signal skew, which eventually accumulates to introduce ambiguity
into the result. A large Wallace tree multiplier could be implemented from logic
which would deliver an answer to new input variables in 200 ns. If new input
values were applied synchronously every 200 ns, the result could also be sampled
every 200 ns, but the result would only be assured of validity for a smaller window, say 50 ns, due to the fact that some paths through the multiplier are faster
than others. Partial results will arrive at the output and jumble the answer until the
slower partial results have arrived and have been taken into account. The only way
to guarantee availability of a result to the next stage in the logic is to synchronize
the outputs by latching them during the valid window with a rank of memory cells
or flip-flops.

The simplicity and relative success of ATG algorithms for combinational circuits has influenced the designers of many major computing systems to restrict the
uses of synchronizing latches to a well-structured set of latches which can be used
in a serial mode for direct access to the combinational logic between latch ranks.
This design style is variously known as LSSD and "scan/set" and is covered in
detail in Chapter 11. Mention is made here to exemplify the relative cost of testing
sequential circuits as opposed to designs which can be treated as sets of combinational subcircuits.

Synchronous Sequential Circuits

In spite of the development of many algorithms which generate stimuli for combinational circuits, few have appeared which address sequential circuits directly.
Instead, there is a general method of adapting the combinational algorithms to

sequential applications. Synchronous circuits present the fewest complications while exemplifying the technique, so they will be treated first.

Using a synchronous circuit allows the flip-flops which make up the sequential parts of the circuit to be segregated easily from the combinational logic. Counters, shift registers, and other complex sequential elements can be reduced to flip-flops and associated combinational logic for determining the next values from the previous ones. The first part of Figure 3.8 shows a circuit which has been segregated. The flip-flops have been reassembled into a single rank of latches. The latches might be clocked by a single clock or a set of enable signals. The clock control signals are not considered in the method; they are only expected to be manipulable between test vectors in some repeatable fashion.

For the purposes of analysis, substitute state variables for the inputs and outputs of the latch rank as shown in Figure 3.8b. The state variable P_0 represents the output of latch bit 0, P_1 represents the output of latch bit 1, and so on. The variables are the previous state of the latch bits. The variables N_0 through N_m are inputs to the corresponding latches and as such represent the next state of the variables. Primary inputs to the circuit are denoted I_0 through I_n, and primary outputs are shown as F_0 through F_q. Note that there must be a one-for-one correspondence between the state variables P and N, but no such correspondence applies to the inputs and outputs.

The block of logic can now be treated combinationally and should be viewed as the circuit in a static state between test steps. Generating a test for the circuit is not quite as simple as in the pure combinational case since the state variables are restricted in their next values to those that can be propagated through the circuit itself. A test for a given fault often requires several test vectors which must be applied in the proper sequence.

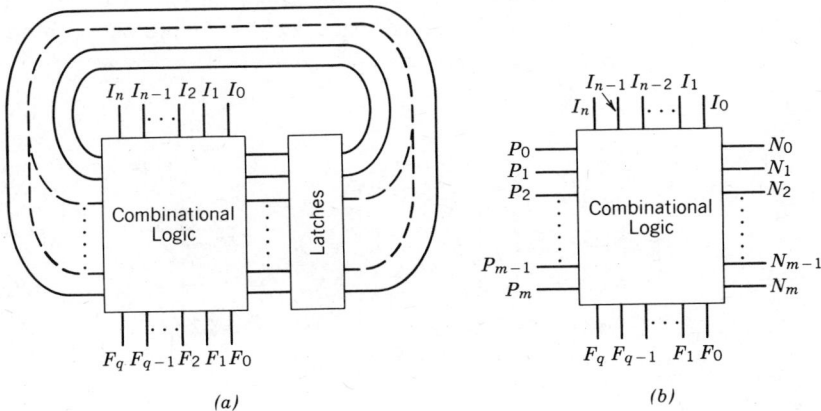

Figure 3.8. The latches of a sequential circuit can be removed, leaving the combinational circuit portion. Each flip-flop is a state variable of the circuit. (a) Latches segregated. (b) State variables substituted.

Suppose a combinational algorithm is applied to detect a fault in the midst of the combinational logic block. Assume the state variables and inputs are all unknown, as though the circuit were just powered up. The test generation alogrithm will establish a path to an output of the circuit from the fault site. The output could be a primary one or a state variable (N). Then the algorithm will attempt to justify the path and activate the fault by selecting input values. Again, the inputs could be primary inputs or state variables (P). The algorithm will reach an impasse if state variables are encountered since they cannot be selected or propagated directly in this time frame.

To negotiate the impasse requires consideration of the previous state of the variables to select inputs and the subsequent state ot propagate fault effects toward primary outputs. To visualize the process, think of the time frames as duplicate circuits (each with the fault being tested in place) connected to each other by the state variables which represent latch values. The three blocks in Figure 3.9 are arranged such that the circuit in Figure 3.9a represents the situation at step $i - 1$, the one in b represents it at step i, and the one in c represents it at the next step ($i + 1$). The logic values at each step are given time to propagate to a stable state during each step so that the steps only occur at clock activation.

In order to select inputs to circuit i from the state variables, backtracing will be needed in circuit $i - 1$. To propagate the outputs from circuit i through state variables, the path sensitization algorithm must be applied to circuit $i + 1$. If either operation reaches the state variables on the outside of the string of blocks, more blocks will have to be used. The blocks can be represented as identical copies of the circuit description in the memory of the computer attempting to solve the problem, or the state variables and inputs may be stored in tables corresponding to their time frame and the same circuit representation can be utilized repeatedly.

In either case the problem is much more complex than the pure combinational logic problem. Not only does the computing time grow rapidly in proportion to the degree of feedback of the circuit, but the space needed to represent the problem

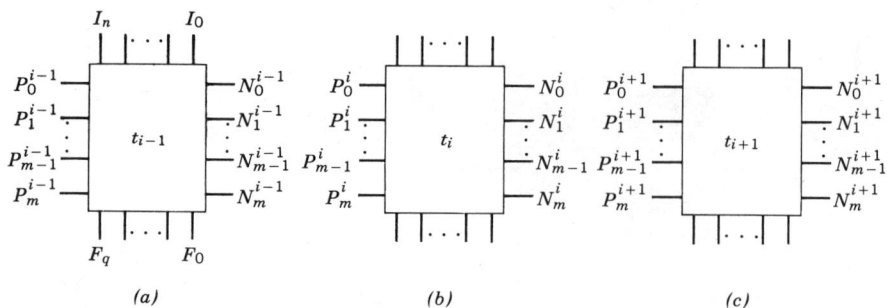

Figure 3.9. Linking the state variables from a given time frame to the next maps time and space together.

in memory can become large. Also, the guarantee of success which characterized combinational algorithms is no longer present. While propagating the fault effects toward the right in the time sequence, requirements for inputs to justify path sensitization are incurred which must be backtraced to the left. Conflicts are often encountered, and the resolution involves more motion in time as well as within each of the given circuit copies.

One of the most successful algorithms for generating sequential circuit tests gains efficiency eliminating the forward propagation of values in time. The extended backtrace (EBT) algorithm [7] chooses a path from the fault site to a primary output using the topological description of the circuit alone. The backtracing operation is then begun as the primary output lines are justified backward in both space and time to sensitize the selected path.

Once the fault site is reached, the goal shifts to achieving the correct logic value for fault activation, but the process remains the same. When a sequential element which must be determined in a previous time frame is encountered, the requirement is noted and the current frame value computation is completed. Then the previous time frame is processed in the same fashion, using the noted requirements as goals. When the process reaches primary inputs, the test sequence is complete. The test vectors are then applied in the reverse order of their generation. Conflicts encountered along the way must be dealt with by discarding partial results back to the last arbitrary choice which was made. The choice is changed, and backtracing is begun again. If necessary, a different topological path choice is made. If no untried choices remain, the fault is untestable.

Particular problems occur when a fault in a feedback path interferes with the determination of a state variable. For such a situation the good circuit becomes known, but the faulted circuit cannot be initialized. Detection of a fault is in question. Of course, binary logic can only assume two values, and if the value of the faulted circuit happens to be the opposite of the value of the good circuit, then the fault will be detected. But if the two values are the same, no detection is possible. Some test generation systems claim probable or possible detection if the situation occurs several times, while other systems will conservatively insist that the fault has not been detected unless the value of the faulted circuit can be made definite.

Advanced heuristic algorithms could be employed to solve the initialization problem in some cases. Since the variable can only assume two values, each value can be asserted in turn, and the result can be evaluated to see if both cases converge to the same solution. Unfortunately, the problem is already complex enough to strain most large computers, and adding this dimension for the thousands of unknowns that may occur singly or in concert overwhelms most test generation tools.

Solving algorithms must also be added to control the clocking of the latches. If the clocking is single phase, the clocking could be left to a postprocessor and assumed to occur after each change in primary inputs. When latches are clocked from several different sources in the same circuit, the sequence can be complex. The restriction of synchronous circuits allows clocking to be handled outside the test generation process.

Asynchronous Sequential Circuits

Controlling clocks within the stimulus generation process is only one of the additional problems encountered when the synchronous restriction is lifted. Hazards and races must be resolved, and the possibility of unstable feedback is encountered. All previous cases could be solved without consideration of propagation delays in the combinational circuitry. Adequate time was merely allowed between latch clocking to assure the settling of all logic states to stable values.

The simplest asynchronous circuit element is the cross-coupled NAND shown in Figure 3.10a. Consisting of two 2-input NAND gates, the circuit is the smallest static memory element in ordinary logic and is often used.

In the truth table of Figure 3.10b the a' and b' values are interepreted as "the value previously held" by outputs a and b, respectively. The result of changing both inputs from 0 to 1 in the same test step is an example of a logic hazard. If a zero delay circuit analysis is performed, the result will be indeterminate. That is, if the entire combinational circuit in which the cross-coupled latch is embedded is treated as gates whose output appears at the instant of application of inputs, then the inputs will change simultaneously no matter where the original stimulus originates. The case of (0, 0) to (1, 1) will therefore always result in a conflict. Before the change, each output will be a 1. After the change, both inputs of both gates will be 1s. The conflict occurs since the outputs are assumed to go to the value 0 immediately. Since only a wire separates the outputs and the feedback inputs, they cannot be both 1 and 0 at the end of the test vector propagation.

Installing a fixed delay in each gate solves the problem for most cases. Since it is the relative delay which is important, and real delays change with temperature, voltage, and other factors, a normalized or unit delay is usually employed. The problem arising from simultaneous change of inputs from (0,0) to (1,1) is not totally resolved since the situation will now cause an oscillation of both outputs with a frequency of 1/unit delay. However, the chances of simultaneous arrival of signals from some distant point is lessened and, moreover, under control of the circuit designer to some degree.

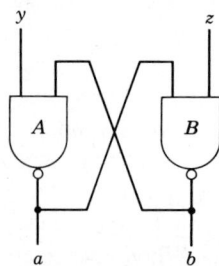

y	z	a	b
0	0	1	1
0	1	1	0
1	0	0	1
1	1	a'	b'

(a) (b)

Figure 3.10. Cross-coupled NAND flip-flop. This is the simplest asynchronous circuit and is often used.

The use of unit delay modeling creates the opportunity for transients generated by races in converging logic paths. In a synchronous circuit such transients were not a concern as long as the transient had disappeared or "settled out" by the time the latch clock occurred. In asynchronous circuits containing subcircuits such as the cross-coupled latch a transient may change the final state of the circuit. If the latch had an input set of $(1, 1)$ and the stored value on the outputs was $(0, 1)$ when a transient going to a 0 and returning to a 1 arrived on input z, the output would be changed to $(1, 0)$. If the unit delay assumption were correctly made, the test generated would accurately represent the circuit, but if the model were inaccurate the result would be worthless.

The stimulus generation algorithms usually rely on simulation of the good circuit to detect races and provide rejection of input vectors which give uncertain results. The basic scheme of backtracing is not affected, but the forward implication may be unreliable if delay is not accounted for in some manner. Simulation is discussed in detail in Chapter 4.

Test Vector Reduction

Sequential circuit tests consist of a series of test vectors, as had been shown. Each test generation procedure could begin with the assumption of a circuit having unknown internal states and unspecified input values. The test sequences generated would then be independent test sequences, each for a single fault. Only the order of test vectors within each subtest would have to be preserved, and the subtests could be arranged in any order to form the overall circuit test sequence. This approach appears sensible as it presents smaller subgoals to the test generation routine, but the penalty for reinitializing the circuit for each fault tested is huge in many sequential circuits. Hundreds of test vectors may well be needed to achieve sufficient initialization, and those will be repeated for each of the thousands of faults detected. The resulting test is extremely inefficient and time consuming to store and perform.

The test vectors can be postprocessed to arrange the subsets in an order that uses the repetitive initialization sequences as a preamble for a number of similar test sequences, but this postprocessing is a complex problem in itself. Instead, the test generation routine for each fault is usually begun with the assumption of state variables as the last routine left them. Some of the variables will need to be altered in most cases, but the overall impact is significantly less than the reinitialization requirement. Maintaining a close coupling between the test generation system and a digital circuit simulator aids the process by supplying the state variables upon request at any stage of the routine.

Feedback from a fault simulator during test generation also reduces the test length by taking advantage of faults detected accidentally while the test for another fault is derived. The act of initalizing the circuit usually results in many faults being detected. Primary outputs, for example, are tested for one stuck fault immediately upon becoming defined. The determination of fault detection by a fault simulator is usually fed back to remove that fault from the list of undetected faults rather than allowing the test generator to remove only faults for which it has successfully derived a specific test sequence.

Selection of the next fault to be tested is now the greatest single influence upon the overall length of the test sequence. Random selection from a list does not take advantage of the present state of the circuit. A counter string may be in a state of (11111111) after the s-a-0 fault for the most significant bit has been tested. If the next fault is the s-a-1 of the active low clear line, the counter will be reset to (00000000). The next fault may be the s-a-0 of the carry out, and the counter must be counted to (11111111) again at a cost of at least 510 input vectors, considering that the clock must be moved to a 0 and then a 1 for each count. Clearly, in the case of sequential circuits, some functional understanding is advantageous.

FUNCTIONAL METHODS

Performance of any of the test vector generation algorithms mentioned previously declines significantly with the size of the circuit to be tested and precipitously if the circuit is highly sequential. The reason for the decline should be obvious, considering the lengthy procedures required to solve circuits with the time frame representation acting as a multiplier to the routine. In addition, the length of the resulting test sequences becomes quite unwieldy. ATG for a sequential circuit with 10,000 gates may take 20 hours of processor time on a mainframe computer, and the resulting test sequence may contain 100,000 vectors. Manual test generation has taken over in complex sequential cases for economic reasons.

An analysis of the differences in strategy between the gate-level automated techniques and the typical manual approach used by a test engineer to derive tests for sequential circuits provides insight into improvements in technique. The test engineer perceives the function of the circuit and associates logic gates to form functional blocks. Usually, the packaging of MSI and LSI circuitry is along functional boundaries, so the perception of functional division is easy. Hierarchical design techniques implemented with CAD and documentation further inform the test engineer as to the intended function of the overall circuit. The advantages enjoyed as a result of cognizance of this knowledge are improved selection of the order of faults to be tested and avoidance of dead-end or conflicting path sensitization and backtracing. The general technique is still one of fault activation and path sensitization, but on a functional level.

ATG programs used in sequential applications are usually augmented with a functional algorithmic or heuristic package to improve performance. Functional blocks of logic such as counters, shift registers, and RAMs are described by solving routines called when the gate-level test generation algorithms encounter them. The solving routines describe the input sequences needed to backtrace through the block, those needed to propagate faults through a block, and the sequences which activate faults within the block. The input values or sequences are passed back to the gate-level algorithms by the functional subroutine and the process continues.

Alternatives chosen within the functional algorithm are optimized only within the block unless the entire test generator operates on the functional level, or unless an overseer portion of the program maintains coordination of the functional blocks. The best way to achieve a (0000) count from the point of view of the functional subroutine faced with a present state of (1110) will probably be to use the clear input. The overall strategy of the circuit is not taken into account. The clear may be inordinately difficult to control, or the use of the clock to count through (1111) may detect additional faults on the carry output. These strategy features are not easily achieved by functional treatment of only portions of the circuit.

Register Transfer Level Description

Hierarchical representations of digital circuitry span several levels from the system architecture or block diagram level to the switch device level (Figure 3.11). The test generation algorithms given for combinational circuits operate at the gate level, although some have been proposed which use the switch-level representation. Augmentation to aid sequential circuit test generation pushes upward into the functional representation level which describes the control and data exchange between blocks usually termed registers. Several RTLs (also called ''hardware description languages'' or HDLs) have been constructed to efficiently describe transactions at this level.

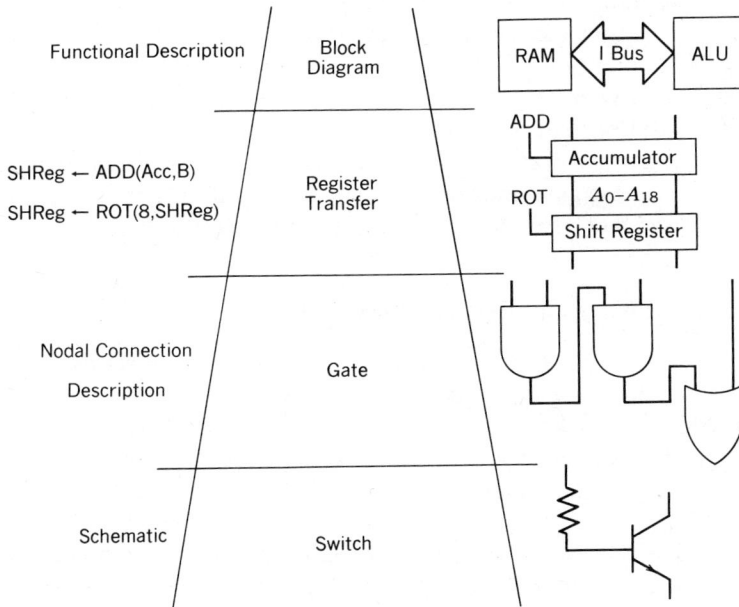

Figure 3.11. Circuits can be represented at several levels of abstraction.

Timing in processes may be expressly specified or implied by the sequence of description. The language AHPL (a hardware programming language) described by Hill and Peterson [8] places individual statements on separate lines and executes them in left-to-right, top-to-bottom order unless branch statements are encountered. Sequentiality is thus implicit. A similar language with Pascal-like control structures has been used in more recent examples by Levendel and Menon [9]. Both a strictly sequential and a parallel operation interpretation are treated in their article. The following examples assume the strictly sequential form which they call procedural. In addition, the left-pointing arrow which denotes transfer of results has been replaced by the Pascal symbol := for ease of typography.

```
If FO Then Acc := INC(Acc) Else Acc := Acc
Maddr := ADD(Page, Index3)
Reg5 := [Maddr
```

The expressions are to be interpreted as follows: If the present value of the Boolean variable F0 is true, increment the contents of the accumulator; otherwise, reload it with the old value. Add the value of the page register to the value of index 3, and put the result in the memory address register. Place the contents of the memory cell pointed to by the value in the memory address register in register 5.

A complete RTL description of a circuit includes definitions for registers and expressions for legal transfers of data between them. Register definition includes the size and name of a register. The transfer expressions describe the data transformations which are permitted between registers and the conditions under which they occur. Control or branching expressions complete the program, which is in itself a description of the control logic needed to make the circuit function. Functions such as decode, count, and shift can be made up of explicit descriptions or described once and thereafter referred to by "function" titles.

Register-Transfer-Level Stimulus Generation

When a circuit is viewed at the register level, the stuck-at fault set used on gate-level representations is not as clearly applicable. Certainly, the inputs and outputs of registers which are on data paths can be faulted in the same way, but the transfer expressions are not as straightforward. The gate-level implementation of the transfer expression can be derived and the gates within it faulted, but that defeats the purpose of RTL description. In the previously referenced paper by Levendel and Menon, a D algorithm extension was described which calculated the fault propagation of functional blocks and suggested faulting schemes for expressions. They conclude with an evaluation of the method of D propagation, saying

"We do not believe that this would be practical for large Computer Hardware Design Language functions. Instead, traversal of the control graphs . . . can be used."

Register-transfer-level test generation consists of steps similar to those of gate-level test generation. A fault mode is selected, the circuit is manipulated into a condition which activates the fault, and the effects of the fault are propagated to an output through a sensitive path. To be thorough in traversing the control graph, each branch of the control description must be taken at least once. That is, both the THEN and the ELSE in the previous description must be exercised. Fault activation has not been achieved in the IF–THEN–ELSE construct unless the results of the THEN branch are distinguishable from the results of the ELSE branch.

To test the expression (IF f THEN XO := A ELSE XO := B), the values of A and B must be different at the point of execution of the expression, and then the value of XO must be transferred to a primary output bus in some form which preserves the distinction between the values of A and B. The expression must be exercised such that A is transferred (equivalent to a stuck-at FALSE test for f) and then such that B is transferred (a stuck-at TRUE test for f).

Arithmetic expressions are not so easily compared to the stuck-at case since the result of a malfunctioning adder might be erroneous in a number of ways. If the expression (Reg5 := Add(A, B)) fails, does Reg5 get the value of A? The activation of an Add fault must presume something about the error to be generated, and a large number of faults could be supposed. In the extreme the adder could be replaced with its gate representation while the remaining circuit is treated as register transfer level. Mixed mode test generation depends upon propagation of fault effects through RTL expressions, but accounts for fault activation on a gate level. Progress has been made toward strict RTL ATG, and test results of some experiments such as those by Khorram [10] are promising. A register-transfer-level test generator derived by Khorram succeeded in generating a test sequence of 189 vectors for the AM2910 microsequencer (ignoring the stack), which detected 99.7% of the 1414 possible stuck-at faults.

Treating the RTL description as a program leads to the possibility of applying software testing techniques. Inductive assertion [11] as applied by Pitchumani and Stabler [12] places assertions or test statements throughout the RTL program which set flags or stop the program execution if verification conditions are not found to hold. The concept is intended to verify the RTL description against formal specifications for the performance of the circuit, not to produce test cases for detecting faults. The idea of using the behavior of the RTL program, which represents a circuit, to provide feedback as to circuit behavior resulting to test sequences is well received as an interactive test generation system rather than a fully automated one. The supposition of theorems about the behavior of the circuit and the verification of proof of those theorems is left to the engineer.

In an RTL interactive environment the test engineer uses knowledge of the circuit specification to propose a theorem of operation and an experiment to investigate. The experimental sequence of input data and commands is directed at the RTL description and an RTL simulation suggests the result. The test engineer is left to verify the result against the circuit specification. If the design has already been verified and there is confidence in the RTL description, the observation of internal actions and data flow during simulation may be useful in determining what portions of the circuit have been tested for what faults during the experiment.

Test generation directly from RTL descriptions will become viable when a method for faulting the description and a method for accounting the fault coverage become viable. The use of artificial intelligence (AI) techniques appears promising. AI systems for theorem proving contain many of the elements used by the design engineer to verify the RTL, and extensions seem reasonable for the proving of fault coverage.

SUMMARY

Stimulus generation consists of a basic algorithm, whether that algorithm is implemented manually or automatically by computers.

1. Identify a fault site and mode (output of gate ZZ s-a-0).
2. Solve the truth table for the driving gate to activate the fault by forcing the opposite value on the node (output of gate $ZZ = 1$).
3. Select a path from the fault site to a primary output and solve the truth table of gates along the path to propagate the effect of the fault. (Sensitize each output to input changes which lie in the path).
4. Backtrace from each of the sensitive path gates and from the fault site driving gate to primary inputs, solving the truth tables for nonconflicting input conditions which satisfy all the gate requirements along the sensitive path.
5. Repeat 3 and 4 repeatedly, making different choices when necessary to attempt to resolve conflicts. If unsuccessful, the fault is untestable.
6. Repeat 1–5 until the entire list of faults has been exhausted.

Ingenious variations of this algorithm have been developed which improve its efficiency and adapt it to a wide variety of circuits. The ever increasing complexity of digital circuits which must be tested strains even the most efficient algorithm implemented in the fastest computer. Testability improvements in design methodology have preserved the capability of ATG techniques in some instances, but no consistently successful routine has been developed to test asynchronous sequential circuits.

The human test engineer is more consistently successful in deriving stimuli for testing complex asynchronous circuits than ATG programs, but the techniques employed are difficult to express in heuristic programs. The recent application of AI techniques appears promising, but the relation of knowledge of the circuit at several levels of description as performed by humans is a difficult task. A well-equipped test engineer sits down to derive stimuli only after becoming familiar with the switch-level characteristics of the logic family, the truth tables of primitive gates used, the function of MSI and LSI devices, and the theory of operation of the circuit as a whole.

REFERENCES

1. C. K. Chin and E. J. McCluskey, "Test Length for Pseudo-Random Testing," in *Proc. 1985 Int. Test Conf.*, IEEE Comput. Soc., pp. 94–99.

2. D. K. Bhavsar and R. W. Heckelman, "Self Testing by Polynomial Division," in *Proc. 1981 Int. Test. Conf.*, IEEE Comput. Soc., pp. 208–216.

3. Y. K. Malaiya, "The Coverage Problem for Random Testing," in *Proc. 1984 Int. Test. Conf.*, IEEE Comput. Soc., pp. 237–245.

4. J. P. Roth, "Diagnosis of Automata Failures: A Calculus and a Method," *IBM J.*, 278–291 (July 1966).

5. P. Goel, "An Implicit Enumeration Algorithm to Generate Tests for Combinational Logic Circuits," *IEEE Trans. Comput.*, **C-30**(3), 215–222 (Mar. 1981).

6. C. W. Cha, W. E. Donath, and F. Özgüner, "9-V Algorithm for Test Pattern Generation of Combinational Digital Circuits," *IEEE Trans. Comput.*, **C-27**(2), 193–200 (Mar. 1978).

7. R. A. Marlett, "EBT: A Comprehensive Test Generation Technique for Highly Sequential Circuits," *Proc. 15th Design Automation Conf.*, June 1978, IEEE Comput. Soc., pp. 332–339.

8. F. J. Hill and G. R. Peterson, *Digital Systems: Hardware Organization and Design*, Wiley, New York, 1973.

9. Y. H. Levendel and P. R. Menon, "Test Generation Algorithms for Computer Hardware Languages," *IEEE Trans. Comput.*, **C-31**(7), 577–588 (July 1982).

10. R. Khorram, "Functional Test Pattern Generation," in *Proc. 1984 Int. Test. Conf.*, IEEE Comput. Soc., pp. 246–249.

11. R. W. Floyd, "Assigning Meaning to Programs," in *Proc. Symp. Appl. Math.*, **19,** Amer. Math. Soc., Providence, RI, 1967, pp. 19–32.

12. V. Pitchumani and E. P. Stabler, "An Inductive Assertion Method for Register Level Design Verification," *IEEE Trans. Comput.*, **C-32**(12), 1073–1080 (Dec. 1983).

CHAPTER FOUR

EXPECTED RESPONSE DETERMINATION

Stimulus generation as covered in the previous chapter is the most difficult portion of creating a digital test sequence, but it is only part of the problem. In order to test a circuit the stimulus sequence is applied, and the actual response is monitored and checked against the expected response. This simple statement presupposes that a convenient source of information about the behavior of the circuit under the effects of the stimuli is available. Determining, recording, and analyzing that information is the subject of this chapter.

Determination can be made using hardware examples of the circuit or software models. Response values can be recorded in their entirety or encoded in a number of ways for compact storage and rapid recall. Analysis is usually a matter of accounting for coverage of faults from a particular class or providing lists of faults yet to be covered for guidance of further test generation.

Before tackling the specifics of response determination, some preliminary thoughts are in order concerning the nature of response. In any test environment only a subset of the response data from a CUT is actually needed. Response could be characterized in voltage levels, currents, time delays, pulsewidths, frequency, phase, or even temperature rise. With occasional exceptions, digital circuits are characterized by voltages interpreted as logic values. The timing relationships of value changes are the second most used information.

Tolerance ranges must be accounted for in any determination of expected response since none of the circuits tested will be exactly nominal or exactly the same as any other. The accuracy of the expected response is the degree to which it correctly separates "good" circuits from "bad" ones. The judgment of "good" and "bad" must be against a predetermined criterion or specification. Functional testing usually demands that "good" circuits always respond with the expected logic values within a fixed window of time after the application of stimuli. Intermittent response or partial response is not acceptable. A calculator which occasionally declares $2 + 2 = 5$ is of no value. The speed with which it finds $2 + 2 = 4$ is a negotiable quality, however.

KNOWN GOOD BOARD METHODS

The most obvious solution to the question ''What does the circuit do when the sequence of stimuli is applied?'' is ''Watch one and find out!'' The ''known good board'' (KGB) technique proceeds in exactly this fashion. A carefully prepared example of the circuit is subjected to the sequence of test vectors, and the outputs are monitored. Some assumptions must be made as to the proper time after each change in inputs to observe the outputs, but otherwise the process is easily automated. Many types of ATE incorporate learn mode routines or provide for real time comparison of outputs from a KGB and the CUT.

Direct Comparison

Using an actual physical example of the circuit for comparison with the suspect circuit during the test has a certain puristic attraction. The logistics of providing duplicate inputs to the two circuits in proper synchronization are somewhat more costly, but several technical difficulties overshadow cost in the implementation of this philosophy.

Unknown Logic Values. Combinational circuits operated in a pseudostatic mode (allowing all logic values to stabilize before sampling) are straightforward examples for KGB test generation. Sequential circuits create several problems, however. When power is first applied to a sequential circuit, some of its internal states may not be defined. These unknown flip-flops or memory cells may be observable to the outside. Unknown values are either logic 1 or logic 0, and there is no means for determining that their present value in a physical circuit is a chance happening. If the test is performed and all outputs are recorded, some of the outputs may be different the next time the circuit is powered up. If a second example of the circuit is tested, its outputs may differ due to chance rather than a defect. ''Good'' circuits may be rejected by the test.

Unknown outputs must be masked from comparison to avoid false failures. Multiple tests could be run with the same KGB or with several examples, noting the differences, but statistically such attempts are doomed for complex circuits. Instead, a separate effort must be made during test generation to initialize the circuit. All expected response recording can be masked until the entire circuit is initialized or the unknown outputs can be determined and masked individually. When the initialization sequence is lengthy, the latter approach is preferable since many faults may have been activated and propagated as known values from portions of the circuit while other parts were unknown. These faults will have to be reactivated and repropagated after the circuit is initialized if all outputs are masked.

For circuits with large RAMs or other deep sequential circuitry the initialization process may be very long. If the test has been partitioned and a separate memory test is to be performed, initializing the block of memory is a waste of time. A few cells can be written, and then the unknown cells can be avoided during the test. If an unknown cell is read, its output can be masked until the address is changed or the cell is written. The necessity to identify unknown outputs is a serious complication to an otherwise simple procedure.

KGB Integrity. A similar "fly in the ointment" is created by the uncertainty of correctness of the KGB. Careful assembly, visual inspection, component testing, and testing in place in a system are techniques used to assure correctness of the example circuit. The accuracy of any of the assurances is a matter of question, but aside from that, the test environment may alter the response, mask a response, or aggravate a transient which invalidates the KGB response at the time of sampling.

The greatest drawback to the KGB method is the lack of fault coverage analysis. Generating a test using the KGB technique for determining the expected response is a one-way process without feedback as to the progress. Completeness of the test must be ascertained by some other method, usually manually. Manual fault detection accounting is tedious, and the estimates kept casually during test generation are almost always overly optimistic. A thorough ATG algorithm is a viable alternative, but as seen in the previous chapter, most ATG algorithms are limited in their rigor by constraints of computer resources or time.

Response Recording

If the KGB is kept as a part of the test equipment suite, expected responses can be freshly obtained at each test. Direct comparison of outputs is easy as long as the masking information for unknown outputs is available. Unfortunately, electronic devices are subject to occasional failure, especially when exercised by repeated insertions into an adapter with abrupt power up and power down cycles. The KGB may become defective, and the test will then erroneously identify CUTs as faulty. Once the KGB is determined to be defective, it must be repaired and retested, but against what standard? A second KGB can be kept as a backup, but if the circuit assemblies are expensive, a considerable cost of inventory can be amassed.

A much more convenient concept is to record the pertinent outputs and use the recording as a standard. Copies can easily be prepared, and backups are inexpensive. When a test is prepared, the expected response values are loaded into a RAM memory from which individual vectors are read in synchronization with the input vector sequence for comparison to actual outputs of the CUT. The permanent recording medium is usually a magnetic disk or tape.

If only the primary output values are recorded, the data storage required is small, and a complete record of discrete values can be kept. Frequently, logic values include X for unknown (masked) values and Z for 3-state values, in addition to 1 and 0. Recording these values requires two binary digits of storage for each output value recorded and thereby doubles the storage requirement. Storage does not become a problem, though, until the circuit becomes very complex with many outputs and a lengthy test sequence or until the values of internal circuit nodes are recorded.

Troubleshooting a circuit as a black box is not efficient, and troubleshooting with only the primary input and output vectors and a schematic is not much better.

Availability of a technique for evaluating the correctness of internal circuit nodes such as IC input and output pins is very important to the isolation of a fault. A probe is often provided as a part of the ATE which performs the test. When held in contact with a node, the probe compares the logic levels measured to a recorded string of values. The data necessary to support internal node analysis are a more serious recording problem. A medium sized PC board with 200 primary outputs may have over a 1000 internal nodes at the IC pin level. Since the test sequence for such a circuit would typically entail 1000 steps, several million bits of data must be stored. When the output data are combined with the input vectors and structured for easy retrieval and interpretation, the test file contains several megabytes of data and would occupy an entire floppy disk.

Transition Counting. In an effort to reduce the data storage requirements, particularly for portable test systems, data compaction is often employed for the internal node values. A simple method used on early systems was transition counting. Rather than record the value at each test step, the node to be examined is connected to a counter, and the final count at the end of the test sequence is recorded. The maximum number of bits required to record the count is $\log_2 N$ where N is the number of test steps in the sequence. The same 1000-test-step example mentioned above would require 10 bits/node or 10,000 bits for internal node storage. Even if primary output values are not compacted, the savings is considerable and the test requires only 100,000 bytes or so to be stored.

Data compaction by transition counting results in a loss of information. Transitions are counted as either 0 to 1 or vice versa, and the X and Z values are masked (counting is prohibited). The counting technique must be modified to differentiate between distinct strings of values which have the same number of transitions. If an internal NAND gate were replaced by an AND gate, the transitions would be identical, although the values would be opposite at any given moment. An enhanced transition counting technique depends upon the sparseness of activity in a typical logic circuit. Less than 20% of the nodes of a typical circuit change state in response to a change in primary inputs. If a counter is incremented at each step in the test sequence and its value is stored in a column corresponding to a node whenever that node transitions, a nearly unique record is formed. The addition of a snapshot of the value of the node at any single point in the sequence makes the record unique. On the average the test will now require 10 bits for each of 200 transitions for 1000 nodes or about 300,000 bytes, including primary input and output values.

Signature Analysis. The use of LFSRs as in Figure 4.1, instead of counters for encoding the values is a further improvement in both compaction and accuracy. The technique was introduced in the portable test equipment arena by the Hewlett Packard Corporation under the name ''signature analysis'' since the contents of the LFSR at the end of the test represent the signature of the node examined. The LFSR is a mechanism that is also used to generate pseudorandom patterns for test generation.

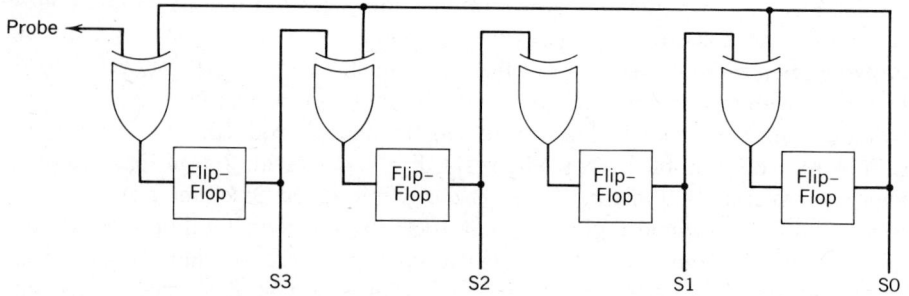

Figure 4.1. LFSR. In an analysis mode of operation the LFSR divides the serial string entering at the left-hand side by a fixed polynomial.

The length of the register and the interconnection of the feedback path determine the characteristic polynomial which is used as a divisor of the serial string input from the probe. By selecting a sufficiently long polynomial a unique result can be obtained for each node's string of values. The parallel output of the register is stored as a binary signature. The signature can be matched with test results as long as the LFSR used to analyze node values during the test is identical in structure. A single number which usually contains about 32 bits represents each of the 1000 nodes in the example. The node value storage has been reduced to 4000 bytes and can even be written on the schematic as a four-hexadecimal-digit number at each circuit node.

Data compaction and the KGB method of expected response determination can result in an inexpensive, portable test technique. The lack of fault coverage feedback during test generation shifts the burden of test generation to the stimulus generator. Confidence in the test quality must be established separately. These factors have resulted in the widespread use of software modeling and simulation rather than the adoption of the KGB method in most cases.

SIMULATION TECHNIQUES

The rules of logic behavior are well documented, and with diligence the outputs of a circuit can be calculated from its input sequence. Computers are known for their diligence. The creation of software to solve for the present state of a logic circuit was a natural progression once the circuits of interest became too complex to solve manually. "Good circuit simulation," or simply "logic simulation," is the process of solving for the logic states (and often the timing relationships) of circuits made up of interconnected predefined logic functions. The task of accounting for fault detection can be added to the process in a number of ways, but the result is always termed "fault simulation."

Logic Simulation

Logic simulation can be performed at several levels of abstraction. Electronic devices characterized as switches can be directly described, and the resulting switch conditions (off and on) can be used to predict the outputs of a circuit without direct reference to the gates or logic functions they make up. For technologies such as MOS this approach is advantageous in the simulation of VLSI chips because the gate boundaries are not well defined. Transistors may be shared by several merged gates.

Circuit Description. Technologies such as TTL and ECL are usually described in terms of simple gate functions which embody only a few of the Boolean operations (AND, OR, NOT) A more generic term for either the switch or gate entities is "primitive," and logic simulation on lower levels of abstraction can be described in terms of the interconnection and interaction of primitives which perform arbitrarily defined logic functions.

The description scheme used in Figure 4.2 is only one of several possible. As long as the description is self-consistent, it can be used. The description can be hand encoded by examining the printed schematic or directly derived from a CAD database by a utility software package. This description refers to primitives by the physical grid location of the component on the subassembly and a uniquely chosen output pin to differentiate between sections in multigate packages. Since the description is given unidirectionally in terms of the inputs to each primitive, primary outputs must be specified by a special case. The construct OUT follows the listing of a primary output, and the driving gate follows OUT in turn.

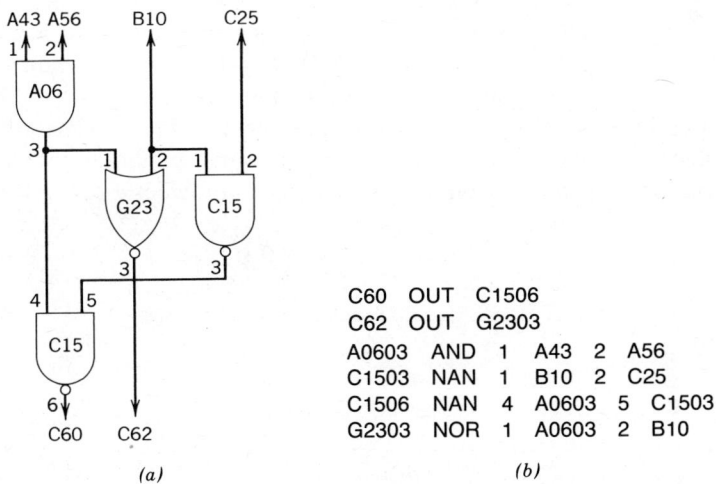

```
C60    OUT   C1506
C62    OUT   G2303
A0603  AND   1   A43   2   A56
C1503  NAN   1   B10   2   C25
C1506  NAN   4   A0603   5   C1503
G2303  NOR   1   A0603   2   B10
```

(a)   (b)

Figure 4.2. Human readable circuit information is converted into machine readable form for simulation. (a) Circuit diagram. (b) Software model description.

The OUT designator takes the place of a primitive name which identifies the function for the other primitives. The primitive name is a reference to an entry in a separate catalog of descriptions, which may be in the form of truth tables or algorithms. Each mention of a primitive name in the description is a separate instantiation of the function. The functions are interconnected by the list of pointers after the function designation. Each pointer pair consists of a unique input designator to the gate being described (in this case a package pin number) and the output designator of the driving gate. Since only one pointer per input is allowed in this description scheme, logic structures such as buses and wired-ORs (wired-ANDs) must be explicitly represented by a primitive in the catalog.

Logic Value Representation. During simulation a table of values is kept in which each gate output is represented. Most simulators use at least 2 bits per output in order to represent 0, 1, X, and Z states. Some simulators use more bits to represent combinations of the basic states, while others identify transitions during a time period as a separate symbol.

Seven static logic states are described by Figure 4.3, making use of degrees of uncertainty that may be present during simulation. For example, if the output of a 3-state buffer gate is enabled but the data input to the gate is unknown, then the output could be described as U rather than X. Although most testers would not take advantage of the information, it is possible to determine that the output enable is not stuck in the inactive state by examining the output value and comparing it against the criterion that the voltage level must be either a legitimate 1 or a legitimate 0. If it is not, then the test has detected a faulty enable.

A further use for extra logic states expresses the "strength" of the basic states. The strength of signals does not come into play in normal TTL gates since each gate is an open feedback amplifier, and once the input threshold has been exceeded, the output is driven to saturation. MOS transistors are sometimes sensitive to the strength of the gate drive, and a weak input may result in a weak output. The most easily envisioned use for strengths is the TTL open collector bus structure in which a pull-up resistor (a weak 1) provides the default level if all attached outputs are high (open). Any output which turns on provides a strong 0 by means

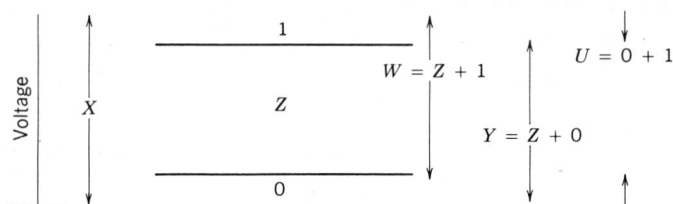

Figure 4.3. Two thresholds form three zones associated with logic states. Four additional states are formed by combining the original three.

of a saturated transistor to ground and overrules the weak 1 of the pull-up. The strength system is thereby used to arbitrate bus conditions.

For explanatory purposes, however, assume that only the four states 0, 1, X, and Z are to be simulated. Since the sample circuit contains no sequential elements, the value X will not be needed if all the inputs are defined, and since there are no buses or tristate drivers, the value Z will not be needed. Begin the simulation by asserting 0s on all inputs of the example in Figure 4.2.

The Simulation Process. A typical simulator software package is divided into two tasks, each of which communicates with the other through a pair of queues (Figure 4.4). The evaluator uses the truth tables or functional descriptions to solve for the output of gates, while the scheduler traces the net of interconnectivity to determine what gates will be affected by the evaluator's updating of a gate's output. Both tasks use database structures which contain the interconnection information for the gates of the circuit. The evaluator alone uses the value tables which hold the logic values of all nodes in the circuit.

The scheduler searches its linked lists to find all gates affected by the inputs, and A0603, G2303, and C1503 are placed in a queue for evaluation. The primitive description for AND is consulted as A0603 is removed from the queue by the evaluator. The description of AND is a truth table, and the output corresponding to (0, 0) is a 0. This value is stored into the value matrix in the position for A0603, and G2303 is removed from the queue for evaluation. The truth table for a NOR function is consulted, and the value of A0603 is looked up in the value table to complete the (0, 0) input set, giving a 1 for an output value. This value is stored in the value table for G2303, and C1503 is removed from the queue for evaluation. The truth table for a NAND results in a 1 being stored for C1503.

The queue is now empty, but the simulation is not complete. As each of the outputs has been entered into the value table, it has also been entered into a schedule queue if the value is different than the previous value for the node. Now the schedule queue is examined, and A0603 is removed for scheduling. The linked

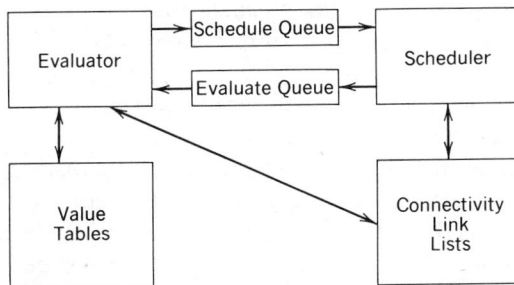

Figure 4.4. The process and data structures of a simulator communicate with one another as indicated by the arrows.

list which describes the connectivity of the circuit is checked to locate all gates which are driven by A0603. C1506 and G2303 are placed in the evaluator queue. The next gate in the scheduler queue is G2303, and it is removed and investigated, resulting in output C62 being placed in the evaluator queue. The last gate in the scheduler queue is examined (C1503), and C1506 is found to be already in the evaluator queue.

The evaluate cycle begins again, and C1506 is drawn from the queue for evaluation. The linked list of connectivity is used to locate its input values from the table (0 from A0603 and 1 from C1503). The truth table for a NAND gives 1 for the output, and this result is entered in the value table and output C1506 is placed in the scheduler queue. G2303 is taken from the evaluator queue, and its value is solved as 1. Since this value is the same as found previously, the value table is left unchanged and the output is not placed in the scheduler queue. The OUT item C62 is removed from the evaluator queue, and its value table entry is updated immediately since its function is a NULL. The evaluate queue is again empty.

The schedule queue contains only C1506, and it is found to lead to the output C60 only. Nevertheless, for the sake of generality the item C60 is placed in the evaluator queue, and the schedule queue is empty. The evaluator is called and finds C60 in its queue. The value table for C60 is updated, and since primary outputs do not affect other gates, the activity has ceased. The simulator waits for an input to change value. When an input changes (i.e., C25) only the gates affected by the change will be evaluated (C1503, C1506, and output C60).

The simulator described does not consider timing or gate delay. Fixed gate delays can be implemented by replacing the evaluator queue with a "timing wheel." A timing wheel is a set of queues, each of which represents a distinct delay time from the present. If the base delay for the simulator is normalized to 1 (usually called a unit delay), then adjacent queues represent events which are scheduled to occur 1 unit delay apart. In ECL logic the unit delay might be 1 ns, while TTL may be simulated in 10-ns increments. The actual value of the delay is arbitrary as long as the relative values given to gates are correct. As a gate is removed from the scheduler queue its fan-out is traced, and the driven gates are placed in the queue which is t queues from the present queue where t is the gate delay associated with the driving gate. When the evaluator is finished with a queue, it selects the next nonempty queue from the timing wheel. When all the queues are empty, the simulation cycle is complete.

The timing wheel usually has a predetermined size. If an event results in an attempt to enter a gate beyond the limit of the wheel, an error is reported which is interpreted as an oscillation of the circuit. As an alternative treatment an unknown value (x) can be asserted for any nodes which are still active beyond the timing wheel limit, and the activity will cease. Information about the circuit state will be lost in this case. The size of the timing wheel must be adjusted to assure that normal operations in a circuit being simulated do not exceed the timing wheel size. The result of the addition of the timing wheel is a "unit delay event driven" simulator. Most simulators associated with fault simulation are unit delay event driven simulators.

Fault Simulation

The primary output sequence and internal node states which are determined by a good circuit simulator are collected for each test step and combined with the input vectors to produce a complete test sequence. The fault coverage of the test sequence is undetermined, however. The prediction and accounting of fault detection adds a major dimension to expected response determination. The most common method for predicting fault coverage is to simulate the effects of each of the possible faults and note whether a difference in primary output state occurs during the test sequence if the fault is present in the circuit.

Serial Fault Simulation. The most straightforward way to perform fault simulation is serial simulation. The good circuit is simulated for all input vectors, and the response is recorded. Then a single fault is injected by modifying the circuit description. An s-a-0 fault is injected by replacing the normal source of signal for the node by a ground connection. Replacing a node driver with an open or a voltage connection injects an s-a-1. Intercircuit shorts can be simulated by adding a wired-OR at the site of the short and driving all fan-outs from either node with the result of the wired-OR function.

Each fault is injected, and the simulation sequence is rerun. During the rerun the primary outputs are compared at each test step with the results of the good circuit simulation. If a difference occurs, the fault is marked as detected. The fault is removed, and another is injected before the next pass. If a single test vector or a series of vectors is added to the sequence, the process must be restarted from the beginning unless the internal states of the good circuit storage elements (assuming it is a sequential circuit) and the internal states of each faulty circuit are saved from the last vector of the old set. This requirement results from the fact that fault effects may be stored and propagated at a later time.

The performance of serial fault simulation in an ordinary computing environment is poor. The number of faults defined for most circuits ranges from 1 to 6 times the number gates in the circuit. The processing time required for good circuit simulation is proportional to the number of gates in the circuit, although event driven simulation performance contains an activity factor. The activity factor is an approximation of the number of gates which change value and are therefore evaluated due to a typical input vector. The activity factor for most circuits ranges from 0.1 to 0.3. The formula for computing time is

$$T = \frac{GFVNA}{S}$$

where G is the number of gates, F is the number of faults, V is the number of input vectors whose effects are simulated, N is the number of instructions in the evaluation and scheduling loop for a single gate, A is the activity factor, and S is the processing speed of the computer in instructions per second. Using typical numbers of 1000 gates, 3000 faults, and 1000 vectors and an activity factor of 0.2, an efficient and speedy computer with 50 instructions in the evaluate–schedule loop and a 5 mips (million instructions per second) processing rate requires 6000 seconds or 1 hour and 40 minutes to complete the simulation. A complex simulation 50,000 gates, 150,000 faults, and 5000 vectors becomes out of the question at 20,000 hours, although similar problems occur in real life.

A simple improvement in the procedure can be made if the simulation of a particular faulty version of the circuit is halted as soon as the fault is detected. The detection of faults is usually weighted toward the beginning of the vector set, so the average simulation run would be less than half as long under this technique. Assuming an average run of 0.3 times the good circuit pass, the times are reduced to 30 minutes for the small circuit and 6000 hours for the large ones.

Parallel Fault Simulation. The serial nature of the process just described is the primary contributing factor to its poor performance. Each pass through the simulator is almost identical to the original good circuit pass. The gate solution routines are the same. Only the data and the order and extent of gate evaluations change as faults alter the results. The data requirements are meager since only a few bits are needed to record the state of a node.

The first improvement made on serial fault simulation took advantage of the width of computer words and the parallelism of the processes involved in each pass. Parallel fault simulation restructured the representation of values as a matrix (Table 4.1) with each row or pair of rows containing the value of a circuit node at any given time during the simulation and each column pertaining to a different circuit. The first column contains the values of the good circuit's nodes, while each of the other columns contains the values of the same nodes under the presence of a unique single fault in the circuit.

The table values given correspond to the input vector $(0,0,0,0)$ for the circuit of Figure 4.2. Only one row is used to represent each circuit node for simplicity of illustration. This restricts the values to 0 or 1 and excludes X and Z, but for a simple combinational circuit the loss is minor. Although most computers used for fault simulation would have more than 16 bits in a word of storage, the size of the example fault block has been chosen for typographical convenience. All the faults in the circuit have not been included, and a second fault block would be used in an actual simulator. The second fault block would repeat the good circuit simulation in the first column and then contain up to 15 other unique single-faulted circuit representations. The good circuit simulation is repeated for convenience in comparison of results.

A review of the primary output rows in the table reveals the detection of several faults by this test vector (A0603 s-a-1, C1506 s-a-0, G2303 s-a-0, C60 s-a-0 and C62 s-a-0). The values in column 3 (C1503 s-a-0) are not identical to those of column 0, but the primary output rows match, so no fault effects have propagated to an observable point.

When the evaluator solves for a new gate value, all columns of the fault block are used in the solution. That is, if input A56 were changed to a 1 and the evaluator

TABLE 4.1. Parallel Fault Simulation Value Table

	0	1	2	3	4	5	6	7	8	9	10	11	12	13	14	15	
	0	0	0	0	0	0	0	0	0	0	0	0	0	0	1	0	A43
	0	0	0	0	0	0	0	0	0	0	0	0	0	0	0	0	A56
	0	0	0	0	0	0	0	0	0	0	0	0	0	0	0	0	B10
	0	0	0	0	0	0	0	0	0	0	0	0	0	0	0	0	C25
	0	0	1	0	0	0	0	0	0	0	0	0	0	0	0	0	A0603
	1	1	1	1	1	1	1	1	1	1	1	1	1	1	1	1	C1503
	1	1	0	1	1	0	1	0	1	1	1	1	1	1	1	1	C1506
	1	1	0	1	1	0	1	1	1	1	1	1	1	1	1	1	G2303
	G	A	0	C	C	C	G	G	G	C	C	C	C	A	A	A	C60 OUT
	O	O	6	5	5	5	2	2	2	6	6	6	2	0	0	0	C62 OUT
	O	O	6	3	3	6	3	3	3	0	0	2		6	6	6	
	D	0	1	0	1	0	0	0	1	0	0	1	1	0	1	0	
s-a-																	

were called to solve for a new value of A0603, then a row of all 1s would be NORed with the row from input A43 and the result would be tested against the A0603 row for differences. In all columns except 14 (A0601 s-a-1) the result is the same as the previous case. Column 2 appears at first glance to be an exception also, but the value 1 in this column is the direct result of a fault and must be restored after every calculation.

Although activity has ceased for the good circuit, the circuit of column 14 is still propagating an event, and the evaluator will place A0603 in the queue for scheduling as it updates the value in all columns (and reasserts the 1 in column 2). The scheduler will place G2303 and C1506 in the evaluate queue, and evaluation will continue. Gates G2303 and C1506 will be evaluated, and G2303 will be given a new value in column 14 only. C1506 is not different in any column, so propagation dies out in that chain. The scheduler and evaluator are called due to the change in G2303, and the value of output C62 is updated. The change in column 14 of the row for output C62 indicates a successful test for A0601 s-a-1. The equivalent fault of input A43 s-a-1 is also detected.

The ideal performance increase achieved by parallel simulation is a factor of $(W - 1)$ where W is the width of the computer word. In reality, the factor is lessened by the fact that simulation of all W circuits in a block must be continued as long as any of the circuits are active. Memory requirements for implementing parallel fault simulation are large but predictable. The memory needed for the value array, for example, is given by the following expression:

$$\text{Memory} = \frac{FG \log_2 (L)}{W - 1} + G \log_2 (L) \quad \text{bits}$$

where F is the number of faults, G is the number of gates, L is the number of possible logic states represented for each node, and W is the width of the computer's memory word. As faults are detected, the tables can be packed to reduce the memory requirements and improve the speed in the same way as serial fault simulation early termination of a pass. The savings are not so marked, though, since the packing process absorbs some time itself and usually is not performed as each fault is detected, but only occasionally.

Using a 32-bit computer with a 5-mips speed, the 1000-gate, 3000-fault, 1000-vector problem will ideally require about 3 min of processing. The actual time is considerably longer because result storage requires I/O operations which are relatively slow. The approximately 1M value tables may not fit into memory at one time, and if this is the case, fault block swapping will require more I/O. The large problem of 50,000 gates requires more than 200 h of processing. Although the large problem is not impractical at this performance level, it is still inconvenient.

Deductive Fault Simulation.

In an attempt to replace the explicit simulation of faulty machines with an implicit method, deductive simulation was introduced in 1972 [1]. The concept of deductive fault simulation is to explicitly simulate the good circuit and then deduce from the good circuit values which fault effects will be activated and propagated.

Once the good circuit simulation is performed, the fault procedure begins at the rank of logic nearest the inputs. At each node a list of fault effects is made on the basis of two criteria. The first criterion looks for fault effects which would originate at the node. If an input were found to be a 1 in the good circuit, then the s-a-0 fault for the node would be listed. The second criterion applies at the next rank and all others as well as to the first test. The second criterion looks for faults which could be propagated to the presently considered node from earlier nodes, given the circuit's present state.

The second test is similar in some ways to the path sensitization concept used in test generation. Examining the 4-input NAND gate of Figure 4.5 as though it were in the midst of a circuit under deductive fault simulation will provide some detail. The good circuit's present values of the input and output nodes are shown alongside the signal lines. A list of fault effects which have been propagated thus far is shown at the terminus of each of the input lines. Each fault effect is shown as a circuit node identifier and a fault effect at that node and does not suggest what effect the fault presence would have at this particular gate. The implication of a fault's presence in the list associated with a line is that the logic state of the line would be reversed from its present value if the fault were present.

The deduction is as follows. Since the output of the gate in Figure 4.5 is presently at a 1, the fault Z4601 s-a-0 should be added to the list associated with its output. The presence of this fault will cause the output to be different without regard for the inputs. No other single fault directly associated with Z46 will cause the output to differ, so the second part of the algorithm is called.

In order for a difference to occur as a result of propagated inputs the input values would all have to be 1. Knowing this, it can be deduced that only those fault effects which singularly cause all inputs to be a 1 will be propagated. Multiple faults are not considered in this example. Since the second and fourth inputs are already 1 due to good circuit effects, the requirement is slightly different. Any fault to be

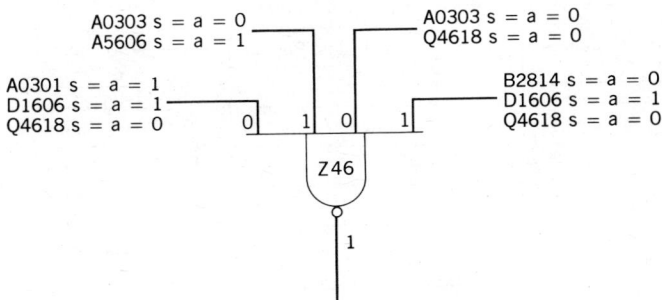

Figure 4.5. A NAND gate with the fault effects for a single evaluation step ready for propagation.

propagated must cause the first and third inputs to become 1 without causing the second or fourth inputs to become 0.

The presence of a fault effect in a list implies that its effect will reverse the line; therefore, the search is for faults which appear in the lists of the first and third inputs, but do not appear in the second or fourth input lists. A0301 s-a-1, A5606 s-a-1, and B2814 s-a-0 appear in only one list each and can therefore be eliminated immediately. D1606 s-a-1 and Q4618 s-a-0 both appear in the lists for the first and third inputs, but Q4618 s-a-0 also appears in the fourth input list. D1606 s-a-1 is added to the list associated with the output node Z4601, but not Q4618 s-a-0. The only remaining fault in any of the lists is A0303 s-a-0, and a quick glance reveals that it appears in only the second and fourth lists (exactly the opposite of what is needed).

The concepts of deductive fault simulation are sound and are often discussed in reference to generalized application to such areas as functional model fault simulation and algorithms for programmable logic fault analysis. The algorithms for deducing the applicability of a fault become considerably more complex when more than two states or other generalizations are included. The similarity in concept to concurrent fault simulation will probably be noticed. Concurrent fault simulation has become one of the most popular schemes and has overshadowed the deductive method.

Concurrent Fault Simulation. Good circuit simulation in the concurrent method remains much the same as in serial and parallel simulation. The data structure for keeping track of fault effects is markedly different. A linked list of fault effects replaces the columns of the fault tables used in parallel simulation. Each logic node in the circuit is represented by a cell or record in the computer's memory. The cell has fields for a node identifier and several pointer fields.

The example in Figure 4.6 combines linked lists which contain the structure of

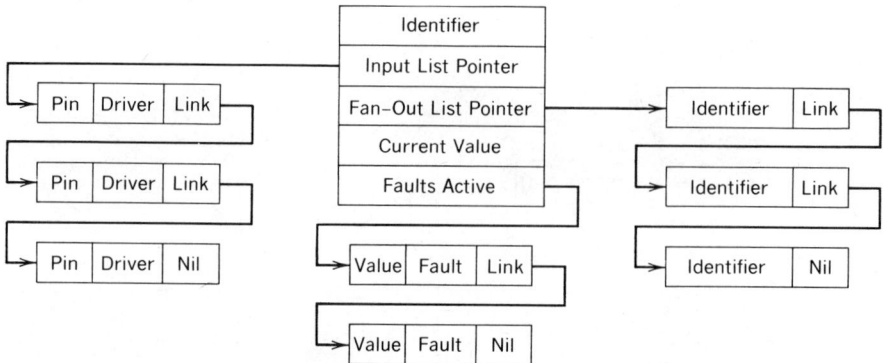

Figure 4.6. Linked list data structures represent gates and fault effects in a concurrent fault simulator.

the circuit with a linked list which holds the fault values for the node. The linked lists may be separately addressed in some implementations. A table of the memory locations corresponding to identifiers directs the scheduler and evaluator to the cell location referred to by their queues. The scheduler is interested in the gates which may be affected by a change in the value of the cell's gate. This information is easily found by tracing the linked list which is pointed to by the "fan-outs" field. Each cell of the linked list has two fields: one for data (a gate identifier) and one for the pointer to the next cell in the list. If the pointer contains the end symbol "NIL," the list end has been reached. Any value which is not a valid cell address can be used as NIL.

When the evaluator locates the cell for evaluation, it first traces the "input" list in a similar fashion and locates all the cells of gates which feed the gate under evaluation. Each gate cell has a field which contains its present good circuit value. These values and the function of the gate under evaluation are used to calculate a new value for the "value" field in the present gate's cell. Of course, the value is compared to the old value before it is written into the cell to determine whether or not a change has occurred. This procedure does no fault simulation up to this point.

To investigate the effects of faults on the gate requires manipulation of the fault effect lists of the input gates as well as the gate under evaluation. If a fault is directly associated with this gate, the effect is considered in the gate solution process directly. For example, if the gate solution results in a new value of 1, but the fault output s-a-0 is in the fault list for this gate, an entry is made in the fault list stemming from the fault pointer field of the cell.

In fact, the fault effect list is reconstructed each time the cell's gate is evaluated. The fault effect list consists of cells which contain three fields. A field is dedicated to the faulted circuit value. Another field contains a description of the fault under whose presence the value holds. (This field usually contains a table address or fault number which refers to the fault rather than actual text describing the fault.) The last field contains the pointer to the next cell in the list or the NIL symbol.

Having taken care of fault effects which originate in the gate under evaluation, the remaining task is to determine what fault effects, if any, propagate through the gate under the present circuit conditions. The list of input gate cells is accessed, each cell in turn for the fault effect values attached to it. The fault effects are applied to the gate under evaluation, and the result value is compared to the good circuit value. Care must be taken to match up the fault effect values if they exist under the presence of the same fault. If such is the case, all effects which exist under the same fault are applied to the gate in a single evaluation. Otherwise, each fault effect which is not matched by the fault effect of another input is applied singly to the gate while the other input values are held at good circuit values. Each time an evaluation results in a value which differs from the good circuit value, the fault effect value with the fault identifier whose input effects were used in the evaluation is added to the end of the fault effect list for the gate under evaluation.

The contents of the linked list concurrent fault simulation data structure shown in Figure 4.7 correspond with the all-0 input vector applied to the circuit of Figure 4.2. Several simplifications have been made, such as the deletion of explicit point-

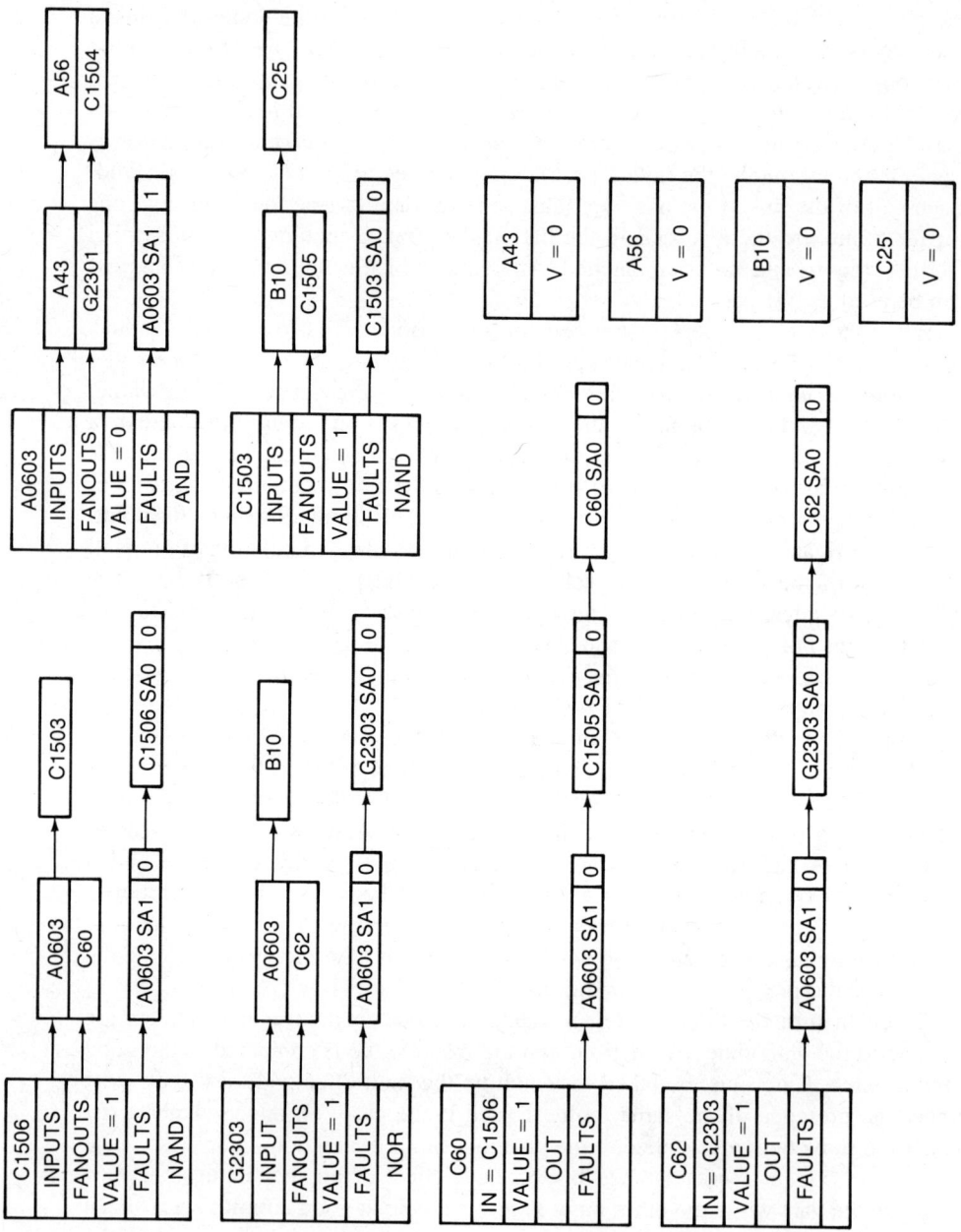

Figure 4.7. Linked list data structure corresponding to the circuit in Figure 4.2 with all-0s input.

ers within the lists themselves. The example is not a particularly efficient representation, only an obvious one. Note the fact that each gate is contained in the input list of another gate, while the fan-out list of the latter contains the former. The database is doubly linked. The cells which hold gate identifiers could be doubly linked to each other if each cell contained a variable number of pointer fields for the input and fan-out sections.

The salient features of the concurrent representation are a reduction in processing and overall memory requirements. The reduction of processing comes from requiring evaluation of only active fault effects. Not only are most faults not active at any given time, but their effects are only felt by a relatively small number of gates. The linked list data structure is not as efficient in memory use as the tables of the parallel method on an item-by-item basis, but the presence of many fewer items to store produces a net reduction in memory use in many applications [2].

The variable nature of the memory requirements is a bane of the method, however. In circuits with high circuit activity the number of fault effects can become large. Sequential circuits aggravate the problem since the fault effect lists must be maintained as long as the fault is present, even if it is latched into a flip-flop. Memory needs cannot be predicted before simulation begins.

Variations. The pressure to reduce processing time and/or memory requirements of fault simulation has given rise to intense analysis of the basic procedures just described. Several interesting variations have come forth. The nature of most variations is to reduce the computation effort for a specific class of problems rather than to improve the basic methods across the board.

As large circuits which require more memory for fault tables or fault lists than is readily available in most computers become frequent problems, the order of the simulation process is often reversed. In the examples the logic simulation and fault effect analysis are performed on the entire circuit for a single test input vector before moving to the next input vector. If the fault effects must be partitioned into fault blocks due to memory limitations, this results in a constant shuffling of data to and from mass storage.

As long as the main memory can hold a complete circuit description, the good circuit simulation can be completed without memory shuffling. The expected response can be recorded from the good circuit simulation for the input vector under consideration, and fault effects whose descriptions fit into memory can be analyzed. Excluding simultaneous multiple fault presence from the calculations, the fault effects from different fault blocks do not interact. It is therefore reasonable to proceed to calculate the results of all input vectors on a single fault block before storing it on disk and retrieving another. The test effectiveness will not be known until all blocks have been processed in any case. This "batch" mode of simulation is effective in many situations due to the computer resource limitations that otherwise limit performance. Unfortunately, the immediate feedback which might be used in an interactive mode to guide the selection of the next test vector is lost. When test vectors are generated by other than interactive manual means, this simple rearrangement of processing is quite effective.

Taking advantage of parallelism in the process, as opposed to the data structure parallelism exploited by the "parallel" method explained earlier, requires some restrictions in the class of problem which can be addressed. A significant class of problem has arisen due to the adoption of the LSSD methodology introduced by IBM Corporation. The technique will be explored in Chapter 11. The important feature with respect to fault simulation is the elimination of sequential elements from the circuit portions to be tested.

In a purely combination circuit several simplifications can be made in the fault analysis and logic simulation tasks. A simple restriction of input values to known logic values and a prohibition of 3-state off states reduces the logic simulation to a 2-state problem. The lack of sequential devices allows the results of all input vectors to be calculated on any logic gate in any order. Simulation can now proceed on a rank by rank basis, calculating a list of gate output values for each gate resulting from the entire test vector set.

Fault effects for each single stuck-at fault are then propagated from the originating circuit node through the circuit. At each node the entire vector set of values is available. If any of the value sets result in the propagation of the fault effects to a primary output, the process is terminated for that fault. This process is termed the parallel pattern single fault propagation (PPSFP) method [3]. The limitation of the PPSFP technique is in the nature of its assumptions. Maintaining a purely combinational design is difficult, and the presence of sequential elements destroys the accuracy of the method. Attempts have been made to generalize the results by relating the effects of violations of the assumptions to the results on a statistical basis.

Statistical Methods. For large data manipulation problems such as fault effect analysis, the power of statistics is attractive. When an overall result, such as the degree of fault coverage for a particular test vector set, is desired, but the individual details, such as precisely which faults have not been detected, are not necessary, the statistical approach can be dramatically cost effective. The limitations of validity for the statistical analyses involved must be observed, however.

The most direct use of statistics is made by selecting a random sample of faults from the total fault set for simulation. A significantly sized sample is used to make up the otherwise normal fault simulation, and the results are generalized proportionally to predict the results of a total fault simulation. This idea is used in two ways. The fault coverage prediction can be used to decide when to stop generating input vectors. It cannot be used to guide the production of input vectors since any attempt to specifically detect a member of the sample of faults invalidates the randomness and statistical value of the sample. When input vectors are selected by random generation methods, the question of when to stop is important and there is no chance of influencing the statistical validity, making the two methods well suited.

Statistical inferencing can also be used to predict the magnitude of the full fault set simulation task. This information can be used for project planning, DFT analysis, and performance comparisons of alternative test generation techniques. Sample size is important to the accuracy of the results. For circuits of 1000–6000 gates a recent study [4] showed the predicted fault coverage was within 2% of actual for a fault sample of 10%. A 1% sample yielded poorer correlation, with predicted and actual fault coverages differing by as much as 7%.

A more complex approach to statistical analysis involves computing probabilities of the individual events which make up a test, fault activation and effect propagation, for each node in light of the input vector set. The prime concepts involved are the notions of controllability and observability of a logic node. These ideas were first used in conjunction with testability analysis, which is discussed in Chapter 11. For convenience the subject will be briefly introduced here.

Controllability of a circuit node is determined by the number of primary inputs which must be manipulated to achieve a 0 or a 1 (combinational controllability) or the number of sequential elements which must be negotiated on the path from primary inputs to the node (sequential controllability).

From the examples of Figure 4.8 the controllability of a gate output can be derived from the controllabilities of its inputs. If the primary inputs of the circuit have a controllability of 1, then each rank of logic attains higher and higher controllability numbers (representing less controllable nodes) as the numbers are added. The combinational controllability of a 1 value on the output of the NAND gate (node C) is the lesser of the controllabilities of 0s on the gate's inputs since only one of the inputs needs to be controlled to achieve the output. The controllability of a 0 on the output of the NAND requires both inputs to be manipulated and thus is the sum of the two controllabilities.

Calculating the observability of each node is more complex. Primary outputs have an observability of 1. For each rank from the outputs back to the inputs the observability is calculated for a gate input by the observability of the gate output and the controllability index of the conditions necessary for sensitizing the output with respect to the particular input.

The observability of input A of the NAND gate in Figure 4.9 is the observability of its output $O(C)$ plus the controllability of the value 1 on the other input $C1(B)$. The value 1 must be achieved on input B to sensitize the gate to the input

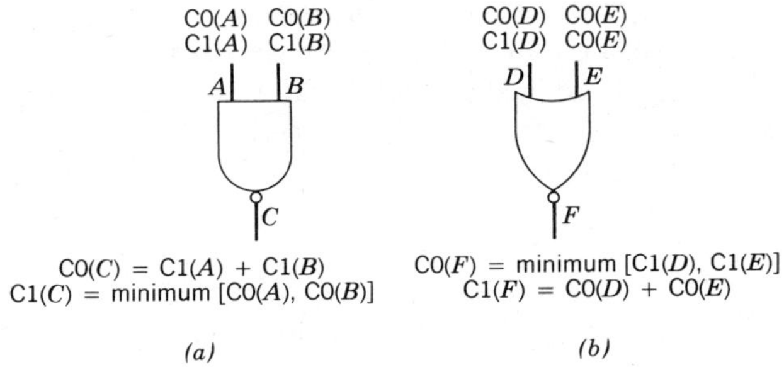

CO(A) CO(B)
C1(A) C1(B)

A| |B

|C

CO(D) CO(E)
C1(D) CO(E)

D| |E

|F

CO(C) = C1(A) + C1(B)
C1(C) = minimum [CO(A), CO(B)]

CO(F) = minimum [C1(D), C1(E)]
C1(F) = CO(D) + CO(E)

(a)

(b)

Figure 4.8. Controllability of a node is dependent on the driving gate's function and the target logic state.

$$O(B) = O(C) + C1(A) \qquad O(E) = O(F) + CO(D)$$

$$O(A) = O(C) + C1(B) \qquad O(D) = O(F) + CO(E)$$

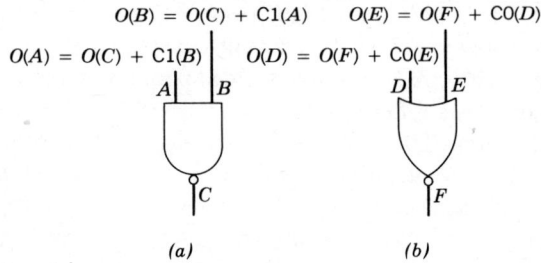

(a)   (b)

Figure 4.9. Observability of a node depends on the driving gate's function and on the controllability of its inputs.

A. Note that the observability figures have no polarity since a sensitive path is sensitive to a change in logic state rather than the value of the logic state.

The controllability and observability values for nodes of a circuit are dependent upon the structure of the circuit and not on the test results. Calculation of the controllabilities and observabilities can be made directly or inferred from the statistics of simulation results, however. If the value of a node is recorded as a 1 for most of the test sequence, it is reasonable to assume that it has a high 1 controllability and a low 0 controllability. Use of the logic simulation results is convenient since it eliminates the need for an explicit calculation of the factors.

Statistical fault coverage determination is a matter of combining the controllability and observability results to estimate the probability of a node having been manipulated into a given state while a path to an output was sensitized. An easily observable circuit (with a low observability index) which has been shown to have had a value of 1 for most of the test was probably tested for a s-a-0 during the test. The use of good circuit results allows individual fault coverage probability calculation, which in turn can be used to guide further generation of test vectors. The statistics used are not dependent on the randomness of input vectors, as was the case in fault sampling.

Sequential circuitry creates additional factors for the controllability and observability calculations and generally lessens the accuracy. For circuits of 100–3000 gates the statistical prediction for fault coverage has been shown to be less than a percent different for combinational examples, but it may be several percent different than full fault simulation for sequential examples [5].

Hierarchical Simulation

Top-down design methodologies and the availability of CAD systems with hierarchical capability have prompted some adaptations to logic and fault simulation. Hierarchical representation refers to the interconnection of several levels of abstraction for the circuit, allowing parts of the circuit to be referred to as a functional description while other parts are represented in gate- or transistor-level descriptions. Simulation of a hierarchically represented circuit requires the capability of solving for expected response of the circuit elements in a behavorial as well as logic gate truth table fashion.

The register blocks of Figure 4.10 are supported by a software algorithm which describes the functions and data flow characteristics of the block. The dotted boxes in the center have been expanded by replacing the behavior description of the adder with its gate-level representation. Output data from the higher-level descriptions must be transferred to the gate-level descriptions and back again.

Hierarchical description methods were created to allow design verification of the circuit or system being designed before the entire design had been reduced to gates. Test generation in the same environment can be accomplished by several approaches. The good circuit simulation portion of the test generation process can be performed at higher levels by software algorithms, but the visibility of internal values is lost.

Accurate functional models must be created without reference to the implementation. Timing features must be explicitly described when they are necessary to the device operation in the system. The hardware description languages available contain constructs to contend with sequentiality and timing. Perhaps the most difficult functional modeling task is the complete description of the block's behavior under circumstances which are unimportant to the design intent. Although these characteristics are not used in the specification of the circuit by the designer, the simulator must be able to solve the circuit in all cases, whether intentional or not. For example, what does the carry–generate output of a 2901 microprocessor slice do when the chip is in the logical operation mode?

Fault Simulation. Fault simulation in conjunction with a hierarchical simulation could also be conducted at the higher levels of abstraction. Functional models would be required to contain descriptions of their behavior under the list of faults which might be associated with the block. Making up an adequate list of possible faults for a functional block is not an easy task either. Both fault effect generation in a faulty block and fault effect propagation through a properly performing block must be included in the description.

A more commonly taken approach involves "flattening" a portion of the circuit to its equivalent gate level representation before beginning fault simulation on that portion. Conventional stuck-at faults can then be applied to the flattened portion of the circuit, and gate-level simulation can be performed for that block. If the block requires interaction with the remainder of the circuit, the data will have to be passed to a higher-level description. Signals will then be handled using the higher-level algorithms for the good circuit case. Only the flattened portion of the circuit is fault simulated; the remainder is good circuit simulated only.

Completing fault simulation in this fashion requires that the circuit be reconfigured several times, and the input vectors must be rerun to collect complete test data for the circuit. In principle, the mixed mode hierarchical simulation just described should be much faster than conventional simulation. Since fault simulation is at least proportional in run time to the square of the number of gates simulated,

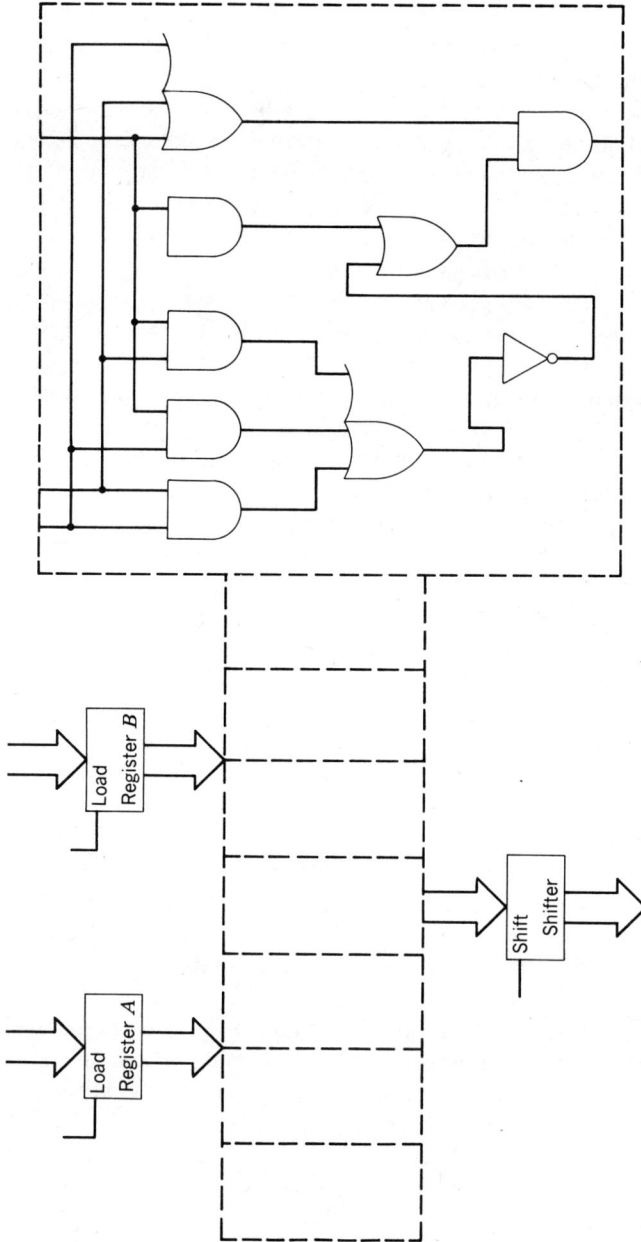

Figure 4.10. Hierarchical circuit representation. Each of the dashed boxes contains a replica of the contents of the rightmost box. Solid boxes have functional representations only.

simulating 7 groups of 1000 gates should be much faster than simulating all 7000 at once by a factor of $49,000,000/7,000,000$ or 7. In reality, the time absorbed in restructuring the problem and in correlating the results reduces the advantage greatly. Smaller partitions and some careful choices or representation techniques have resulted in improvements for some problems of 60 to 1 [6].

Hardware Acceleration

Logic simulation and particularly fault simulation are computationally intensive problems. Although simulation has long been accepted as the most accurate technique for determining expected response and fault coverage, other methods have been developed and used to avoid the expensive computer resources needed in fault simulation. New algorithms provide some relief, but none has proven overwhelmingly successful.

An analysis of the computing resources needed and the nature of the operations performed in fault simulation reveals a poor fit to the traditional von Neumann architecture. SISD machines bog down amidst the sea of nearly identical data tables and repetitive operations of parallel fault simulation. The block-oriented memory management methods struggle with swelling and shrinking linked lists of concurrent simulation. The massive number of simple logic operations needed is not expedited by the addition of floating point accelerators.

Creation of high-speed specialized hardware is generally centered around identification of processes which can be performed in parallel. In fault simulation there are three easily identified sources of parallelism.

1. Activity in a logic circuit takes place on many parallel paths at the same time.
2. All the gates of a rank of logic can be solved simultaneously without interaction. The algorithms which make up simulation are simple and applied repetitively to different parts of the circuit. They also are made up of distinct processes which can be performed in parallel, such as evaluation and scheduling.
3. Fault effects create many possible values which may exist on a logic node at the same time.

The first characteristic leads to the use of parallel processors. SIMD machines are properly configured to perform the simple operations on many gates simultaneously. The hypercube "Connection Machine" developed by the Massachusetts Institute of Technology is an example of such a machine which has been applied to the simulation problem [7]. The IBM Yorktown machine [8] uses conventional multiprocessor concepts to take advantage of the same parallelism while pipelining within each processor to take advantage of the parallel nature of the algorithm steps themselves.

Both of the previous examples were designed primarily as logic simulation machines for good circuit simulation only. Fault simulation was performed on the Yorktown Simulation Engine by serial fault simulation. The speed improvements proved to be insufficient to offset the disadvantages of serial fault simulation, however.

Use of a wide data path and memory structure in a prototype parallel fault simulation engine developed by a Sperry team including the author proved to give a speed improvement over conventional computer parallel fault simulation even though the processor speed of the simulation engine was slower than that of the general purpose machine. A combination of several features including pipelining, parallelism, and hardware implementation of the concurrent fault simulation algorithm enable the Zycad Fault Evaluator [9] to achieve a two orders of magnitude improvement over mainframe computers executing the concurrent algorithm.

The sheer size of data structures and the magnitude of a problem which grows exponentially relative to the complexity of circuits creates a challenge to any and all approaches. Constant improvement of method and tools is required to maintain the status quo in the fault simulation arena.

SUMMARY

A wide range of techniques and tools is available to determine the expected response of a CUT and to determine the thoroughness of the test. The costs and results of the techniques vary widely also shown in Table 4.2, which suggests that care should be taken in selecting the proper technique for a particular application.

The selection of an expected response method is closely paired with the stimulus generation decision. If the application is the self test environment where an internal generator will create random patterns, and conservation of data storage and hard-

TABLE 4.2 Expected Response Alternatives

Technique	Cost	Coverage	Feedback	Comments
KGB	Low	No information	None	Must maintain KGB
Recorded (signature analysis, etc.)	Low	No information	None	Portable
Logic simulation	Medium	No information	None	Accurate
Parallel fault simulation	High	% detected	Fault list	Slow, memory hog
Concurrent fault simulation	High	% detected	Fault list	Faster, but unpredictable memory use
Statistical	Medium	Approximate %	Some versions	
Hierarchical	High	% detected	Fault list	Complex support
Accelerated concurrent	High	% detected	Fault list	Big Capital Investment, but fast running

ware size is a premium, the LFSR will serve well as both the pattern generator, and with slight modifications, a signature analysis monitor. Secondary fault grading will be required to assure coverage, however, and the statistical simulation technique is a reasonable choice. If the tests are to be used in a high-reliability manufacturing or field testing location, the accuracy and fault coverage proof of fault simulation is well suited. The patterns can be supplied to the simulator by an ATG program supplemented by well directed manual generation to assure optimum fault detection. The final input and output vectors will probably be applied by high-speed ATE with relatively large data storage capability. For high-volume test generation operation an accelerator may be an economically viable alternative.

REFERENCES

1. D. B. Armstrong, "A Deductive Method for Simulating Faults in Logic Circuits," *IEEE Trans. Comput.*, **C-21**(5), 464–471 (May 1972).

2. K. Son, "Fault Simulation with the Parallel Value List Algorithm," *VLSI Syst. Des.*, **VI**(12), 36–43 (Dec. 1985).

3. J. A. Waicukauski, E. B. Eichelberger, D. O. Forlenza, E. Lindbloom, and T. McCarthy, "A Statistical Calculation of Fault Detection Probabilities by Fast Fault Simulation," in *Proc. 1985 Int. Test. Conf.*, IEEE Comput. Soc., pp. 779–783.

4. P. Goel, C.-Li. Huang, "Statistical Fault Sampling and Full Fault Simulation," in *Proc. 1985 Int. Test Conf.*, IEEE Comput. Soc., pp. 801–802.

5. S. K. Jain and V. D. Agrawal, "STAFAN: An Alternative to Fault Simulation," in *Proc. 21st Des. Automat. Conf.*, IEEE, 1984, pp. 18–23.

6. W. A. Rodgers and J. A. Abraham, "CHIEFS: A Concurrent, Hierarchical, and Extensible Fault Simulator," in *Proc. 1985 Int. Test Conf.*, IEEE Comput. Soc., pp. 710–716.

7. G. A. Kramer, "Test Generation Algorithms: Overcoming the von Neumann Bottleneck," in *Proc. 1983 IEEE ATPG Workshop*, pp. 86–95.

8. G. F. Pfister, "The Yorktown Simulation Engine: Introduction," in *Proc. 19th Des. Automat. Conf.*, June 1982, pp. 51–54.

9. R. Landry, "Zycad Announces Fault Evaluator," *IEEE Des. Test Comput.*, **2**(2), 97 (Apr. 1985).

CHAPTER FIVE

CIRCUIT AND FAULT MODELING

Electronics is explainable. The magic often brushed off or attributed to noise is actually a rational physical effect whose cause is simply too obscure or complex to be worth the observer's investigation. If the "noise" creates a failure, the depth of understanding may be increased by an intensive set of measurements or experiments, and the actual cause of the problem will be found. The thoroughness of the explanation of a circuit's characteristics reflects the usefulness of the information. The representation of a circuit by schematics or prose technical description is a humanly readable model of the circuit which conveys sufficient information for human use while remaining tractable for the human mind.

Circuit modeling in the sense usually discussed refers to machine-readable representations of the circuit which contain enough information to allow a computer program to predict the circuit's output in response to a set of stimuli. The accuracy of the prediction is directly proportional to the information conveyed in the model. The computation difficulty, and hence the time required to solve for the present circuit outputs, is directly proportional to the information about the circuit which the program is expected to take into account. A model of a circuit is appropriate to the task if it is detailed enough to give sufficiently accurate answers, but terse enough to minimize computation time.

If purists are upset with the idea of approximating a circuit with a model, the simple example of Ohm's law should provide convincing evidence of the occasional necessity of the concept. Impedance is commonly modeled as a device parameter which appears in formulas as a constant proportionality of current voltage. In the real world, effects such as temperature and skin effect change the constant to a dependent variable. A very thorough Ohm's law would contain temperature coefficients and the frequency of voltage or current being supplied. Since this would be cumbersome for most calculations, a simplified model is used.

In fact, digital circuits are modeled in several different ways in response to the needs of the tools to be used. During a "top-down" design process a digital system

Gajski–Kuhn "Y" Diagram

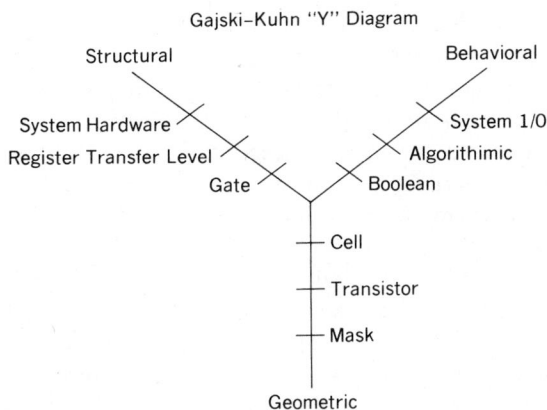

Figure 5.1. The Gajski–Kuhn "Y" diagram relates the circuit representations technique graphically. (© 1986 IEEE.)

and its circuits are constantly remodeled from a particular description to a less abstract one until a physical realization is achieved. The Gajski–Kuhn "Y" diagram [1] seen in Figure 5.1 suggests that there are actually two top-level descriptions from which the design may begin. The structural model of the system specifies the "black boxes" which make up the system and the buses or signals which connect them. On the other hand, the behavior model describes the processes and the information they will exchange without explicit reference to the partitioning of the circuit.

As the circuit definition is forced downward from the abstract to the detailed realization, the structural path passes through a model based on the registers (blocks of storage elements), data paths between the registers, and the allowed transfers of information along those paths. This is the register transfer-level description, which is now often referred to as the HDL description. Futher refinement results in remodeling the circuit into logic gates having primitive functions like NAND, NOR, EXCLUSIVE-OR, and so on.

From the behavioral system description the path downward passes through a software-like description of the functions of each circuit partition, which is termed an algorithmic model. At this level each of the system functions is described as conditional actions resulting in information exchange and control interaction with other modules of the system. The final refinement in this case models the circuit functions as Boolean logic descriptions. The behavior modeling method has found its best application in the conversion of microcode into gate array VLSI devices.

A common model of a complex digital circuit such as a PC board containing a mixture of standard and custom ICs is a hybrid of structural and behavior models. Most of the circuit may be modeled as gates, but pockets of logic may be described as Boolean equations or microcode algorithms. Test tools such as fault simulators often require the homogenization of circuit description into gates only. Boolean or algorithmic models must then be converted to gate realizations. Not all tools require such conversion, however. Some tools allow behavior and structural descriptions to be mixed, and some permit hierarchical modeling in which sections of the model may be described at differing levels of abstraction.

Models on the geometric or physical description arm of the diagram have a one-to-one correlation to the realization of the circuit. This permits physical characteristics to be accurately represented, but the complexity quickly becomes a concern for test tools.

Although all of the levels of circuit representation are useful in some domain, most test tools use gate-level models. Sufficient use is also made of transistor (switch), algorithmic, Boolean, and register-transfer- (HDL) level models to warrant some discussion. Each of the alternatives will be described in enough detail to allow the reader to recognize and differentiate the models, while gate level models will be discussed at length.

GATE-LEVEL MODELS

Gate-level circuit representation is historically derived, and its popularity results from the availability of gate-level tools and the broad base of familiarity among users. Digital ICs, which caused the computer industry to mushroom, were at first straightforward embodiments of Boolean functions. Circuit designers quickly abandoned the tedium of describing the transistors which made up each Boolean gate and substituted symbols for each function. Since any logic circuit can be described from a few types of logic functions (actually, only two are needed), the gate-level description is easy to create and easy to understand. The gate-level model of a circuit conveys the function while holding an identifiable correlation to the components which make up the physical circuit.

Primitives

Gate-level models of digital circuits are networks of interconnected devices called gates. Each gate has an associated functional equation or truth table which provides the relationship between the gate's inputs and its outputs at any given time. The historical definition of a gate has been broadened to include any element which has a single bit output and a function which can be described in a truth table. This includes combinational functions and simple sequential elements such as 1-bit storage cells or flip-flops.

In addition to the traditional AND, OR, NAND, NOR, and inverter gates, there are hybrid versions in which one input and a traditional gate is inverted while the other is left uninverted. Hybrid gates were not provided in the SSI ICs of commercial suppliers, but are sometimes fabricated as part of custom VLSI circuits. The hybrid variations such as those shown in Figure 5.2, result in the use of fewer gates for complex functions and more closely represent the actual implementation for some circuits.

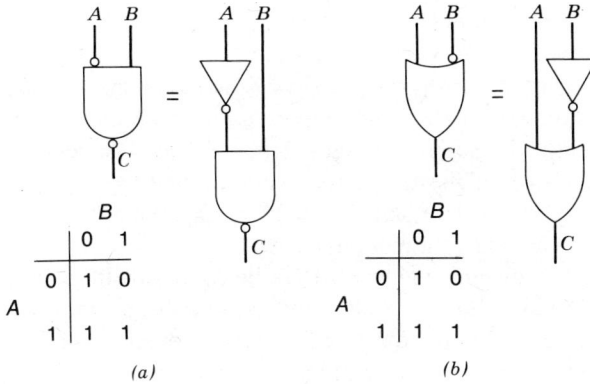

Figure 5.2. Hybrid function gates and their standard gate equivalents.

Bus Elements. Primitives are also created to represent the action of bus elements and bus drivers. The bus element most commonly used is the "wired-OR" or "wired-AND" function, which requires an additional model when 3-state logic is involved. Each bus driver must include a primitive which interprets the "enable" signal and provides a representation of the "off" or "Z" state when the driver is disabled. The bus primitive is then tasked with resolving the inputs from the drivers into a present bus state. In the actual circuit the resolution is a matter of the physical characteristics of the drivers and the wire which makes up the bus.

Passive elements such as pull-up resistors may affect the resolution and are therefore also modeled as primitives. Nonlogic influences and the time-dependent nature of the signal on an undriven bus complicate the bus primitive. A typical bus primitive for TTL circuitry and its corresponding driver primitive are shown in the example of Figure 5.3.

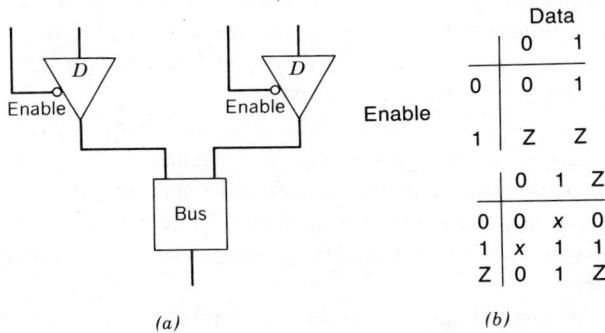

Figure 5.3. (a) Three-state drivers and a connective bus element. (b) The truth tables delineate the functions of each gate type.

Bus elements are the foci of signal arbitration associated with the strength concept of digital modeling. In strength models the values of a logic signal (0, 1, x) are modified by a strength. A power supply or ground connection has the strongest signal, while an ordinary gate is the highest strength which changes its value but is overcome by a power supply strength signal. Passive elements such as pull-up resistors have the third highest strength, while an undriven bus line's capacitive storage constitutes the weakest signal. The hierarchy makes bus resolution simpler and represents the actions of a real circuit well.

Signal strength and signal value must not be confused. In models which do not use strengths explicitly, the Z, or off, state is often discussed as though it were a signal level between the 0 and the 1, but this is an oversimplification of the situation. The Z state is actually an undriven strength. In TTL circuits the I_{IH} of the gates fed by a node tend to charge the node to the threshold value of about 1.7 V, leading to the mistaken impression concerning the Z state.

If the bus has previously been charged to a logic 1 state, the slow discharge due to leakage in the driven gates will give an exponential decay in the signal voltage toward the threshold level. For a few microseconds the signal will remain a valid 1 before becoming neither a 1 nor a 0. A more accurate representation of the situation for an undriven signal strength is to leave its value unchanged for a few microseconds of modeled time and then change its value to an x or unknown state.

The presence of a pull-up resistor connected to the bus would change the behavior. If the bus signal had been a 1 before the drivers were disabled, the signal would stay at a 1 indefinitely instead of floating toward the threshold. If the previous signal were a 0, the bus would quickly change to a 1 when the drivers were disabled. By design, the pull-up resistor is sized to charge the bus in a period of time that is short in comparison to other circuit events, and therefore the signal can be considered to switch immediately to a 1.

In essence, the model works as though a gate which has a permanent output of 1 were connected to the bus when no other gates are driving the bus. Modeling the pull-up as a gate which is on when all other drivers are disabled is a viable technique in a modeling environment which does not use explicit strengths. The second circuit of Figure 5.4 shows this technique for a TTL bus.

If strengths are modeled explicitly, a permanent logic 1 with a strength less than an active gate should be installed as an input to the bus element. If any driver is active, its value will take precedence over the pull-up value without the necessity of additional gates to turn the pull-up off. The strength of the pull-up is modeled as sufficient to overcome the undriven capacitive state of the bus, and the time-delayed unknown outcome is thus avoided. A complete mapping of the outcomes of bus arbitration is the strength modeling system described previously is shown in Table 5.1.

The table entries which are followed by ''!'' are actually violations of the specifications for most TTL logic devices. The outcomes shown in these cases may be presented or an alarm may be triggered to alert the user of the model of an illegal condition on the bus. Multiple resistive values can be present at the same time,

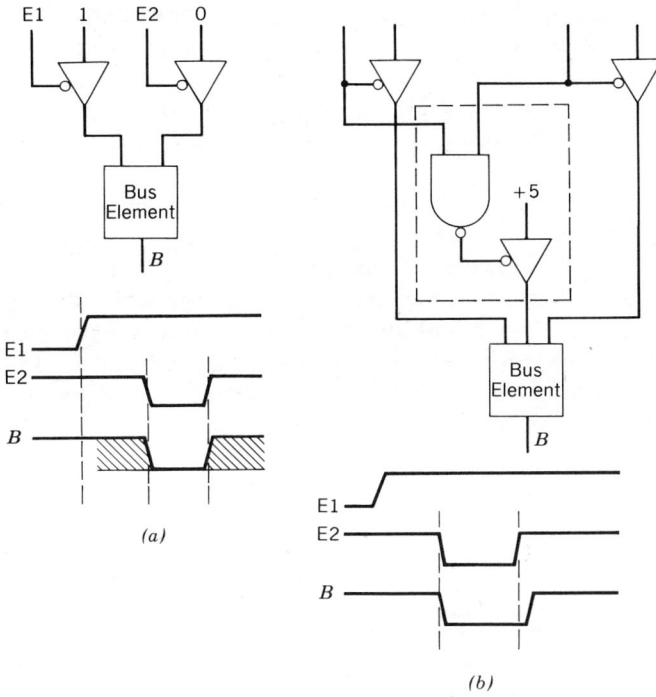

Figure 5.4. Models and resulting waveforms for TTL bus elements. (a) Without pull-up. (b) With pull-up.

although pulling the bus up and down simultaneously results in an unknown since the evaluation of actual resistive values and solution of the circuit equations to determine the actual state are beyond the scope of a digital model.

Macros

The primitives described (gates and bus elements) are sufficient to model digital circuits implemented in bipolar transistor technologies (TTL). These gate-level

TABLE 5.1.   TTL Bus Arbitration Using Strengths

		Active		Resistive		Undriven
		0	1	0	1	(Z)
Active	0	0!	x!	0	0	0
	1	x!	1!	1	1	1
Resistive	0	0	1	0	x	0
	1	0	1	x	1	1
Undriven		(Z)	0	1	0	1	x

models are also useful in modeling MOS digital circuits, although some effects must be ignored or accounted for by special constructs. In either case, however, constructing a model for an ordinary VLSI chip or a PC board results in a sea of gates, which gives little insight into the function of the circuit without tedious analysis. The repetitive specification of each primitive in large circuits is also tedious.

Larger structures of a circuit which occur repeatedly can be constructed for primitives once and thereafter copied for each remaining instantiation. The individual primitives in each new use of the structure must be carefully relabeled to avoid duplication, but their functions and interconnections within the copied group can remain unchanged. A model structure used in this way is often called a "macro." Unique labels for each gate can be maintained by attaching a label for each instantiation of the macro as a prefix or suffix to the gate labels within it.

If the model paradigm supports the nesting of macro structures, MSI and LSI devices such as the shift register shown in Figure 5.5 can be easily constructed. Each flip-flop in the shift register is a macro of NAND gates, and the entire shift

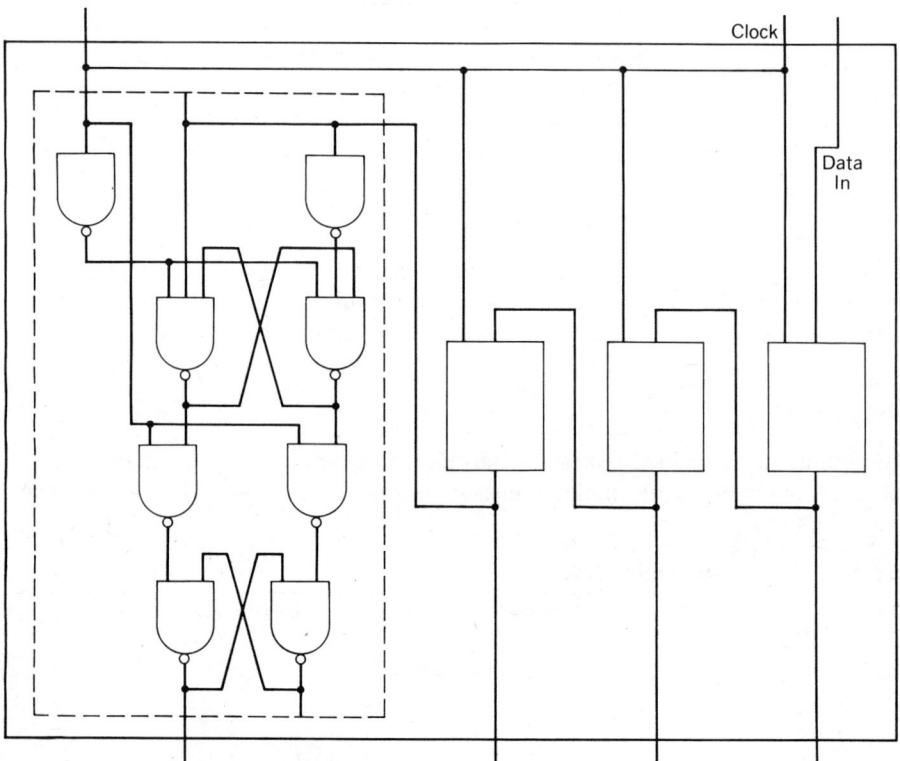

Figure 5.5. Using nested macrostructures, a shift register is composed of flip-flops, which are in turn constructed from gate primitives.

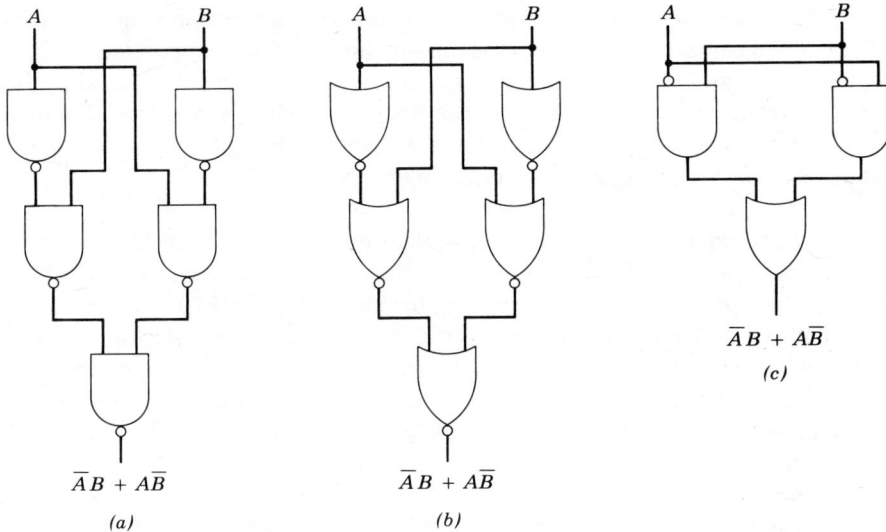

$\overline{A}B + A\overline{B}$

(c)

$\overline{A}B + A\overline{B}$

(a)

$\overline{A}B + A\overline{B}$

(b)

Figure 5.6. Three models of the exclusive-OR function with differing structures.

register is a macro of flip-flops. In a complex circuit, a giant micro such as a universal asynchronous receiver–transmitter (UART) might be in turn constructed using the shift register macro. Although the user would undoubtedly think of the circuit as a block which exhibits the characteristics of a UART by responding appropriately when stimulated with UART instructions and data, the tools which use the model would see a mass of NAND gates or other primitives.

The equivalency of the model and the real device is limited to the effects observed on the outside of the model. Unless the precise makeup of the circuit being modeled is known and rigidly followed (rarely the case), nodes within the model may not correspond to those of the real circuit. In Figure 5.6 the three models of the exclusive-OR function use different primitives to achieve the same external characteristics. Restrictions of the modeling method may result in any of the three being used, or the exclusive-OR function may itself be a primitive. The aim is to create a model which sufficiently corresponds to the real circuit in behavior while minimizing the number of primitives needed.

Timing

The behavior of digital circuits is characterized by Boolean equations and by the timing relationships between inputs and outputs. If a circuit is completely combinational, modeling it without regard for the timing will produce a logically correct output, once the transients have settled out. Sequential circuits cannot be modeled so casually.

Modeling the timing aspects of a circuit adds a new dimension with new difficulties. Adequate accuracy in the modeling of timing depends upon the applications intended for the model. In static functional testing, a rough approximation which is consistent enough to predict transients or glitches is adequate. If a high speed test is to be performed, sophisticated models which include rise and fall timing or delay modeling of circuit path lengths may be needed.

Unit Delay Models. Propagation delay is the time between the application of an input change and the appearance of a new result at the output to a gate or circuit. Not only does propagation delay vary with technology (TTL, ECL, etc.), temperature, and power supply voltage, it also varies from device to device as the semiconductor manufacturing process varies. The latter variation has been reduced greatly as integration levels which place large circuits on the same die are now commonplace.

For a logic technology which makes use of well-defined gates of a variety of functions such as NAND, NOR, AND, OR, and so on, each gate's propagation time can be characterized and normalized such that a NAND gate has a delay of 1 unit of time, for example. Unless the transistors used to form a gate were sluggish or a greater number of them were used on a complex gate function, most gates would exhibit normalized delays of close to 1. The difference would not be apparent unless the designer used many ranks of logic to manipulate signals without resynchronizing them to a clock. Since this practice is not common due to concerns for temperature-induced timing variations within a circuit, the unit delay method is usable in many cases.

As long as the propagation delays of all gates in the circuit do not vary greatly, relative to one another, the circuit operation will be logically correct. A circuit path which is 4 gates long will always beat a 5-gate circuit path in such a situation (ignoring loading effects) even if the gate delays are doubled by a temperature or voltage change. A simple method of modeling the delay gives each gate a delay of 1. If a gate's inputs change value, its outputs do not change until 1 unit of time later. The real value of the unit may be in microseconds, nanoseconds, or picoseconds, but the model does not need to consider measured values.

Unit delay modeling greatly simplifies the simulation task. Minor modifications to the scheme are usually required to improve the accuracy of modeling for such devices as bus elements. If the bus capacitance is not considered, the bus device (which is actually only a wire) is usually modeled as a 0 delay gate. The presence of gates whose outputs change immediately upon a change in input requires special handling for the simulator which uses the model.

Inaccuracy in the model timing most often appears as failure to resolve hazards correctly and/or the generation of extraneous transients. In combinational circuits these transient effects can be ignored if sufficient settling time is allowed before sampling the outputs. If the transients enter the clock input of a flip-flop or similar element during simulation of a sequential circuit, the state may be changed and the circuit values may differ from the model until the next correct clock pulse is encountered.

Scaled Time. Improvements to the unit delay timing model can be made by allowing the delay of each gate instantiation to be adjusted. Individual gate types from the model catalog can be given proportional values of delay, and fine tuning to compensate for circuit routing and conductors is performed only if necessary at individual gates which exhibit transient problems. The propagation times modeled are chosen to scale the actual values of the gates.

Complex timing relationships can be represented well with separate delay times for each direction of gate output transition. Separate delays allow the modeling of rise and fall times, although the modeled change in output value is made in a single step instead of an analog ramp of voltage usually associated with rise and fall (see Figure 5.7).

Models which are to be used for design verification simulation in addition to fault simulation for test generation benefit use scaled timing in worst case analysis investigations. The effects of tolerances such as device variation, temperature, and voltage swings can be approximated by skewing the delays of some or all of the gates of a model. Thorough analysis requires the ability to vary the delay of a single instantiation of a gate without affecting the others which are based on the same model. The large number of gates in most circuits prevents most designs from being exhaustively worst cased, but statistical (Monte Carlo) selection of individual gate delays within their limits has proven effective.

Pulses. In either case of timing control (unit delay or scaled timing), there are occasions for the representation of lengthy pulses. Asynchronous circuits frequently contain monostable multivibrators (one-shots) which produce a timed pulse when triggered by a transition of one or more inputs. If the one-shot pulsewidth is of the order of a few gate delays, a simple gate model as shown in Figure 5.8 will suffice. Note that a few gates can be used to provide pulses of either direction in response to positive transition triggers, negative transition triggers, or both.

Longer pulses become cumbersome unless a specifically adjustable delay is provided in the model paradigm. A gate which performs no function except to delay a signal is usually provided, and its delay must be specified for each instantiation in the circuit being modeled. The only limitations to the pulse lengths produced

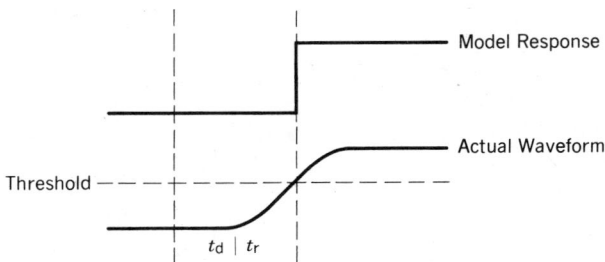

Figure 5.7. Modeling rise time represents only the delay before the threshold is crossed.

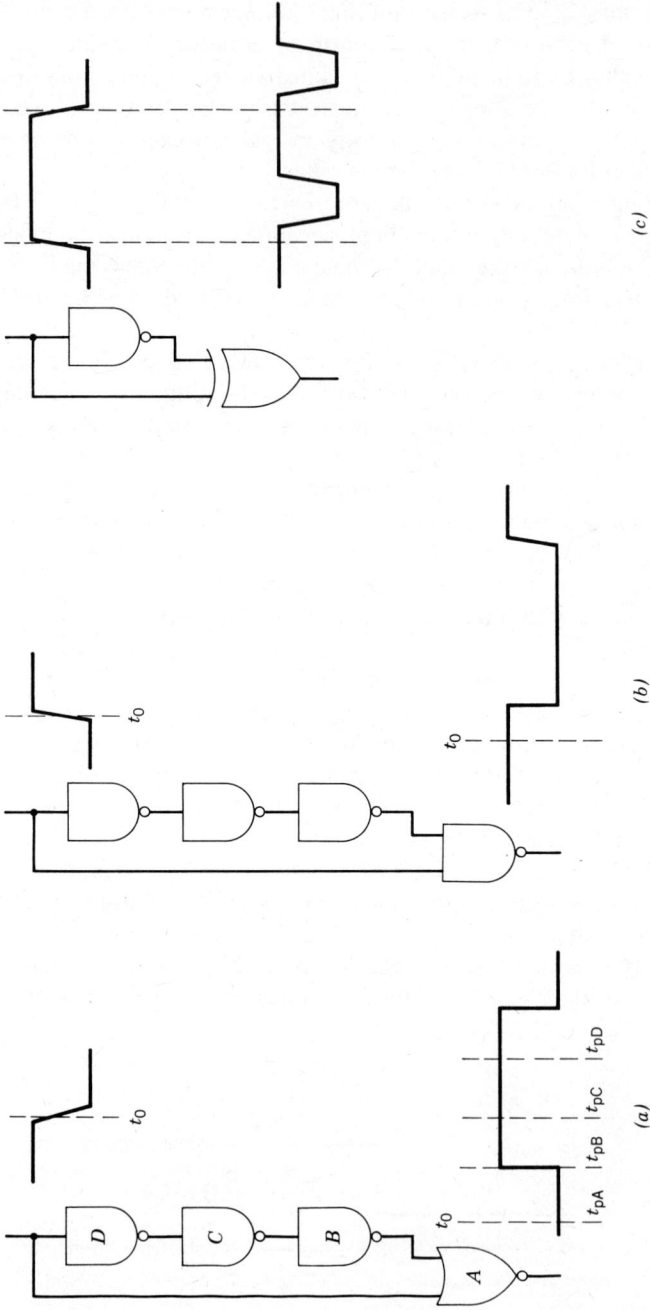

Figure 5.8. Gate models for various edge-triggered one-shots.

in this way are the ability of the simulator or other tool and the patience of the user. A 1-s delay may be used in the actual circuit, but processing a simulation time slice for each 10 ns of that second would be terribly wasteful unless other circuit activities were to happen along the way. If the rest of the circuit settles in 1 μs, use a slightly longer delay (1.2 μs) for the model, and verify the actual timing as a measurement on the ATE.

Free-running oscillators can also be modeled by gates, but the result is not often usable due to the problems of controlling and monitoring simulation of such a structure. Synchronization is the first difficulty since a free running oscillator may be either a 0 or a 1 when the simulation begins. Unless the model contains a reset or synchronizing method, the simulator must assume the output is unknown at any given time. Specific commands may be included in the simulator to identify the initial state at some defined point in the process.

Modeling and simulating the synchronization does not make it happen in the real world. A matching synchronization must be achieved at the tester in order to use the simulation results for testing a real circuit. Test step application must be maintained in synchronization throughout the test sequence also. If the ATE has the capability to lock onto the circuit's oscillator, this is possible, but it is seldom easy and therefore usually avoided.

Simulators with simple timing are often event driven. Event driven simulators begin evaluating the circuit model when input values are changed and continue until the nodal states are stable everywhere in the circuit. Such a scheme will not work for models with continuous oscillations, and additional information such as the sample interval and input strobing interval must be provided. Synchronization must be achieved between the simulator evaluations and the model, as well as between the ATE and the simulation results.

Hazard Detection. The modeling of transient events has two sides because device inputs have associated timing requirements such as setup, hold, and minimum pulsewidth. Not only must the generation of transients be correctly predicted, but the resolution of their effect on the circuit elements to which they are propagated must be achieved. If a 1-gate delay pulse is generated by an inappropriate input change sequence at a NAND gate and the gate drives the clock input of a flip-flop, will the data in the flip-flop be held? In a conservative modeling and simulation system such events may be detected and a warning or error may result. Other simulators may assume a result and leave the verification of correctness to the test engineer and the ATE.

Error detection gates are included in some modeling schemes to monitor the changes in certain device inputs. A transient detection gate might be attached to the clock and data lines of a flip-flop, for example. The gate would simply create an error message if both inputs were to change in the same evaluation time frame. The assumption made by this technique is that the setup and/or hold time of the data input with respect to the clock input is significant to the operation of the flip-flop, but the values of these requirements are less than the granularity with which simulation is to be performed.

Faults

At the gate level of modeling, the classic stuck-at faults are most appropriate. As discussed in Chapter 1, defective input or output transistors of gates usually produce the effect of clamping the logic level of a node to a 1 or a 0. If a chip output enable becomes inoperative, or both the output transistors of a TTL gate become open-circuited, the output will seem to be stuck at a Z or off condition. A gate modeling paradigm which supports bus structures and their 3-state drivers can represent the drivers as permanently off and thus s-a-Z.

The correlation of gate level faults to failure mechanisms is an important concern in gate modeling. At the periphery of packages, the stuck-at failure mechanisms prevail for the common pc board test conditions. Solder shorts, circuit path opens, and misplaced components result in signal characteristics which appear as logically stuck nodes. Within the chips the correlation weakens, as the actual circuit implementation and the model may not correlate in some areas. Devices which consist internally of a large collection of gates can be represented accurately with stuck-at faults at the their interconnections.

Devices such as memories exhibit additional failure modes and often consist of devices which do not resemble gates. Correlation of faults between a real IC which contains transistor–capacitor cells and a model which uses a classic 6-gate cross-coupled flip-flop is marginal. Gates with specific functions such as single-bit memory latches are supported by many software simulation schemes, lending a somewhat more accurate fault representation for memories. Alternate models for large memories in the form of algorithms or direct substitution of the simulating computer's memory resources for the simulated circuit's cells are more efficient than gate models, but they leave the fault accounting and fault effect prediction to software algorithms which are disconnected from the model.

The greatest weakness in gate modeling of IC faults occurs in the MOS technologies. Gate functions are often merged in MOS designs. A gate model may represent the function, but the nodes between the gates do not always correlate to points in the circuit, as evidenced by the simple circuit of Figure 5.9. Frequent use of the gates of MOS transistors for storage of signal values between clocked stages introduces the need for more phantom gates to correctly represent the behavior of the MOS circuit. The bidirectionality of MOS transfer gates further complicates the representation since standard logic gates are unidirectional. The bus element concepts applied in TTL circuits are useful in modeling transfer gates, but the implementations must be made either as special function gates (if allowed by the software) or as a several-gate macro. Faulting is difficult to apply in either case.

SWITCH LEVEL MODELS

Progress toward the goal of achieving greater accuracy in the representation of logic circuits produced in MOS technology was made with the introduction of

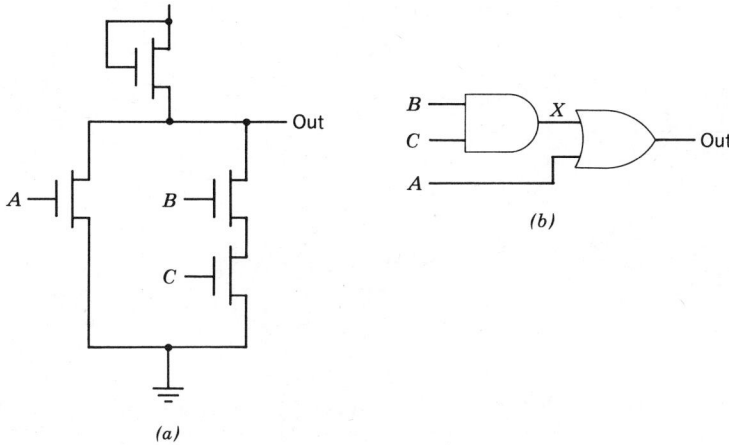

Figure 5.9. (a) MOS merged gate realization. (b) The equivalent gate-level model. Point X cannot be located in the transistor realization.

switch-level modeling and simulation [2]. The basis of switch-level models is the large swing in impedance of a MOS transistor between the off and on states. The action is, for logical purposes, that of a switch which is either open (off) or closed (on). Allowing the basic modeling element to represent the transistor instead of the gate relieves the confusion in the merged gate circuits of common MOS ICs.

Strengths

The MOS transistor is not a simple switch in reality, and the differences have been used to produce desired effects in logic circuits. Transistors are produced in differing dimensions to provide specific conductance strengths. Switch-level simulators were improved to include the relative strengths and to include the capacitive storage effects which dominate MOS circuits due to the high input impedances of the transistors. A typical use of the strength characteristics of MOS buffer gates is the flip-flop circuit shown in Figure 5.10. Direct inverted feedback latches the circuit to a value and seems disastrous to those unfamiliar wih MOS technology. Operation of the circuit depends on the dominance of one of the inverters (in this case inverter A). Data flow is in the direction of the dominant inverter. An additional requirement for operation is the dominance of the driving device connected to the input at the left-hand side over the nondominant inverter (inverter B). To achieve dominance, the transistors of an inverter are made larger than the other inverter, giving them a greater conductance and greater strength. A model which cannot express and resolve strengths cannot model the flip-flop accurately.

Early switch-level simulators modeled in three strengths: active (switching devices), resistive (pull-up devices), and capacitive (undriven nodes). Later simu-

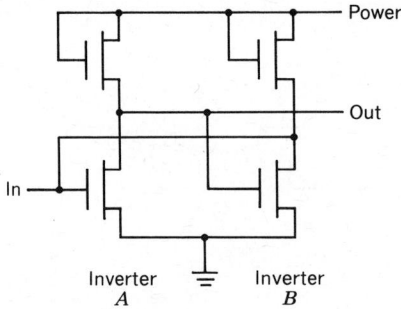

Figure 5.10. MOS flip-flop which depends on the relative strengths of transistors for proper function.

lators [3] calculate individual strengths for each transistor size and for each node capacitance to give very realistic representation. Design verification benefits from this accuracy, but the fault representations become more complex.

Faults

For switch-level models the traditional "stuck-at" faults of s-a-1 and s-a-0 must be augmented by a "stuck-at open" (s-a-op). The need for an s-a-op fault class stems from the capacitive storage and the use of transfer gates which connect circuit nodes together without performing logical operations on the signals. In the transfer gate case if the connected nodes all have drivers, the values for both buses are resolved by strength comparisons for the active drivers when they are connected. Failure of the transfer gate would not result in a fixed logic value for either bus, but might change the values on either, depending on the conditions.

As a simple example of the s-a-op fault, consider the circuit in Figure 5.11. The pairs of MOS transistors in parallel form a transfer gate, and their gates are driven in concert such that both transistors are on or off at once. The pair is needed to assure that the signal voltage does not affect the path impedance by reducing

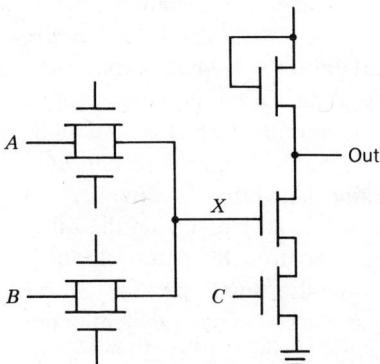

Figure 5.11. MOS transmission gates feeding and NAND gate.

the differential bias of gate to drain. The transistors which make up the pair are complementary (one has a *P*-type channel while the other has an *N*-type channel). The gates are therefore driven oppositely, and one of the transistors will be maintained in the on condition (if desired) no matter what signal voltage exists on the channels.

Assume that one of the transistors of the upper transfer gate (which connects *A* to node *X*) is open. Further, assume that when the signal at *A* is a 1, the remaining transistor of the pair will not be turned on due to insufficient gate drain bias. The result will be an inability to transfer a 1 from *A* to *X*, while a 0 will be transferred properly. The situation is further complicated by the capacitive storage of node *X* (primarily a result of the gate capacitance of the devices it drives). The s-a-op fault does not force the node to a 0; it merely inhibits changing the node from a 0 to a 1 via input *A*. If the node has previously been charged to a 1, it will remain a 1 for a much longer time than the usual operating cycle of the circuit. Testing for the s-a-op condition requires setting up the opposite condition on the driven node and then using the suspect gate to reverse the state. Moreover, if the model does not support the concepts of strength, the situation cannot be represented. At least one method has been introduced for remodeling circuitry such that s-a-op faults can be equivalently represented by conventional stuck-at faults [4], but many gates are needed to represent a single MOS gate accordingly. The added complexity limits the usefulness of this technique.

Other strength faults can be defined in terms of the transfer characteristics of transistors under abnormal conditions, but the situation becomes complex and difficult to resolve quickly. If the gate of a transistor is driven by a weak or floating signal instead of a strong one due to an s-a-op fault, the signal passed will be weakened if it passes at all. Predicting the effects of cascaded transistors operating under these levels is intractable for practically sized circuits.

In general, the use of switch-level simulation is limited to ICs. Circuits as large as those found on pc boards contain too many elements to be tractable as switch-level models. Instead, the pressure has been to replace gate-level modeling with a higher-level representation to reduce the complexity at some loss of representation accuracy or richness.

HARDWARE DESCRIPTIONS

Neither the concept of hardware description languages nor its use in fault simulation is new. Early HDLs were termed register-transfer-level descriptions and resembled APL [5] or other languages of the 1960s, but the method used to describe circuit structures and actions is the same as the ADA based VHSIC hardware description language (VHDL) of the 1980s [6].

Describing the data flow and control actions of a circuit by software algorithms has been a part of fault simulation for almost as long. Algorithmic models of regular structures such as RAM or ROM were quickly recognized for their efficiency of host computer memory and computing power in comparison to gate models of those circuits. The rise of PLAs with their Boolean descriptions has extended the use of algorithmic or functional models.

Functional Device Models

Notice the progression of digital functions from software, through firmware, to specialized hardware, and the relative equivalence of the various expressions of structure and behavior is evident. ICs which perform the complex calculations necessary for graphic display control are now commonplace, but have only recently supplanted microprocessors executing complex software. The process which replaced software with hardware was a translation of the software description of the data flow and data transformations needed to perform the tasks. The original software assumed the existence of a structure corresponding to the microprocessor. That structure was replaced by a streamlined version when the specialized ICs were produced.

The essence of functional modeling and of HDLs is a two-part description of the digital circuit. The structure of the circuit is described, and then the functions (alternatively called behavior) of the circuit are delineated in terms of the structure. HDL descriptions of a circuit do not necessarily constitute a functional model. Additional information concerning the physical attributes of a circuit (i.e., packaging information) is sometimes included in the HDL, but is not necessary for a functional model. The HDL description may appear to be a program, but a translation is often required to allow it to be used as a subroutine to a test generation fault simulation program. Compilers for HDLs are aimed at reducing the description to a physical implementation rather than converting it to executable code. The similarities between functional models and HDL circuit descriptions are strong enough to encourage use of HDLs for functional modeling.

Structural Description. In the early programming language of AHPL, the physical structure was implied by its use in describing the actions of the circuit. Later HDLs such as VHDL and Zeus [7] follow the practice of structured software languages and require the explicit declaration of all signals and entities prior to their use in the process definition section. Further structuring is encouraged by the capability of linking modules which contain definitions of parts of the circuit into an overall program. Complex hierarchical descriptions can be evolved in the same way "macros" lead to large models in the gate modeling scheme. The bottom level of description need not be a gate or primitive in this case.

Devices of arbitrary complexity are coded as entities, components, or packages in various languages, but in each case the code corresponding to an entity lists the available interface connections or ports. Ports are often used to describe a symbolic or bus-oriented connection which may interface several related bits in a single reference. If a port is defined as a word of 16 bits, it may be referred to later as a single value rather than a bit at a time.

The example circuit of Figure 5.12 includes both sequential and combinational components with various degrees of functional complexity. The data and control

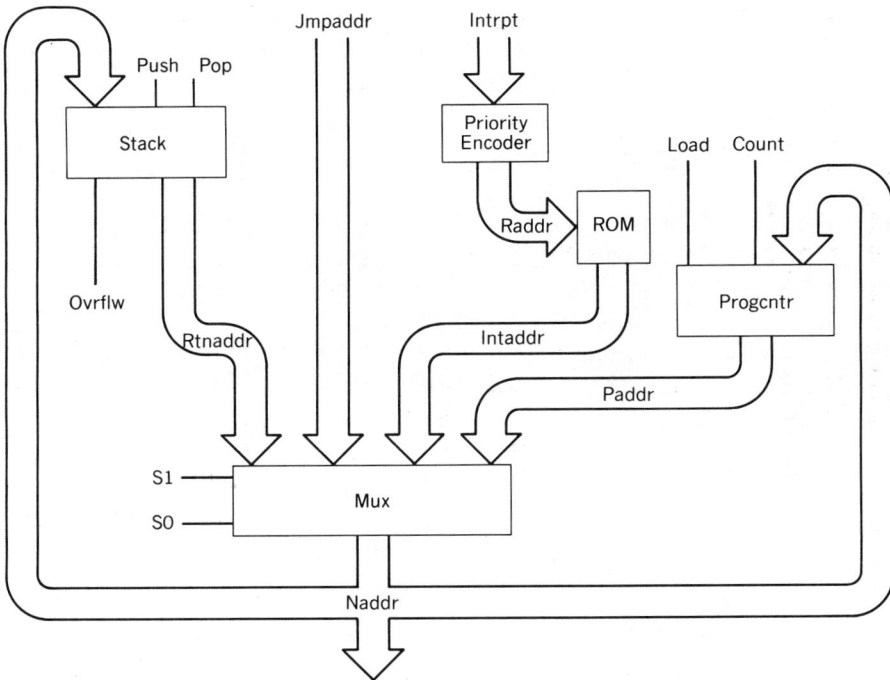

Figure 5.12. Next Address Generator block diagram.

paths have all been given names, but a name may represent a single signal or a bus. This "next address controller" is obviously intended to be part of a larger control circuit, but it will suffice as an example. The same description techniques which allow the use of packages such as Stack within the description of Naddrgen can be used to embed the result into a larger description of the controller.

The HDL code which follows is not intended to correspond to any particular language rigorously. It will resemble the newer languages in form, but some simplification has been adopted to present a clearer picture of the concepts without the entanglement of syntax. The construction of a description of the example begins with a declaration of the Naddrgen module as a single block with its interface to the outside world defined in parameter fields.

```
ENTITY Naddrgen(INPORTS(Push,Pop,Ld,Cnt,S0,S1,
            Intrpt,Jmpadd,
            OTPORTS(Naddr,Ovrflw)).

SIGNALS
    BEGIN
        Push,Pop,Ld,Cnt,S0,S1,Ovrflw : BIT;
```

```
            Intrpt : VECTOR(0 to 31);
            Jmpadd,Naddr,Intaddr,Paddr,
                Rtnaddr : VECTOR(0 to 23);
            Raddr : VECTOR(0 to 7);
        END;

    COMPONENT Subrstk : Stack(INPORTS(Naddr,Push,
        Pop),OTPORTS(Rtnaddr,Ovrflw));
    COMPONENT Ivrom : Rom8×24(INPORTS(Raddr),
        OTPORTS(Intaddr),DATA(Ijmps));
    COMPONENT Progcntr : Cntr24(INPORTS(Cnt,Ld,
        Naddr),OTPORTS(Paddr));
    FUNCTION Priority_Encode(Invector : VECTOR(0 to 31))
        CONSTANT Mask = Hex(80000000)
        BEGIN
            FOR bit = 32 to 1 DO
                BEGIN
                    IF (Invector AND Mask) = 0 THEN
                        Invector = Rot(Invector,1) ELSE
                        RETURN(bit);
                ENDFOR;
            RETURN(0);
        END;

    PROCESS
        BEGIN
            CALL SubrStk;
            IF Ovrflw = 1 THEN error;
            CALL Progcntr;
            Raddr = Priority_Encode(Intrpt);
            CALL Ivrom;
            CASE Concat(S1,S0) OF
                0 : Naddr = Paddr;
                1 : Naddr = Intaddr;
                2 : Naddr = Jmpaddr;
                3 : Naddr = Rtnaddr;
            ENDCASE;
        END.
```

The "signals" block identifies all variables, both internal and at the circuit perimeter, and defines their type for ease of compilation. Component definitions are used to relate local names for subcircuits and their port signals with global circuit definitions elsewhere. Later calls to these component names result in a subroutine style evaluation of the subcircuits using the actual parameters passed as arguments.

Following structured programming style, the functions or subroutines used in the main process are defined prior to their calls. Finally, the process which describes the actions of the circuit as a whole is given. Although the statements are ordered for correct results in a sequential evaluation sequence, they could be easily rearranged into parallel code sections. Explicit timing has been omitted in this case, but most HDLs contain facilities for specifying the delay associated with an evaluation or the application of a signal. Commands for synchronization, conditional timed execution, and other timing operations are part of most HDLs, but have been omitted for simplicity.

Fault Models in HDLs. It is easy to see how a program can simulate the actions and data flow of a good circuit. Accuracy of the simulation is judged by the outputs of the simulation in response to input vectors. Timing and response to transients or unusual conditions are typically the shortcomings in a good circuit simulation by an HDL, but careful construction of the model can usually alleviate any difficulties in critical areas of concern. Accuracy of results under faulted circuit conditions is quite another matter.

Fault simulation using gate-level models consists of modifying the model slightly to correspond to a circuit with a defect such as a short or open on one of the nodes connecting the gates. The physical correspondence is removed by another level of abstraction in the HDL models, and it is not easy to identify the effects of a short or an open upon the model. In a data path a short or open is equivalent to inserting a masking operation in the process to set or clear a given bit (for single faults) before the data are used. Control path faults are more difficult.

With both the schematic and the HDL description available the correspondence of some faults can be derived. The potential stuck-at faults associated with the mux in Figure 5.12 are s-a-0 and s-a-1 faults for each of the 24 outputs (Naddr), another 24 of each for each of the input ports (Rtnaddr, Jmpaddr, Intaddr, and Paddr), both faults for each of the control lines ($S0$ and S). The mux function is modeled by the "case" statement of the HDL. If only the case statement is considered when the asking the question "What happens if there is a fault in the construct?" then predictions of faulty behavior will be difficult. In light of the correspondence with the physical mux structure and the knowledge of its potential faults for the single stuck at class, the solution is readily perceived.

To begin with, each bit of the $S1$, $S0$ concatenation might be stuck at either extreme. In each instance of a single fault for these control lines, the number of cases available will be reduced from four to two. If $S0$ is s-a-0, then the cases of 0 and 2 will be unaffected, while case 1 will actually activate the case 0 statement and case 3 will activate the case 2 statement. If the case structure is modified to appear as follows, it will emulate the fault behavior correctly.

```
CASE Concat(S1,S0) OF
    0,1 : Naddr = Paddr;
    2,3 : Naddr = Jmpaddr
ENDCASE;
```

Data line fault effect injection causes any one bit of any operand or result to become s-a-0 or s-a-1. If an input bit is stuck, the bad value will be transferred to the output (Naddr) only if that input port has been selected by the appropriate case. This corresponds correctly to the real circuit's behavior and illustrates path sensitization once again.

A more confusing example is the faulting of the counter. The code representing the counter is not shown, but only referred to with a call statement in the example. The code representing the counter is easy to derive and makes use of a stored value of the current count in addition to the data and control inputs. Considering only a functional description for the counter, the fault effects are not obvious. If the counter is at a count of 3 and it fails on the next clock to advance to 4, what does it do instead? Does it stay at 3, reset to 0, or jump to some other value, such as 6? A knowledge of the physical device and the stuck-at fault leads to the conclusion that most probably the counter will return to 0 since the failure that would stop a transition from 3 to 4 is an s-a-0 bit in the third least significant position. A fault in the Cnt line would never have allowed the count to reach 3 unless the value 3 were loaded into the counter originally.

On the whole the counter behavior under single stuck-at faults will be quite peculiar. The counter will change values in response to the Cnt signal in most single fault cases (except Cnt line failures), but the values will depend on the fault position and type as well as the data loaded into the counter at the start of the count sequence. A further consideration in modeling is the separation between the internal count value and the outputs of the package. If the outputs are buffered from the internal mechanism, a stuck output will result in observed holes in the count, but the sequence will remain unchanged, (i.e., 0, 1, 2, 3, 0, 1, 2, 3, 8, 9, 10). Altering the HDL code to correctly track the fault effects of all the possible counter faults is a complex task.

The difficulty in simulating the fault effects with HDLs has limited their use to indirect modeling. A circuit description using HDL is compiled into a gate-level realization as though it were to be fabricated, and the resulting gate-level model is used with conventional faults applied.

Behavioral Modeling

Behavioral representation of a circuit is an extension of HDL modeling at a higher level of abstraction. Sometimes termed "algorithmic modeling," it contains a process description similar to the HDL example, but usually written in a popular language such as FORTRAN or Pascal. The structure definitions are only needed to allow compilation of the routines. Programs that represent models are compact and efficient, but less readable than the HDL equivalents. Control of modeling and verification of correctness are more difficult in conventional languages since the structure of the circuit is not explicit.

The only concern in behavioral modeling is that the correct output is obtained for each input condition or vector. Timing aspects can be simulated with careful construction of the routines, but conventional languages do not contain aids which define the timing relationships between instructions. The internal structure of the circuit being simulated is lost, with state variables becoming merely register or memory locations in the host processor.

Fault effect propagation through a behavioral model is straightforward. The faulty inputs are evaluated in the same way the good circuit inputs are evaluated, and the resulting outputs are labeled accordingly. Fault simulation within the behavioral model is not attempted since the relationship between physical circuitry and the models routines is not preserved. Functional faulting may be approached in the same way as the HDL case, but results are more difficult to obtain.

Boolean Models

The class of logic devices known as Programmable Array Logic (PAL) lends itself to a simple Boolean representation. Within the PAL, the combinational logic structure is limited to a two-level AND/OR implementation. This correlates to the classical two-level Boolean equation which appears as the OR of a set of AND terms. Each AND term is made up of true and complemented circuit inputs. Outputs of the device are derived by solving Boolean equations such as:

$$F5 = (I3.I6^*.I8) + (I0^*.I1^*.I2) + (I6.I7.I8)$$

where $F5$ refers to the fifth output function and $I6^*$ is the complement of the sixth input value. The plus sign is interpreted as OR and the dot as AND.

PALS commonly use a few simple extensions of the boolean scheme to control the direction of bidirectional bus pins on the chip. A Boolean equation's output is always a simple true (1) or false (0), and the condition is used to answer a question such as "Is pin 5 an output?" Use of the input $I5$ in another equation would depend on the answer to the bus direction equation. The value of the pin would be externally supplied if it were an input, and the result of the solution of $F5$ if it were an output. In the latter case a feedback path is established within the device model.

A popular type of PAL includes a register for storing values from the previous clock cycle. The values of the register are available as feedback values to influence the outputs along with current input variable values. The model for this PAL type is a classical state machine. Just as with the classic state machine, circuits of arbitrary complexity can be represented as boolean equations which include stored state variables. Each state variable must have an associated equation which predicts its next state. The following equations describe a PAL of this sort.

$$F1 = (I2^* \cdot I6 \cdot R2^* \cdot R3) + (I1 \cdot I4 \cdot I6 \cdot R3^*) + (I6 \cdot R2^*)$$
$$F5 = (I4^* \cdot R3) + (I6 \cdot R2 \cdot R3)$$
$$R2 = (I1^* \cdot I6) + (R3)$$
$$R3 = (I4 \cdot I6^* \cdot R3) + (I2 \cdot I4 \cdot R3)$$

The appearance of an R variable on the right hand side of the equation represents the present value of the variable, while the use on the left-hand side indicates the value to be latched into the variable at the next clock.

Clock phases are usually implicit to the model's use. When the clocks are also included in the Boolean descriptions, the models become an expression of generalized circuits. Gate delay timing needed to resolve conflicts is not explicit in Boolean equations. Overall circuit timing must be achieved by structuring the equations with intermediate variables so that, solving them in event order, one equation level at a time will result in correct operation. That is, when $I3$ changes, all equations which directly employ $I3$ are solved using old values of feedback variables. Then all the output variables from that rank are updated at once, and the next rank is evaluated.

Faults in Boolean Models. Inserting faults in a Boolean model is straightforward for the "stuck-at" fault set. To solve for the effects of variable $I3$ s-a-0, simply replace $I3$ with a 0 in all equations. Inputs and outputs of the circuit can be faulted with this simple approach, but the internal structure faults need a further faulting mechanism. In an equivalent physical circuit the outputs of the AND gate which form each term and the corresponding OR gate inputs of the second stage can also be faulty. These faults are inserted by substituting a 0 or 1 for one of the parenthetical expressions in an equation.

In actual practice the PALs are manufactured as complete arrays with all possible terms input to each equation. The user then programs the device by blowing or removing the unwanted terms. A side effect of this approach is the possibility of the presence of variables in the equations which should have been removed. Thorough faulting for these defects would involve solving for all possible equations involving all combinations of variables and comparing the outcomes with the intended equation. In fact, it is possible to predict a more likely set of misprogramming faults and to reduce the cases significantly with proper analysis [8, 9].

Hierarchical Modeling

In order to use hierarchical simulation techniques models efficiently, circuits must be modeled on several levels. The gate-level or switch-level models will be used for the circuit portion which is of most interest during a given portion of a simulation. The behavioral or HDL models of the remaining circuit portions will be used to reduce the computation required. An appropriate use of hierarchical models in test generation involves full fault simulation of the gate-level models in the center of focus, while the remainder of the circuit is simulated with good circuit techniques applied with higher level models.

Fault effects can be propagated through the higher-level models to reach primary outputs without injecting faults into those portions of the circuit. Later passes of the test generation will replace a different portion with gate-level models and return the previous section to higher levels of abstraction. A complete test consists of a concatenation of the portions for each focus.

Hierarchical modeling is not a distinct scheme, but a method of use for models of several levels to capture the benefits of detailed modeling without the computation of simulating the entire circuit at a detailed level. The associated overhead of creating multilevel models and shuffling the circuit representation as the focus moves from subcircuit to subcircuit can be great enough to offset the computation of a fully detailed representation.

SUMMARY

Test generation by examination of an actual circuit in response to stimuli on a bench was driven to obsolescence years ago by the complexity of digital circuits. All test generation methods now in popular use employ a model of the circuit at one level or another. The information content of the model and the difficulty of using it to compute good circuit responses and fault effects are conflicting properties which create a tradeoff for the test engineer.

There is no correct way to model a circuit, but rather there is a range of modeling schemes which represent the circuit at differing levels of abstraction. Analog modeling of the Ebers–Moll variety is far too involved for digital circuitry, and the simplifying assumptions of switching mode make it unnecessary. Switch-level simulation is instead the most detailed modeling scheme in frequent use for digital circuits. Gate-level, register-transfer (HDL) level, and behavioral-level models contain successively less information and are successively more compact and easily used by computer simulation software.

Switch-level simulation is popular in MOS applications due to its ability to represent merged gate functions accurately. The algorithms for solving switch-level representations of circuits have been expanded to include effects such as undriven circuit nodes and wired bus logic by the application of signal strength concepts. Switch elements can be faulted as stuck open and stuck closed to augment the nodal s-a-0 and s-a-1 faults in accurately predicting faulty circuit behavior. The transfer gate which connects two nodes without forcing any logic state to either is better represented with the stuck at open fault definition and the strength representation.

Gate-level models are perhaps the oldest, having been used to represent DTL and TTL digital circuits since the predominance of SSI. The solution of a gate-level model is a simple truth table lookup. The stuck-at class of faults was created in conjunction with the gate level model, and the propagation characteristics of faults in this realm are well documented in the D algorithm and its derivatives. Solution of gate-level models is simple and quick, but the advance of VLSI has created circuits with such massive numbers of gates that the aggregate solution is a major computing task.

HDLs such as VHDL were developed to deal with the complexity of representing a circuit during the design process, but the same models can be used to reduce the computational tasks in test generation. Although accurate fault simulation may require a mixed level simulation in which a portion of the circuit is modeled in gates while the remainder is described in an HDL, these higher-level models provide a technique for handling extremely large circuits which might otherwise be unmanageable.

Obtaining models in any format is often a tedious task of reverse engineering which necessitates lengthy cut-and-try sessions. Lack of documentation of the entire behavior of circuits is the biggest hurdle to be overcome. The behavior of circuit portions which may be unused in a given operation is of no concern to the designer, but may be critical to test generation. Extreme care must be used in acquiring and applying digital circuit models during test generation.

REFERENCES

1. A. Dewey and A. Gadient, "VHDL Motivation," *IEEE Des. Test Comput.*, **3**(2), 15 (Apr. 1986).
2. R. E. Bryant, "Switch Level Modeling of MOS Digital Circuits," in *Proc. ICCC Conf.*, IEEE, 1982, pp. 68–71.
3. G. G. Schrooten, "SWITCH, A Switch Level Logic Simulator with Unique Element Models," in *Dig. Papers IEEE Int. Conf. Comput.-Aided Des.*, 1984, IEEE, pp. 245–247.
4. S. A. Al-Arian and D. P. Agrawal, "CMOS Fault Testing: Multiple Faults in Combinational Circuits, Single Faults in Sequential Circuits," in *Proc. 1984 Int. Test Conf.*, IEEE Comput. Soc., pp. 218–223.
5. F. J. Hill and G. R. Peterson, *Digital Systems: Hardware Organization and Design*, Wiley, 1973, pp. 103–128.
6. R. Lipsett, E. Marschber, and M. Shahdad, "VHDL—The Language," *IEEE Des. Test Comput.*, **3**(2), 28–40 (Apr. 1986).
7. S. M. German and K. J. Lieberherr, "Zeus: A Language for Expressing Algorithms in Hardware," *IEEE Comput.*, **18**(2), 55–65 (Feb. 1985).
8. V. K. Agrawal, "Multiple Fault Detection in Programmable Logic Arrays," *IEEE Trans. Comput.*, **C-29**(6), 518–522 (June 1980).
9. C. W. Cha, "A Testing Strategy for PLA's," in *Proc. 15th Des. Automat. Conf.*, IEEE, 1978, pp. 326–334.

CHAPTER SIX

AUTOMATIC TEST EQUIPMENT

The complexity of digital circuitry has not only complicated the process of deriving adequate test sequences, it has caused a revolution in the means for applying and observing the tests. Automatic Test Equipment (ATE) has become a major factor in the design, manufacture, and service of digital electronics. The size, cost, and capabilities vary with the application. For purposes of discussion, in this text digital ATE will be defined as "a programmable system which controls and monitors digital electronic subsystems to determine their operability." This is a relatively broad definition which might include portable and built-in ATE as well as the stand-alone type of ATE found in factories. The primary limitation is the purpose of the equipment, which is restricted to the testing of digital subsystems such as ICs, PC boards, or functional modules, but does not include power supplies, analog electronic assemblies, or large networks. This restriction in the definition is needed to contain the subject to a single chapter. ATE exists for other areas, but it is not generally called digital ATE.

This chapter will be concerned with the architectural and functional aspects of digital ATE, leaving details of the DUT interface and the ATE programming languages to other chapters. Emphasis will be placed on the application of ATE to the production environment. A typical use is the functional screening of PC boards after assembly.

Illustrations of architectures, capabilities, and applications of equipment are to serve as examples and are not to be viewed as endorsement of a particular brand or approach to testing. It is important, rather, to become aware of the alternatives in ATE and the potential applications of each configuration and style before making a decision to adopt a testing method and the associated equipment.

THE DEVELOPMENT OF ATE

At the time of the birth of computers and throughout the 1950s, digital subassemblies were usually not explicitly tested until they were incorporated in the final machine. Each assembly step might have been followed by a close visual inspection of the workmanship, but the proof of operation was left to the programmer who tested the computer or the technician who checked out the product. Electron tubes and transistors were essentially well understood from an analog vantage point and were tested in that vein even when their application was switching.

With the advent of ICs, digital circuitry began to be viewed as an assemblage of Boolean gates, and the intermediate level of testing which concerns itself with correct logical function arose. Digital thinking replaced the signal generators with simple switches and the oscilloscopes with lights. The components themselves were still tested with analog instruments to determine switching speed, drive capability, and so on, but the PC assemblies contained too many components to check each input and output individually. The relationship between inputs and outputs had become too complex for analog test methods. Switches and lights were inexpensively assembled on a test panel and connected to the CUT as in Figure 6.1.

The test operator connected the switch and light panel and any power supplies required and then proceeded to manipulate the switches according to a printed list of patterns, observing the pattern of lights at each step. The typical test sequence was only a dozen or so steps. Typical circuits had only a few outputs and 10–20 inputs. Diagnostic notes may have been written in the margin or on the schematic of the CUT. Troubleshooting procedures were left to the technician.

Automation entered the picture as the circuits gained complexity to over 30 inputs and outputs and test sequences grew to 100 steps. The switches were replaced with relays, and the lights were replaced with threshold comparators. Custom-designed testers with dedicated sequencers to drive the switches and check the comparator outputs proved too costly since each circuit configuration to be tested needed a separate tester. Numerically controlled machine techniques were

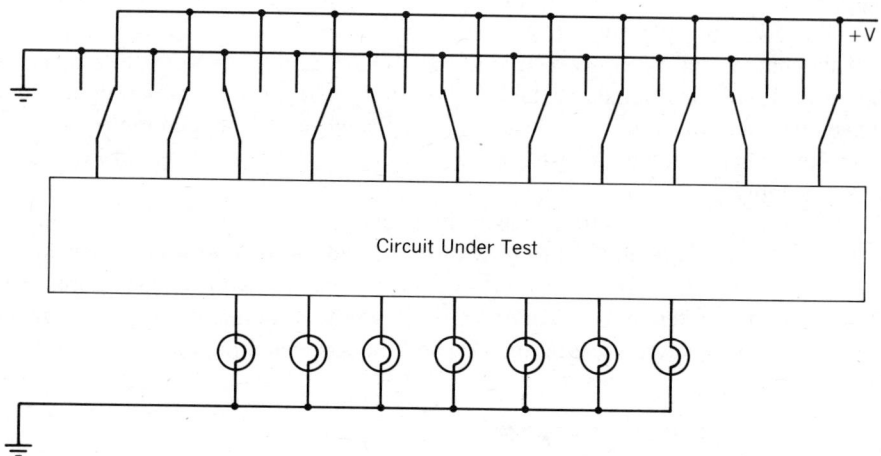

Figure 6.1. Manually operated test panel diagram.

successfully applied, and paper tape driven ATE was born. An interface adapter which accomplished the task of connecting circuit inputs to tester relays and circuit outputs to tester comparators was created for each circuit to be tested, and a separate paper tape program for each CUT provided the test sequence.

The test method that had emerged was static functional testing. Since the test vector application rate using relays was only about 10 steps/s, even the resistor–transistor logic and diode–transistor logic of the era was well settled into a static state before the outputs could be sampled and the next vector applied. The term functional in this usage refers to the technique of accessing the circuit primarily through the connectors which are later used to communicate with other logic circuits in system operation. Since these connectors are usually placed around the periphery of the PC board, the term functional is sometimes replaced by the phrase "edge connector" testing.

Further increases in circuit complexity and a desire to speed up the test process led to the substitution of electronic driver circuits for the relays and direct computer control for the paper tape. The drivers used were logic circuits similar to those being tested, but were usually augmented with high-current outputs. Protection circuitry was often added to keep defects such as shorted inputs in the CUT from damaging the test equipment. Operational amplifiers made excellent choices for both the driver and receiver functions. Receivers are buffer circuits which isolated the comparators from the CUT and reduced the loading on the latter.

Providing both a driver and a receiver for each connection to the CUT became common practice (Figure 6.2). The driver circuit is connected through a relay which isolates it when the CUT pin assigned to the pair is an output. The relay does not switch at test speeds but only during "setup" of the test configuration. The output of the comparator can be used to assure correct driver operation when the CUT pin is an input.

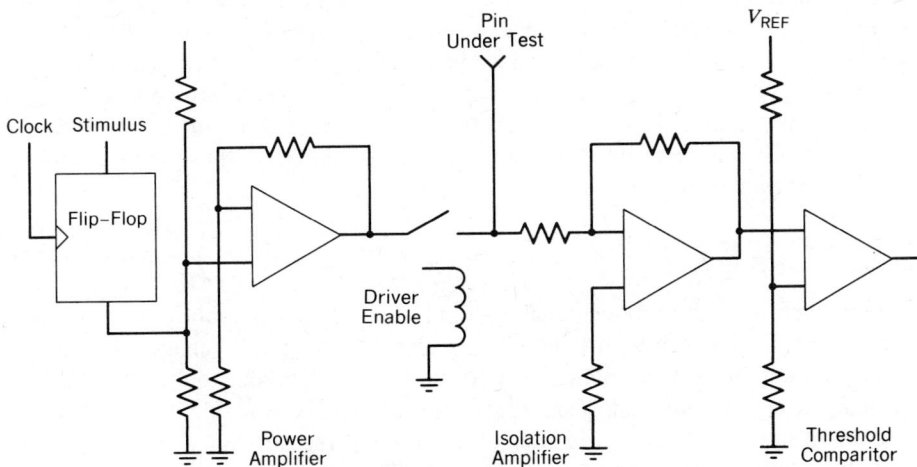

Figure 6.2. ATE pin electronics including driver and receiver circuits for a single pin.

The computers available in the early 1960s were large and general-purpose, leading to the connection of the tester as a peripheral. The electronics in the test equipment typically consisted of a computer interface and several ranks of holding registers as well as the driver–receiver electronics. Circuitry for controlling power supplies and reporting failures to the operator was incorporated in the more sophisticated models.

In 1965 the first commercially produced ATE for IC testing was introduced by Fairchild [1]. The Fairchild model 4000 was originally programmed via paper tape, but later models were converted to direct computer control. Companies such as Teradyne with its model J259 and Texas Instruments with the model 553 began selling ATE at about the same time, providing a base for the present day competitive ATE business. The important transition which occurred in the later 1960s integrated the computer into the test equipment it controlled, leading to specialized architectures and languages for control.

ATE ARCHITECTURE

Selection of appropriate ATE for an application is a difficult task due to the many factors to be considered, the variety of equipment available, and the large sums of capital required in many cases. It is imperative that the test engineer acquire a firm understanding of the minimum requirements of the application and of how the general classifications of ATE might meet the requirements. Details concerning particular brands within a general classification can be acquired from a representative of the manufacturer. Far too many models exist, however, to begin interviewing candidate suppliers without a well-developed idea of which ones are likely to satisfy the minimum requirements.

Several distinct system architectures have evolved for ATE, stemming from the needs of users such as IC manufacturers, hybrid (digital and analog) PC board manufacturers, and field service repair installations. Speed, accuracy, flexibility, portability, or low cost may be optimized in a given system, but no system achieves them all. A typical architecture can only be discussed within the context of the digital testing application. The general order of consideration will be from those systems which use centralized general-purpose computers to those which embed computers deeply into the test system.

Instrument Bus ATE

Manufacturers who produce analog–digital hybrid subassemblies or who have a mixed production line with low volumes of any one type of subassembly often find instrument bus ATE systems to be the most economical due to their flexibility. A great variety of both analog and digital instruments is available with bus interface control. Using them is a matter of assembling the necessary instruments to provide stimulus or analyze response for the CUTs to be tested and then interfacing the control ports to a computer and the instrument I/Os to the CUTs.

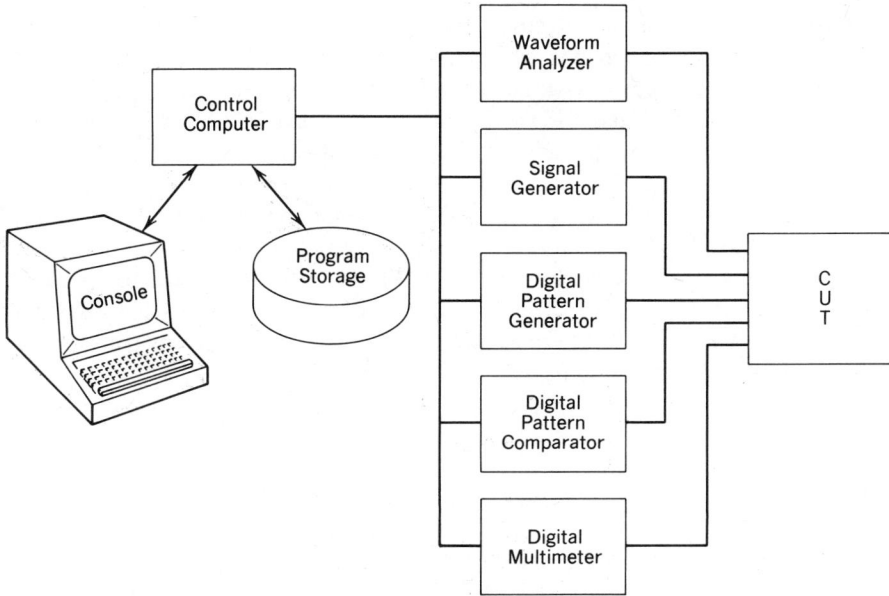

Figure 6.3. Diagram of an instrument bus ATE system.

The control computer provides operator interface through a console and uses ordinary peripherals such as disk storage, line printer, and so on. The bus used to communicate with and control instruments may be one of several in electrical characteristics and protocol. The control computer considers the bus to be an I/O port (see Figure 6.3).

IEEE Standard 488 defines an instrument bus protocol which is the most pervasive among instrument makers and system integrators. The bus protocol standard defines the general-purpose instrument bus (GPIB) which is based on a bus structure developed by Hewlett-Packard under the name HP-IB. Physically, the bus is implemented on a 24-pin connector parallel cable which can be connected in daisy chain or star arrangements between multiple controllers and instruments. Usually, a single controller is used with a number of instruments. Pin assignments, physical characteristics, and command details can be obtained from the IEEE standard. Specific programming sequences depend upon the instruments to be used since the functionality is not part of the standard. Once the bus protocol has established communication between a controller and an instrument, the data passed could be setup codes for the instrument, data to be used by the instrument, or results of a measurement from the instrument.

Each device monitors the bus, and each has a unique address by which it identifies when it is being requested. In addition, each device is classified as a LISTENER, a TALKER, or a CONTROLLER. The action taken is defined by address, classification, and the state of dedicated control lines on the bus. The

CONTROLLER sets up the test configuration by asserting an attention signal followed by a series of commands which activate TALKERS and LISTENERS. A LISTENER can only accept data, and a TALKER can only transmit data. Devices can be both TALKERS and LISTENERS, but only one TALKER can be given the bus at a time. Following setup, the CONTROLLER starts the process and relinquishes the bus by removing the attention signal. The CONTROLLER may assume the role of a TALKER or LISTENER temporarily until the process is finished. Data exchange takes place over an 8 bit parallel bus.

Instrument manufacturers who produce IEEE-488 instruments usually produce one or more controller devices. These are mini- or microcomputers with the IEEE-488 interface attached and dedicated protocol and test management software installed in ROM. Interface cards are available for personal computers, and I/O port options are available for most major brands of minicomputers as well. Software drivers are installed to allow test system control to be mixed with other computational functions.

Moving the IEEE-488 bus into the personal computer fold reduces the price of a controller. The specialized controller computers can be replaced by the mass-produced, general-purpose computer by installing a peripheral driver and a software handler. Support for the personal computer is much greater in both hardware and software maintenance. Postprocessing of data obtained from the instrumentation can take advantage of the many utility software packages available. Peripherals such as printers and plotters are easily adapted. Mass storage in the form of disk drives is less expensive than the larger minicomputer versions. All the mass market factors of the personal computer add up to a friendly environment for the system developer and the end user [2].

In inexpensive instrument bus systems the RS-232 serial bus serves as the instrument control and communication path. Such a system might use a personal computer or a single-board microcomputer as a controller. The protocols used by RS-232 instruments are not standardized, but several suppliers are available. The speed restrictions of the medium limit its use to areas which minimize bus traffic. Therefore, the digital test arena has seen little use of this configuration. Field test applications use this method by contacting a centralized controller and test database over telephone lines. The instruments in the field can be attached to the phone via an inexpensive Modem, and the test can be carried out by remote control.

The popularity of remote operation has prompted the manufacture of protocol conversion modules which convert RS-232 data to IEEE-488 commands and data. No standard set of serial representations for IEEE 488 exist, however, and the makers of converters specify commands in their own high level languages.

Other system standards have been adopted for application by the military. The modular ATE (MATE) has established a complex bus protocol to allow testing of a wide variety of electronic subassemblies ranging from power supplies to radar transmitters. The instruments used in MATE are usually ATE systems in themselves.

Application of the instrument bus architecture to digital testing is limited somewhat by the lack of general-purpose digital stimulus generators capable of generating long input vectors. Complex circuits usually have hundreds of input signals and hundreds of output signals, all of which must be manipulated simultaneously. The instruments available for signal generation usually generate waveforms rather than parallel digital vectors. Accurate timing from pulse generators and complex waveforms from arbitrary waveform generators are useful in pseudorandom test sequences or for serial digital device testing such as in USARTs.

Broadside pattern generation is usually left to a custom-designed stored pattern generator consisting of RAM loadable directly from the controller. Direct loading bypasses the bus and relieves the bottleneck that otherwise restricts test loading speed as hundreds of thousands of bytes of digital test patterns flow over the rather slow 8-bit bus. The expected response data are loaded into another section of RAM by direct means and used by a rank of comparators connected to the CUT outputs. The test application can be synchronized to other instruments using the instrument bus. The instrument bus approach is, however, corrupted, and separate software and hardware are required to test in this way.

Whether or not the test application instruments remain purely IEEE-488 controlled, the CUT interface is a somewhat messy problem. If each interface pin of the CUT is to be connected to only one instrument during the entire test, a hardwired adapter is usually used to connect the CUT and the nonbus I/O lines of the instruments. Since hybrid boards are often tested using bus architecture schemes, the wiring from instruments to the interface jacks are usually coaxial cables, and an effort is made to provide adapters which preserve the signal integrity. The typical adapter is therefore bulky and mechanically well constructed with high-reliability connectors between the tester and the adapter.

When the CUT demands changes in configuration during the test, as when a signal is to be treated digitally for part of the test and measured in an analog fashion for another part, the interconnection becomes difficult. Relays or electronic switches such as MOSFETS can be controlled by the IEEE-488 bus, but at additional cost in both signal degradation and dollars. Many instrument bus testers offer building blocks which provide signal matrices on the tester side of the interface adapter.

Matrix Architecture

The open matrix architecture is used in purely digital testers also. The matrix is made up of columns of stimulus–response signal lines from the tester and rows of CUT interface lines (Figure 6.4). The characteristics of the stimulus–response lines vary in sets and are connected by adapter wiring or programmable relays to the appropriate CUT interface pins for each variety of CUT to be tested.

A typical arrangement might provide 512 fixed voltage or driver–receivers, 128 programmable voltage driver–receivers, and 16 measurement lines. The driver–receiver circuits are usually arranged in sets of 8, 16, and so on, on PC cards called ''pin cards'' since they provide signals for individual pins of the CUT in-

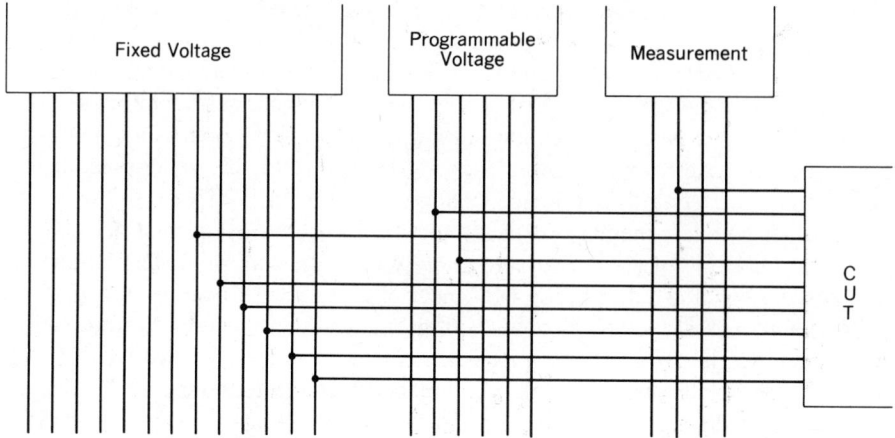

Figure 6.4. Matrix ATE pin electronics connections. CUT pins are wired to ATE pins according to function.

terface connectors. The fixed voltage lines would be settable with a screwdriver or jumpers to levels such as drive 1 = 4.2 V, drive 0 = 0.2 V, receive 1 = 2.4 V, and receive 0 = 0.8 V for testing TTL logic. This arrangement provides some compensation for adapter noise and allows the receivers to be used to assure correct driver operation.

Setting the voltages to the specification given in TTL data books is theoretically acceptable, but usually results in trouble due to the cable loading involved in matrix systems. Programmable voltage pin cards can be adjusted by the test program, usually only in sets of 8, 16, 64, or some multiple of 2. These lines allow easy testing of CUTs with mixed logic levels. Interface circuits often have TTL voltages at most signals with voltages such as +12 to −12 at the I/O port. Commands in the tester control language set the drive and receive levels similar to those in Figure 6.5. In some cases the setup of levels can only take place at the beginning of a test sequence, while other testers can change levels within a test sequence. The added size and cost of programmable voltage pin cards is usually significant, resulting in the use of mixed selections of fixed and programmable models.

The most complex, and therefore the most costly, pin cards are capable of measuring parameters such as pulsewidth, delays, and voltages of signals from the

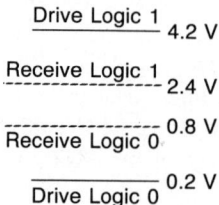

Figure 6.5. Four thresholds for driving and receiving logic signals must be properly nested to avoid conflicts.

CUT. Testing asynchronous ouputs such as one-shots or variable outputs such as D–A converter signals require measurement. The inclusion of a few pin cards with measurement capability is adequate for most digital testing at the PC board level. IC testing may require a full complement.

The matrix wiring performed in the adapter to connect the correct capabilities to the correct tester pin cards is exemplified in Figure 6.6. The CUT is an RS-232 interface card with handshake signals requiring TTL on the computer side of the interface and programmable interface signals for the serial port. Several irritations involved the use of ATE become evident when looking at Figure 6.6. The wiring of the adapter is complex and noise prone. The designations of ATE pin card lines seldom match those of the CUT connector. Even if all the signal lines of the CUT are TTL, the assignment is not linear due to the power and ground connections which are usually interspersed throughout the CUT connector. Tables of correspondence between pin cards, CUT lines, and test program signal names are a normal part of ATE programming for matrix systems.

Relay or electronic multiplexing of the matrix connections is usually not employed in functional testing due to the cost of the N^2 nature. For in-circuit testing, however, the number of pin cards may be small (a few dozen), while the CUT connections are many. Multiplexing the pin electronics is more appropriate since

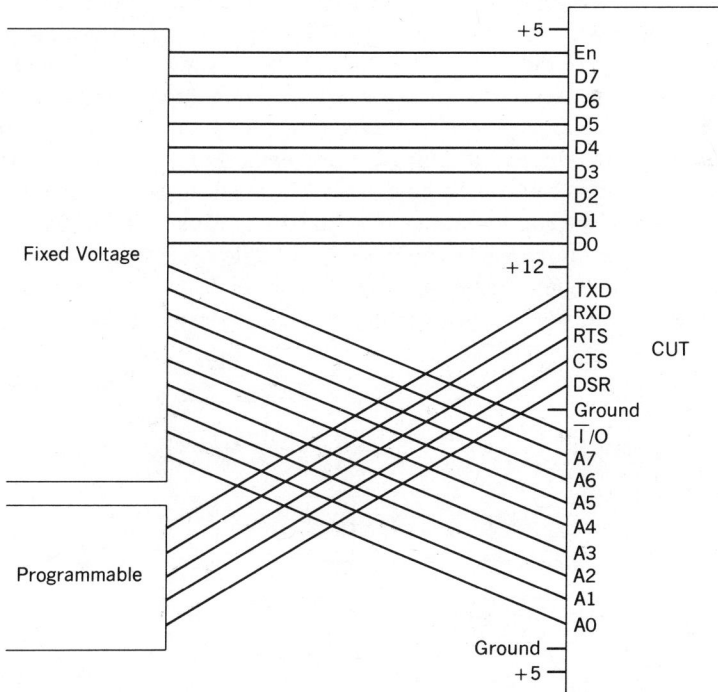

Figure 6.6. Pin electronics allocation by voltage range in a matrix ATE system.

the in-circuit method does not use all the pins at the same time, as does functional testing. Intelligent selection of multiplexing arrangements can achieve a mux ratio of 10:1, saving the cost of additional pin electronics cards [3].

The control side of the pin electronics for matrix systems is usually a high-speed parallel bus. The internal bus of the controlling microcomputer or minicomputer is often used. Even at the fastest computer bus speeds the bandwidth is not sufficient for high-speed testing of CUTs, so buffer memories are installed on the pin electronics cards. The control computer loads the test vectors into the buffers and then triggers a test clock which sequences through the test. If a mismatch of expected response occurs, the control computer can be interrupted and the test halted, or the results can be stored in the pin electronics buffers and reported only at the end of the test. The former is useful for troubleshooting, and the latter is normal for screening CUTs into good and defective bins quickly.

Actually, there are up to four buffers in most pin cards, depending on their capability (Figure 6.7). Once the buffers are loaded and the test sequence triggered, the pin cards must operate autonomously from the control computer since the data bandwidth of the link between them is too slow in any case. Therefore, the stimulus vectors and the expected response vectors must both be stored in buffers. The remaining two buffers are not so obvious in function.

The inhibit buffer comes into play when the CUT line connected to the particular pin in question is bidirectional and switches during the test sequence. The data buses of most microcomputers change direction from one instruction to the next. If the tester is to provide instructions to the microprocessor and check its response during the next half of the instruction cycle, the pin card assigned must function as a driver and then quickly switch to a receiver. The inhibit buffer output is used to switch the driver circuit on and off. Synchronizing the reading of all the buffers allows the driver to be active during one test vector and off during the next. The microprocessor cycle would then be treated as two test vectors, and clock synchronization would have to be arranged on this basis for a successful test. Testers with complex cycle timing can switch during a single cycle using a stimulus time frame and a response time frame in the same test vector.

In some testers the number of buffers is reduced to three by using the output of

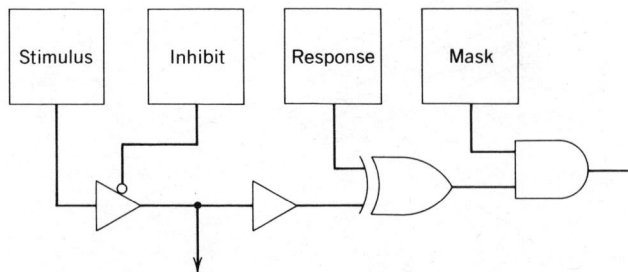

Figure 6.7. Buffer memories hold digital test information for simultaneous retrieval in this pin electronics diagram.

the inhibit buffer to direct the output of a ''logic value'' buffer to be used as either stimulus or response. The assumption made in this case is that a tester pin can either drive or receive, but not both. Alternatively, the same logic value can be used for both drive and receive on all tests. The former method prohibits the use of the receivers as self-check devices for the drivers. The latter restriction may be a problem if the driver is used as a circuit pull-up for an open collector output of the CUT. The situation ''drive a 1, receive a 0'' is common in this case.

During the application of the first few vectors (or perhaps the first few hundred in complex circuits) some of the outputs of a sequential circuit are likely to be unknown. Since the contents of the expected response buffer are a single binary digit, only a 1 or a 0 can be represented, and there is an excellent chance that some of the unknown outputs will disagree with either case if it chosen randomly. The mask buffer is read in synchronization with the expected response buffer and used to disable the error output from the comparator when the expected output is actually unknown.

All four buffers could be placed in a single RAM memory using a 4-bit data path since the read clocks are usually all in synchronization. Alternatively, a single bit memory could be read at a rate 4 times the test vector rate as long as the outputs are latched and properly strobed. The elimination of glitches and erroneous error stops depends on correct application of these signals, as shown in Figure 6.8.

To be assured of eliminating bus contention, the inhibit signal must be moved toward turning the driver off before changing the driver value, but toward turning it on after changing its value. The mask and response values can be changed at any time as long as they are settled before the error strobe is opened. An error strobe is necessary to allow settling time of the CUT. The mask signal can be used to inhibit the error strobe for the same effect as blocking the comparator output.

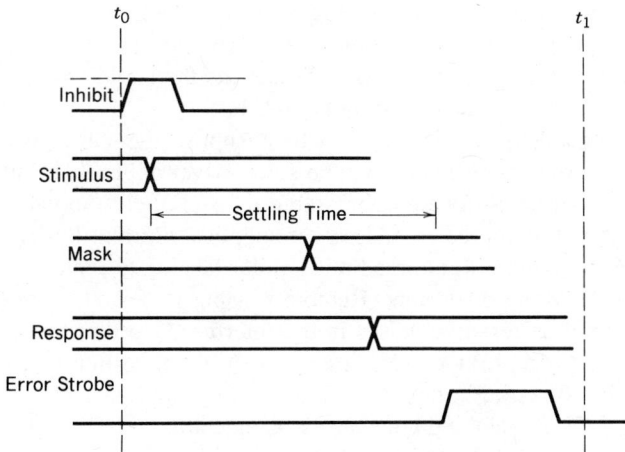

Figure 6.8. Timing diagram for a single test step.

Cycle time of the test is measured from one application of stimuli to another. The circuit response time allowed for is somewhat less, however.

Matrix systems are limited in speed by the necessity of constructing adapters which contain relatively long wires. No matter how carefully the pin cards are arranged, the worst case line length spans the entire width of the test head for some CUTs. In addition, the construction and storage of separate test adapters for each variety of CUT become costly for installations which must test more than a few types of circuits. Field installations and factories which have small volumes of many different circuits suffer from this drawback.

Programmable Pin Electronics

As the density of electronics improves, the capability of pin electronics also increases. The matrix problems have been counteracted by a more flexible (although generally more expensive) style of pin card. The programmable function pin card referred to presently should not be confused with the ''programmable voltage'' pin card just discussed since the subject at hand is a device which can supply power voltage or ground and perform some types of meansurement on the CUT signals as well as interact logically.

Several techniques have been employed to create programmable function pin cards. A common form uses a number of power and reference buses which are distributed to all pin cards. Relays or electronic switches on each card are controlled by configuration registers and in turn connect voltages to the CUT pin. The actual voltages present on the buses are programmed separately by means of D–A converters or digitally controlled power supplies.

The relays in the illustration of Figure 6.9 could be replaced by VFET switches, but the function remains the same. Before power is applied to the buses the switches are set which select whether a power bus, ground bus, or logical driver–receiver is to be connected to each CUT pin. The power supplies are programmed, and the logic family switches are thrown. The logic family selection sets up the values to be driven or compared against by connecting a set of rails (buses) to the driver–receiver pair. the voltages on the rails represent the thresholds for a logic family, such as TTL or ECL, which is used on the CUT.

The programmable power supplies which supply the CUT or define logic thresholds for the driver–receivers may be screwdriver adjustable, but usually are under software control. Software addressable registers hold the values which are D–A converted and amplified by the power supplies. Current limits for the CUT power supplies are often programmed in a similar fashion to protect the ATE from damage due to shorted power nets. Remote sensing is desirable for CUT power supplies to counteract resistive losses in the interface. Parallel lines may be provided in the adapter to provide sensing as close to the final interface as possible.

Programmable function in pin cards reduces the complexity of the adapter to a linear wiring of CUT matching connectors. Connector A1, pin 1 of the CUT can be wired to ATE pin 1; CUT pin A1, pin 2 can be wired to ATE pin 2, and so

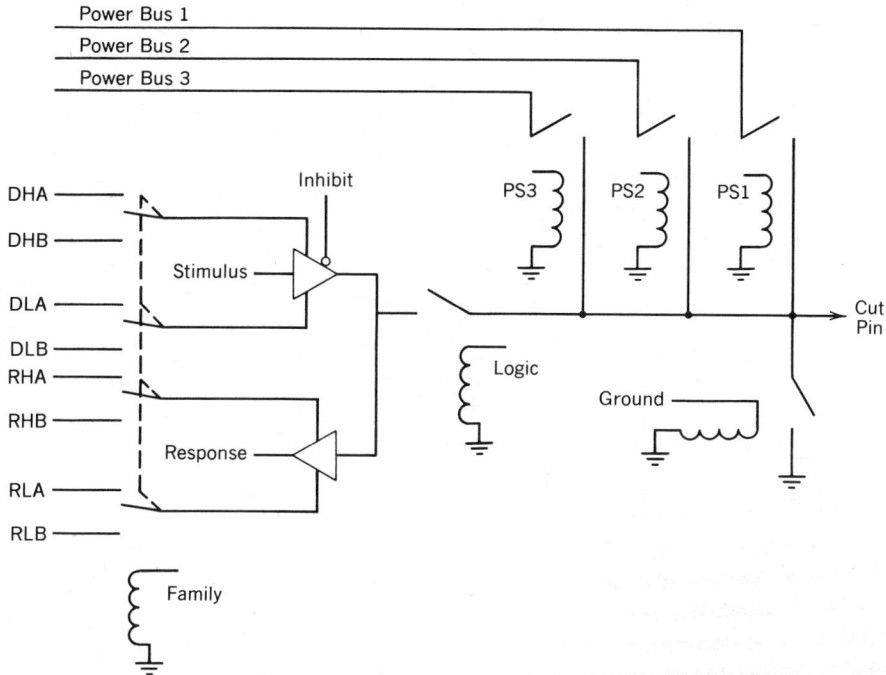

Figure 6.9. Relay-switched fully programmable pin electronics circuit.

on. Simpler wiring in turn shortens the path between the driver–receivers and the CUT, thereby reducing induced noise and loading on the connecting wires.

Termination and loading of the CUT outputs can be critical to some designs and may be specified in the test requirements. Provisions for the addition of resistors are commonly made on the adapter and may in some systems by included in the pin electronics. Relay-switched resistors of common values for loading logic families such as TTL and ECL might be included, for example. In cases where the connecting wires from the tester to the CUT are unavoidably long, series resistors are sometimes added to slow rise times and reduce noise.

Achieving accurate loading is difficult since the actual load seen by the CUT is a combination of the distributed impedance of the connecting wires, the input impedance of the receiver, the off-state output impedance of the driver, attached loading resistors, and any other gates on the CUT net driven by the output gate. Several of the factors are frequency dependent. In the case of TTL and similar technologies the impedances vary with the logic state of the driving and receiving logic gates (Figure 6.10). Specifying a load properly requires a careful analysis of these factors and the tolerance from one tester to the next. Fixed resistive loads are becoming unpopular due to the inaccuracies of application from one output to the next.

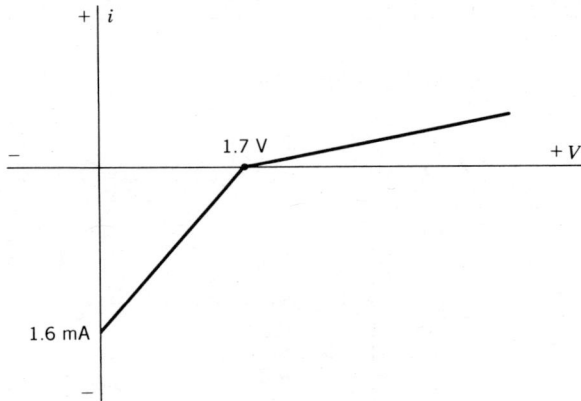

Figure 6.10. Approximate load line for a standard TTL input.

In place of resistors and relays the driver in the pin electronics circuit has been improved to take on the loading functions. Control circuitry of the driver can be modified to produce a constant current or a constant voltage. In the latter case the tester is forcing a logic level, as described earlier. When instructed to force a constant current, the driver represents a programmable continously adjustable load for the CUT output pin connected. A feedback loop in the driver can also allow resistive loads to be simulated as a current forced which is proportional to the voltage present. Nonlinear relationships can even be used to model true TTL loads, although the control mechanism and the programming become so complex as to be impracticably expensive in most cases.

Distributed Computer Systems

As pin electronics becomes more and more programmable the point at which each CUT pin is controlled and monitored by a separate microcomputer is passed (Figure 6.11). The availability and familiarity of 8-bit microprocessors has been used by many ATE manufacturers to create computerized pin electronics with each byte of CUT pins handled by a separate chip. The tasks of coordinating and timing the application of input vectors and the comparison of expected outputs can be easily cared for by a microprocessor. The buffer storage associated with each pin is attached to the microprocessor as data memory. The drivers and receivers become I/O channels, and a separate program memory defines the actions to be taken during test.

The major advantage of distributing processors occurs when memories are mixed into the logic of the CUT. As Chapter 10 reveals, test vectors for memory structures are well known algorithms. Conventional logic test generators are very inefficient at testing memories, on the other hand. The microprocessor-controlled pins of a tester can generate directly the input vectors needed by using memory test

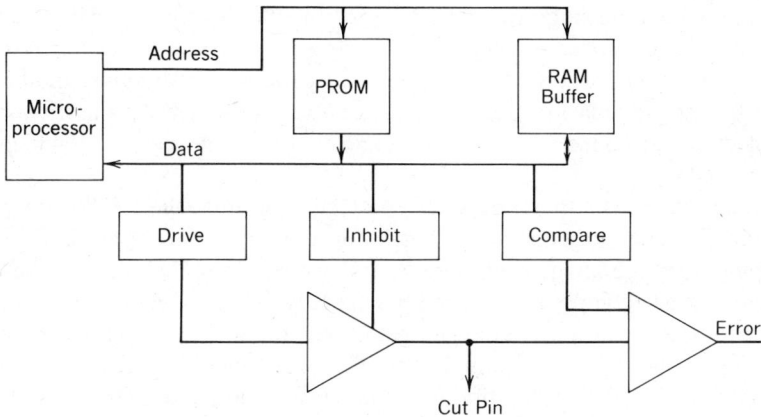

Figure 6.11. The diagram for a single pin electronics circuit includes a dedicated microcomputer and support chips in a distributed computer architecture.

algorithms which have been previously loaded into their individual instruction memories. Pins which correspond to address bits can be instructed to work together as a counter, while pins connected to CUT date lines complement their contents each cycle. Other, more complicated schemes can be worked out to read and store the contents of PROMS in a CUT for comparison with later units.

ATE with mixed capabilities, such as a complement of fixed voltage pins and a small number of programmable voltage pins, are quite common. A mix of microprogrammable pins with fixed voltage pins is sometimes available, although access to all the pins must be through a matrix architecture. When programmable function pins are implemented, the capabilities are uniform from pin to pin. Otherwise, the advantages of simple adapter wiring could not be realized.

OPERATING MODES

In addition to the variety of architectures presented by ATE, the modes of operation differ from one system to the next and from time to time within a given system. The execution of a test sequence for a particular CUT may require switching from one mode to another to bring capabilities of the test system to bear on individual partitions of the CUT. Static functional testing may be adequate for most of the logic, and the ease of troubleshooting afforded by this technique is reason enough to use it when applicable. But the CUT may have dynamic circuits which pass through several important states after only one input vector has been applied or whose outputs disappear a short time after they are asserted. The sequence of input vectors required to trip a condition or activate a state in the CUT may be time dependent, requiring the presentation of vectors within a window of time. Any of these conditions require careful timing of the stimulus and response functions.

Dynamic functional testing provides such control either explicitly or implicitly by running at sufficient speed to synchronize with the CUT. Both static and dynamic variants of functional testing depend on stored patterns for stimulus, inhibit, expected response, and masking, and both variants access the CUT primarily through the connections provided for later communication with the system in which it is to be installed.

In-circuit testing is a major difference in technique since the CUT is not tested as one circuit. The components that make up the circuit are tested individually even though they are already installed in the printed wiring board, thus the name "in-circuit." Each component is tested functionally, and the board is usually tested for point-to-point continuity along its paths, but the overall functionality of the board is not explicitly tested. Complexity of the test generation process is lowered significantly by this physical partitioning, but physical limitations also apply. Either stored patterns or algorithmically generated patterns may be used within this method.

Some subassemblies to be tested are well structured and thus suited to algorithmic test generation. RAMs and ROMs are examples. Logical partitioning or physical partitioning can often be used to separate such circuits from the surrounding logic and allow algorithmic testing to be used in conjunction with functional or in-circuit techniques. Once the algorithms have been generated, they can be expanded to stored patterns, but the compactness of the method is lost. Many testers make special provision for understanding and carrying out algorithms directly from the compact programming form.

The capability of ATE pin electronics to control and measure both current and voltage has led to the availability of automated parametric testing. Functional stored patterns force the circuit into the correct state for measurement, and the parametric mode is then activated. Although the differing modes of test application are often used in conjunction with one another and are somewhat intertwined, they will be discussed individually.

Static Functional Testing

Early digital testers were controlled by paper tape applied tests in the static functional mode. Sometimes called pseudostatic testing more accurately, the key point is the time allowed between application of input vectors. As long as the time between stimulus applications is greater than 10 times the normal clock period for the CUT, the test can be considered static. All propagation times, delay line pulses, and one-shot pulses must have been settled before the expected response is considered valid in any case.

The test generation method used to create the input vectors and the method used to determine the expected response must also take the static assumption into account. Free-running oscillators or clocks are the most common obstacle in the static method. They must be blocked or disconnected during testing. If either the good circuit or a faulted circuit becomes oscillatory during test generation, special steps must be taken by the test generator or simulator. If the good circuit becomes oscillatory, the input vector or sequence must be rejected. If a faulty circuit becomes oscillatory, the result must not be used to detect the fault.

If the criteria for static testing are strictly observed, the result will be a test which is insensitive to the speed of the tester as long as it does not exceed a given limit. The time between input vector applications could be a millisecond, minute, or an hour. Troubleshooting in this environment is greatly simplified since the technician can use a voltmeter to compare logic values at any test step to the expected values.

Timing cannot be ignored, however. In addition to determining overall minimum settling time, test generation must account for the possible generation of transients and occasional time-dependent operations. Transients can be generated internally or by input vector skew. Internally generated transients are problems of design, but input vector skew is a test problem. When more than one input signal is changed in a single test step, the possibility of a logic hazard exists. Such a hazard causes a transient or pulse when a gate is moved through an intermediate logic state on the way from an initial state to a final state and the intermediate state has a different output than the other two states.

The minimum time required for the presence of an input condition to cause a reaction in IC logic is on the order of a nanosecond. Unwanted transients could thus be avoided by assuring that all signals in a given input vector arrived at their respective destinations within a few hundred picoseconds of one another. Not only does this condition imply costly pin electronics, but it requires very well controlled transmission characteristics in the adapter wiring. The only practical way to avoid problems entirely is to move only one input signal at a time and compute the expected response on that basis. Usually, simulation models include some approximation of timing and warn the test engineer when a transient may be generated by an input vector.

If a primary output of the circuit is a one-shot which has no other feedback to the circuit, it cannot be tested by strictly static means. If the input vector is applied and sufficient time is allowed for the entire circuit to assume a stable state, then by definition the output will always be sampled in the unfired state. Static functional testers are sometimes equipped with a transition detector attached to the receiver which can be invoked to solve this problem. A simple edge-triggered flip-flop which is cleared at the start of each test step suffices. The tester must be informed during setup to look for pulses rather than levels on this pin. If the pulse is mixed with a level in the circuit, the mode of the test pin will need to be changed from time to time during the test.

An alternative method of detecting one-shot outputs requires intervention of the test engineer into the sample timing of the expected response circuit. The procedure results in waveforms, shown in Figure 6.12. At the beginning of test step t_0 an input is asserted which fires the one-shot after a propagation delay t_p. At some time greater than t_p but less than $t_p + t_w$ (the width of the one-shot pulse), the output is sampled. A dummy test step with no change in inputs is inserted at t_1,

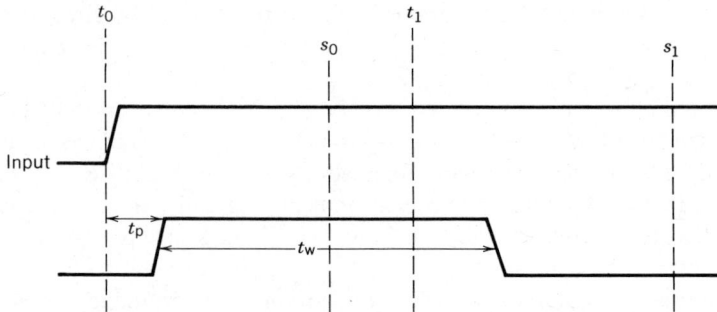

Figure 6.12. Timing diagram for one-shot output tests using a static functional tester.

and the output is sampled again after the one-shot is sure to have settled. Normal static testing can then resume.

Some circuits contain lengthy one-shots or delay line signals which are intended to bracket other operations. The time-out signal in an I/O controller is a prime example. Strictly static testing of a block transfer sequence could be impossible if the entire sequence has to be completed while the time-out one-shot is active. Once the input vector which activates the one-shot is applied, the rules of static testing require waiting until the output has settled back to its stable state, resulting in a time-out condition. The model used to determine expected outputs has to be altered to assume that the state of the one-shot is bistable. The test engineer must then assure that all vectors required within the fired state are accomplished rapidly before the one-shot time has elapsed. Then the bistable model is reset to simulate time-out, and a wait is placed in the actual test sequence to allow the one-shot to settle. A short dynamic segment has just been inserted into an otherwise static test sequence.

Dynamic Functional Testing

Contending with the difficulties just mentioned, and the desire to perform tests as closely as possible to the operating environment, has created the technique of dynamic functional testing. Sometimes called real-time testing, the actual speed is not necessarily that of the intended use of the circuit since testing technology is seldom ahead of other areas of circuit design. The architecture of the ATE is not greatly different, although at higher speeds the programmable pin electronics style is favored due to its inherently shorter adapter lead lengths.

Multiphase clocking and programmable timing of events in the test cycle are characteristics of dynamic testing. The input signals may be associated with one of several clocks, and outputs may be sampled after a programmable settling time has occurred. Data inputs might be asserted at phase 2 with bus control signals at phase 1 and a register clock at phase 4. A common programming arrangement for test clock phases allows the specification of rising and falling edges in terms of master clock pulses where the master clock is much faster than the test rate.

Arriving at the proper timing for signal application is a matter of uncovering and encoding the specifications which relate the signals of the CUT. The example of Figure 6.13 includes only a single IC, but the ideas are easily extensible.

The data inputs D_0 through D_3 have been selected as "return to complement" mode with the leading edge on phase 1 and the trailing edge on phase 4. The data being applied on this test step are $(1, 0, 1, 1)$, which requires the inputs to be set to the complement until the leading edge time phase been reached and to be reset to the complement at the occurrence of the trailing edge time. The time between the leading and trailing edges of the inputs is the sum of the setup and hold times for data on the D flip-flops. The clock leading edge is placed at phase 3 to divide the time into its setup and hold times properly. Thus, both of these criteria are tested during each step.

The clock pulse is formed by defining a "return to zero" (RZ) mode of operation with the leading edge on phase 3 as mentioned before and the trailing edge on phase 5. The latter is selected to give a clock pulsewidth corresponding to the

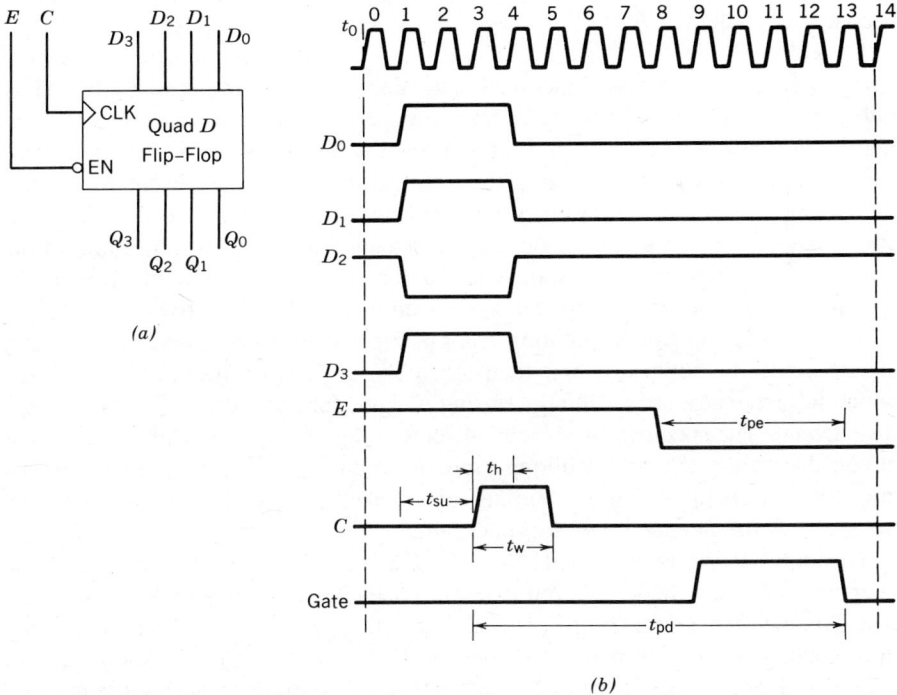

Figure 6.13. Application of the test stimulus and response according to the diagram in (b) tests several timing specifications of the circuit in (a). (a) Example circuit. (b) Pin electronics clock phasing.

minimum guaranteed by the system. The leading edge of the clock is also the beginning of the data propagation time through the chip. The test receiver gate is closed after the propagation time to prohibit slow parts from being accepted. Even the output enable activation time has been delayed in this case to test the propagation time from enable to outputs.

The example given illustrates the ideal case of dynamic testing on a very simple example. Complex circuits are not tested so thoroughly due to the impracticality of deriving the test specifications and the difficulty in arranging test conditions which measure them. When two NAND gates are used in series, the tolerance of the overall circuit is the sum of the individual tolerances in the worst case (i.e., if each NAND has a propagation time of 3–5 ns, then the pair has a propagation time of 6–10 ns). Actually, the distribution of most characteristics is statistically normal, and the circuit times usually fall on the classic bell-shaped curve.

Testing the propagation times of circuits with dozens or hundreds of elements in the paths from input to output would involve considerable statistical analysis. The resulting test specifications would probably bear little resemblance to what is actually expected of the circuit. Statistical limits other than worst cases are more realistic in their relationship to the application of the circuit, but the risk of finding a defective propagation time on a path made up of "in-specification" components becomes a problem.

Consider the problem of troubleshooting a circuit which exhibits a late-arriving output signal in a dynamic functional test. Verifying the failure as a late arrival rather than a stuck level requires an oscilloscope synchronized to the tester. Tracing the failure requires analyzing the expected delay times for each of the gates along the path as well as analyzing the logic action taking place at internal nodes during the test step. If a gate can be implicated, it can be replaced with an in-specification part. If a tolerance buildup has resulted in the failure, the gates of the path will have to be replaced somewhat randomly to tune the circuit. The task is not an easy one, and the components replaced may not be defective.

The difficulty in troubleshooting has made the dynamic test less popular as a testing mode for initial test of circuit assembly. It is sometimes applied to units which have already been debugged using a static functional test as a high grade final screen. The specifications used are seldom tighter than the worst cases of the circuit due to the tolerance buildup problem in any case. Very often, the inputs are applied in parallel and the outputs are checked after a time which represents the speed of the intended application or some significant fraction thereof.

Dynamic testing is also used when the CUT itself is dynamic and cannot be operated in a static mode. Microprocessors, memory management ICs, and interface circuits often contain time dependent circuits. MOS devices are often operated in a dynamic mode. Even if the devices are capable of static operation, lack of DFT may leave clocks in the circuit which cannot be disabled during the test.

If the circuits are asynchronous in nature but contain elements which utilize charge storage such as memories or MOS ICs, the ATE can be used in the same fashion as static testing (providing all inputs including clocks and sampling outputs after an appropriate delay). When the clocks in the CUT cannot be interrupted, the ATE must be slaved to the CUT clocks, and the test mode becomes synchronous. The tester may be required to base its multiphase clock on a single CUT clock input, or signals such as "valid address" or "bus enable" may be combined to define the stimulus application and response sampling times.

The extreme case of synchronous ATE application is the bus-oriented "in-circuit emulator." A connection is made to the bus or buses which control and pass data in a circuit such as a single-board computer, and the controlling IC (usually the microprocessor) is disabled. The ATE assumes the role of controller and exercises the remainder of the circuit by imitation. The imitation includes all timing and enable signals and requires the tester to respond to circuit stimuli as well as issue its own. Generating a test in this environment is similar to writing a diagnostic program. The ATG algorithms do not perform well since the level of organization necessary in the test sequence is beyond their realm.

Emulation instruments are microprocessor-controlled programmable testers in a compact form. User-friendly menu displays and a keypad or keyboard enable the test engineer to program a test sequence in the language of the IC being emulated. Further automation is provided in most cases by an IEEE-488 or RS-232 port through which the program can be loaded and results returned. A major ATE system can use such a device as an instrument in the same way a counter or a voltmeter might be used.

Portable stand-alone emulators are usually equipped with option modules, enabling the control of several distinct emulations one at a time from the same programmable unit. Each module is an I/O channel, and the switching is by software control up to a maximum number of available slots or channels. The modules are sometimes called "personality" modules since the hardware of the module is responsible for timing and voltage matching with the device to be emulated.

ICs are not the limit of emulation possibilities. Standard protocol I/O buses themselves can be tested by programming the emulator to imitate a matching transmitter and/or receiver. Test messages can then be exchanged, including parity, timing, and start/stop bits. Extra routines are often included in the software of bus protocol emulators to create errors in order to check the error detection of the unit under test (UUT). Noise generators, voltage offset injectors, and other hardware tolerance checking hardware may be included for testing worst case limits of operation.

In-Circuit Testing

The difficulty of troubleshooting complex circuits, even in the static mode, is considerable. It requires knowledge of logic and familiarity with the particular actions of the test sequence. Structures such as feedback loops and long sequential strings require rerunning the test many times and correlating the observed states at various test steps. Test generation is costly for similar reasons when circuits become complex. DFT efforts attempt to improve the controllability and observability of internal circuit nodes to reduce the problem, but most DFT programs are compromises of design requirements versus test cost. DFT will be discussed in a later chapter. Its mention here is to relate such efforts to a significant development in test technology aimed at the same problems.

In-circuit testing is an adaptation of both function and architecture of ATE, although the largest innovation is in the function. The pin electronics is modified in an in-circuit tester to substantially increase its current drive capability. Additionally, the CUT interface adapter is changed from an edge connector contact fixture to a "bed-of-nails" fixture. The name is a reference to the eastern bed of nails used by mystics to demonstrate mental dominance over pain. The fixture consists of a planar cabinet with spring-loaded pins protruding from the top. The pins are arranged accurately with respect to guideposts so as to contact all internal nodes of the circuit board. Contact is made at solder connections of component leads on the underside of the PC board. The board is held against the pins by a vacuum or by mechanical pressure to assure reliable connection.

If the contacted points were used only to monitor the nodes while input vectors were applied to primary inputs, the troubleshooting problems would be greatly reduced and the test generation problems would be somewhat reduced since fault activation, but not fault effect propagation, would be necessary to test for stuck-at faults. The in-circuit concept provides the additional capability to drive portions of the circuit through the bed of nails as well. High-current drivers in the ATE override the normal driver in the circuit nodes which input to the DUT. Each component can then be tested independently of the rest of the circuit, but nevertheless while physically installed in the circuit.

The test sequence for the entire subassembly is a series of test subroutines which define a single IC's input nails, define its output nails, and then execute bursts of test vectors which are directed at testing that IC. Inputs and outputs are selected from a field of contacts which includes at least one nail for each net, as shown in Figure 6.14. Since each component is tested independently, fault isolation to a

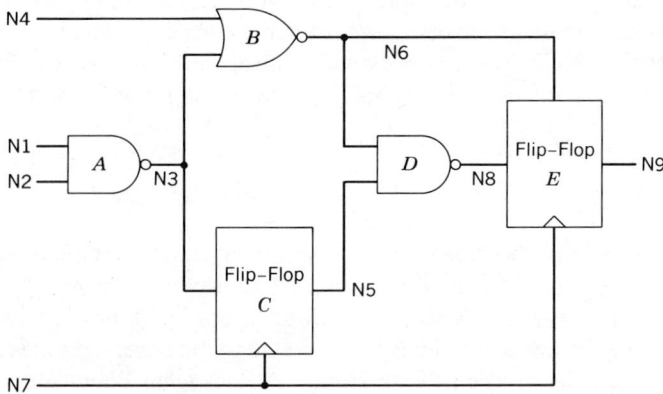

Figure 6.14. Contact points called "nail locations" are shown on each net of the circuit diagram.

single component is automatically achieved. The input vector subset needed to control the component being tested is applied in a burst. The "backdriving" of CUT gate outputs by the ATE drivers must be limited in duration in order to prevent thermal damage to the IC. Overcoming a conducting transistor in the lower case of a TTL totem pole output can require currents of 150–400 mA to maintain 4.0 V [4].

Power dissipation in the IC rises greatly during the burst. The total energy must be limited to prevent damaging an internal lead wire or destroying a junction. Limiting the energy is achieved by limiting the duration of signal application. Unfortunately, calculating the energy absorption limits of digital ICs is not straightforward. Studies of the thermal characteristics of semiconductor junctions in bipolar technologies [5] have established models for determining backdriving limits, and experimental results indicate that these technologies are not easily damaged by the practice. MOS transistor technologies show signs of being more fragile, although safe practices can be established [6]. The number of outputs from a single chip being backdriven and the circuit configuration of connected nodes both play a role in determining what current will be required to successfully backdrive and how much power can be applied simultaneously. The burst length is limited to less than a second and usually less than 200 ms. With a test vector application rate of less than 1 vector/μs, the useful test sequence length is limited to less than 2000. Complex components such as microprocessors may have to be tested in several segments.

The general mode of operation for in-circuit testing is a concatenation of tests. Before applying power to the CUT, the nets which make up the circuit are usually tested for internet shorts. Since the minimum nail allocation for thorough testing of components is one nail per net, a net isolation test is an easy and natural first step. After shorts have been eliminated, the components are tested one at a time using bursts of test vectors that treat each component as a CUT. These tests can be generated by the techniques of the earlier chapters if the components are digital parts such as latches, gates, microprocessors, and so on.

Bus structures are treated separately since the "logic" of a bus is actually the interaction of several components. Subroutines provided by the ATE manufacturer exercise the bus through manipulation of all its drivers in concert. Measuring routines determine that the bus is isolated when all drivers are off and track the current on the bus to find a "stuck-on" driver. Similar measurement subroutines can make analog measurements of resistance and capacitance to verify the correctness of pull-up and filtering components associated with digital circuits.

The in-circuit test structure is a significant improvement over functional test programs for the reduction of test development time. Since the test is a concatenation of subtests which may be taken in any order and are independent, changing a test section to compensate for a change in the UUT design has only an effect proportional to the change magnitude. In functional testing even a small change often causes the entire process of test generation to begin again.

Achieving a physical connection between ATE nails and CUT nodes is difficult in some geometries such as surface-mounted device modules. Circuit boards which have been hermetically sealed with a coating for protection in hostile environments cannot be tested after coating. Physical factors of DFT, such as maintaining a grid of standard PC layout coordinates to assure that contact nails can be assigned, drilled, and placed automatically, become the overriding concern in the in-circuit test method. Even with direct access to all circuit nodes, the in-circuit technique is not immune to testability problems. The ATE driver cannot be made a perfect current source. The connections between ATE and CUT add impedance and result in imperfect transient response when the stimuli pulses are applied. An active clock driver which has high-current output to drive a large net on the CUT is pictured in Figure 6.15 as it is connected to the ATE driver which is to hold a logic 0 for the test vector being applied. Although the tester driver has a high-current capability itself, the real circuit includes distributed resistance and capacitance on the adapter and within the CUT. The driver is to hold a constant voltage, but the feedback is unable to respond quickly enough to the current changes as the clock driver switches. The actual voltage maintained on the node will have ringing correspondence to each clock transition. If the intervening connection impedance is high enough, or the feedback circuit is slow enough, the pulses will be of sufficient magnitude to reach the logic threshold and trigger the clock inputs of the CUT, invalidating the test. Good testability practices pertain to all modes of testing [7].

The bed-of-nails technology originated in the continuity testers, which are still used to check bare PC boards (before adding the components) for correct wiring. In continuity testing a very thorough bed of nails is used to contact every terminus of each net, and a simple switching matrix connects an automated ohmmeter across each pair of terminals. The continuity found on a KGB is recorded, giving a pattern which tests for connection within nets and isolation between nets. New boards of the same type are quickly scanned and compared to this example.

The one-nail-per-net distribution of the in-circuit tester is not sufficient to per-

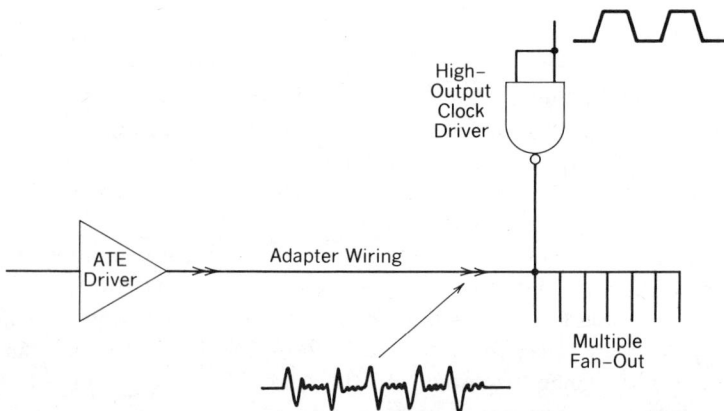

Figure 6.15. Single in-circuit nail with wiring to a net under test. Noise arises due to the wiring impedance.

form connectivity tests within a net (it could be augmented if desired), but isolation tests between nets are routinely run prior to functional testing. The isolation tests locate some solder shorts immediately and protect the power supplies from excess loading.

SUMMARY

The range of ATE available provides many tradeoffs to be made in choosing a system for an application. Each area of ATE is progressing with time, and it is therefore impossible to provide a detailed comparison of manufacturers or models. General characteristics of architecture and operating mode, as shown in Table 6.1, serve as a guide, however. The troubleshooting column refers to the ease of isolating a bad component or other defect once the go/no go test has detected a failure. A lower rating reflects the need for more training and knowledge on the part of the troubleshooting technician to perform fault isolation. The cost range typically begins at about $40,000 and exceeds $2 million for high-speed multicapability systems. Speed also varies over a wide range with static and instrument bus systems performing dozens or hundreds of test steps per second and high-speed dynamic testers operating in the 50–100 MHz area.

A typical manufacturing flowchart includes possible application areas for several different types of ATE. The tradeoff is often not a matter of which test concept to use, but where to use each one. Consider the manufacture of a minicomputer which has a high-speed serial bus interface.

In the manufacturing flowchart of Figure 6.16 the single lines which connect each symbol represent a volume flow of many different types of components or assemblies. Storage facilities used for staging have been omitted for simplicity. The receiving department routes PC boards (bare) and components to the appropriate test station. Printed circuit boards are tested for possible opens and shorts and sent to the assembly area to be populated. Components are tested using a high-

TABLE 6.1. Comparison of ATE Types

ATE Type	Cost	Speed	Troubleshooting	Comments
Instrument bus (portable)	Low–medium	Slow	Poor	Hybrid capabilities
Instrument bus (rack)	Medium–high	Slow–medium	Poor	Hybrid capabilities
Static functional	Low–medium	Med	Good	
Dynamic functional	Medium–high	Fast	Fair	
In-circuit	Low–medium	Medium	Excellent	Cannot test coated circuit boards

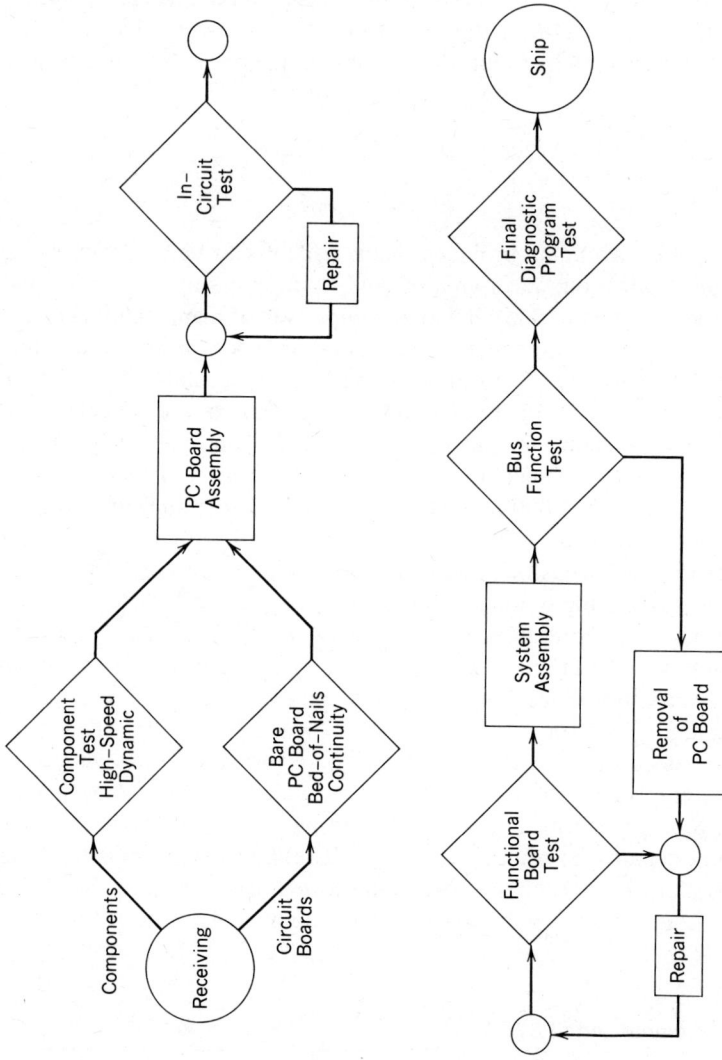

Figure 6.16. Flowchart of a manufacturing and test process which uses several types of ATE.

speed functional tester with parametric capability. The supplier's specifications sheet and the actual design requirements of the intended use guide the test conditions.

After assembly the populated and soldered PC boards are tested quickly on an in-circuit tester to eliminate assembly errors (missing ICs, reversed components, etc.). A functional test assures that the boards operate correctly as complete assemblies before they are installed in the final system. Before the entire system is tested using diagnostic programs, specialized test equipment such as a bus emulator is used to efficiently test specific aspects of the system which are difficult to test or diagnose using software. After final acceptance the system is shipped.

The example given assumes that the nature of the product and the volumes involved in the manufacturing process justify the use of a complete suite of test equipment. Subsets of the ATE shown in this case may be sufficient for other problems. Product configuration determines which forms of ATE will be efficient. Functional testing was supplanted in many cases by in-circuit testing during the 1975–1985 time frame. But recent trends toward the use of leadless chip carrier style components and ceramic substrates or double-sided PC boards complicate the use of the bed-of-nails adapter. A resurgence of functional testing is the result [8].

Further discussion of the selection and placement of test technologies within the manufacturing and other scenarios will be presented in Chapter 12.

REFERENCES

1. T. J. Healy, *Automatic Testing and Evaluation of Digital Integrated Circuits*, Reston, Reston, VA, 1981, pp. XIV–XV.

2. J. Purvis III, "The Personal Computer as Instrument Controller," *Test Measurement World*, **6**(2), 42 (Feb. 1986).

3. M. S. Hoffman and J. F. Wrinn, "Channel Card Architecture for Multimode Board Test Systems," in *Proc. 1984 Int. Test Conf.*, IEEE Comput. Soc., pp. 589–579.

4. *Test Programming 227X Circuit Board Test Systems*, Gen Rad, Inc., Boston, MA, 1982, pp. 6–89.

5. J. M. McPhee, "The Effects of Backdriving Integrated Circuits—An Accurate Electro-Thermal Model," in *Proc. 1985 Int. Test Conf.*, IEEE Comput. Soc., pp. 518–522.

6. F. H. Hielscher and J. C. Pagano, "Backdrive Stress-Testing of CMOS Gate Array Circuits," in *Proc. 1985 Int. Test Conf.*, IEEE Comput. Soc., pp. 523–533.

7. S. Caplow, "Conquering Testability Problems by Combining In-Circuit and Functional Techniques," in *Proc. 1984 Int. Test Conf.*, IEEE Comput. Soc., pp. 581–588.

8. S. F. Schieber, "Full Circle: The Re-Emergence of Functional Testing," *Test Measurement World*, **6**(4), 20–31 (Apr. 1986).

DEVICE-UNDER-TEST INTERFACE

Perhaps nothing seems more mundane than the connecting of an electronic device (circuit) to an automatic test system. The first impression is "Just hook it up!" Few articles of test equipment are as poorly designed on the average or result in so much consternation as the device-under-test (DUT) interface. A carefully designed and constructed DUT interface, or device adapter as it is sometimes called, is transparent to the ATE and to the test operator. A poorly designed or constructed one masks faults and causes good DUTs to be discarded. Either case is costly and frustrating.

The importance of the interface adapter is often underestimated. If the test is to be performed in a number of different locations more or less simultaneously, the cost of producing the adapters can be considerable ($500–10,000 each). Moreover, the success of the tests which may have been costly to create in their own right, may be jeopardized.

Whether the DUT interface is termed an interface adapter, device adapter, or a fixture, it incorporates several functions into an electromechanical assembly, and all the functions must be considered to assure success. Providing power and connecting signals between the ATE and DUT are the functions most easily conjured up. Physical support and environmental control are quite important in some applications.

INTERFACE DESIGN GOALS

Application parameters guide the design. The configuration of the DUT, the configuration and capabilities of the ATE, and the testing site determine the optimum approach to use. The adapter built in the laboratory for check-out of the prototype would probably be unreliable if it were subjected to day-after-day repeated use in the factory or constant packing, unpacking, and transportation in the field. The durable adapter designed for testing several different devices in the factory may be too heavy or bulky for the field.

The DUT interface is often a multipart assembly with "personality boards" or switches which are replaced or altered to permit connection of more than one DUT to the ATE without duplication of the entire interface assembly. This strategy works well for testing ICs which share the same package style. The physical holding mechanism, the DUT connector, and the ATE connector are the same for each member of the 24 pin DIP family, but the point-to-point wiring of chip pins to tester channels varies from part number to part number. Jumper blocks can match up the DUT and ATE pins. If some pins require termination, loading, or buffering, the personality boards may be actual daughter boards with passive or active components on them.

Individual adapter designs vary widely, even for different devices tested using the same ATE. General guidelines for interface design are more useful than individual details as long as engineering judgement is used in their application to individual problems.

Transparency. The DUT should appear as if it were installed in the product or final application. In this way the test environment impact on test results will be minimized. When the DUT is to be used in more than one product, a compromise must be effected.

Simplicity. A complicated interface device is usually more costly to build and maintain. Complications of use such as the installation and removal procedures should also be minimized since a small cost in time per use may add up to a large life-cycle cost.

Durability. Selection of mechanical components and connectors is critical to providing a test adapter which requires little or no maintenance. The reliability of the test is dependent upon the reliability of the equipment used, including the adapter.

Safety. Logic testing is usually conducted at low voltages and safety is therefore often neglected. The safety included here is in a broad sense, including DUT safety from electrostatic discharge, overvoltage, or overtemperature, as well as operator safety from direct connection between high current terminals and a watch, ring, or other jewelry.

The design tradeoffs made among these goals depend upon the application. Before providing several examples of interface designs for specific application, the common factors dealt with on every interface adapter will be discussed.

ELECTRICAL CHARACTERISTICS

Both power for the DUT and signals pertaining to the test inputs and outputs are usually conducted through the interface adapter. Power may be in the milliwatt range for testing ICs or in the watts range for pc cards. Multiple voltage may be required, and in some instances the ground returns must be kept isolated from one another. The simple principles of dc power wiring seem trivial, but they are so often not followed thoroughly in interface construction that they bear restating.

Power Wiring

The programmability of power sources in ATE leads to the belief that the voltage which is requested will be present on the requested pin under all circumstances. Although the equivalent circuit for the power wire from ATE to DUT shown in Figure 7.1 is needlessly complex for most cases, it serves to remind the test engineer of the potential effects that can be generated by transient power drain changes [1]. The effects from the characteristic complex impedance of the wire are minimal if the wire is short and the transients are kept small. In the ideal case the current is constant and the voltage at the DUT is simply the programmed voltage minus the ohmic drop over the wire. Current requirements for the DUT can be several amperes for a large pc assembly, and the resulting voltage loss can place the circuits at the edge of their operating range instead of at the programmed nominal. Moreover, the effect is doubled if the ground wiring also produces an uncompensated voltage drop. Consider a pc card which requires 5 A power at 5.0 V. Assume the adapter contains 1 ft of wire between the ATE power bus and each of the four DUT power pins and four ground return pins. If the adapter is wired entirely with AWG 28 signal wire, which may be the case if the adapter is universal and the power pins are rearranged on each DUT type tested, the voltage seen between power and ground buses on the pc card will be 4.83 V. This is dangerously close to the 5% margin for operability of the TTL components. Replacing the power leads with AWG 18 wire results in a voltage at the card of 4.94 V and a better margin for noise.

Compensation for such losses and for low-frequency variations in load voltage due to changes in the load current is ordinarily achieved by remote sensing. The output voltage feedback lines of the power supply regulator are often accessible at the terminal block for use in this way. These remote sense lines should be connected as closely as possible to the load using two wires independent of the load

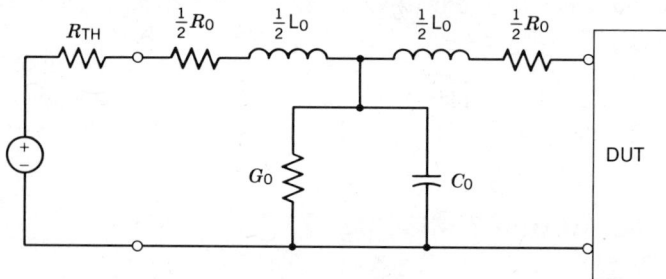

Figure 7.1. Equivalent circuit for DUT interface power wiring.

current flow. The four wires thus used result in the term four-wire or four-terminal power supply connection. The sense lines which lead to the control circuit of the power supply carry little current, and the resistive effects are therefore negligible. The power supply may provide 5.2 V at its output terminals to assure 5.0 V at the remote sensing terminals.

The power supply control circuit has a delay or lag associated with its response to a change in load current. This lag is aggravated by the inductance and capacitance of the power distribution wires and the sense lines themselves. The sum of the delays may create oscillations in the power supply voltage when the DUT load current changes rapidly. Overshoot on power up is a result of this lag, but it can cause voltage variations when a bus on the DUT is switched or similar massive changes occur. Undamped oscillations (often mistakenly called "ripple" and associated with ac–dc conversion) can be set up if the line lengths reach several feet.

Dynamic response of the entire power circuit can be improved by the placement of significant capacitance near the DUT. The frequency composition of current variations can be very broad. High-frequency transients are filtered by ceramic capacitors, and lower-frequency components are cleaned up by aluminum or tantalum electrolytic capacitors. A typical power supply connection is shown in Figure 7.2 for a pc board with an average current draw of 5 A and a peak of 7 A.

The wire gauges shown are for example only. Using four-wire connection lessens the need for large wire in the power circuit, but the largest size convenient should be used to reduce the heating. Wire size is usually limited by the connectors used at either end and the need for flexibility and light weight. Wire size in the sense loop is not critical and can be the same used for signal wires in the adapter. The two 100-Ω resistors bridging between the sense and output wires of the power supply are safety devices. If a sense line were to become open-circuited due to a break or a bad connection, the power supply regulator would cease to sense a voltage at the load and the output voltage would be raised to compensate. Further compensation would ensue since no effect would be sensed until the limits of the

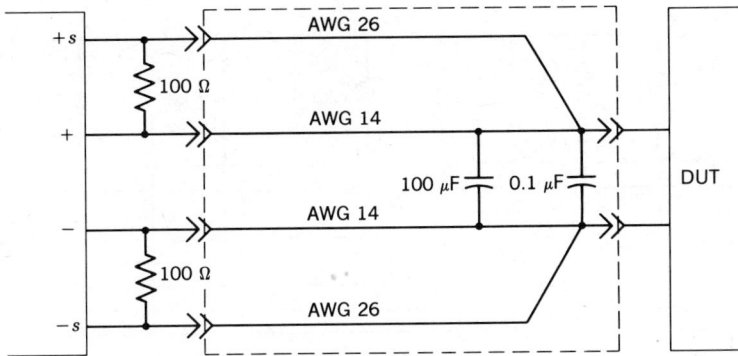

Figure 7.2. DUT power wiring diagram utilizing remote voltage sensing to compensate for resistive loss.

power supply are reached. The 100-Ω resistors prevent the runaway by feeding back the output to the sense in such a case. Their resistance is much larger than that of the sense loop to prevent degrading the performance of the four-wire connection when it is working properly.

Additional protection for the DUT is often provided by the power supply manufacturer in the form of an independent overvoltage circuit which monitors the output and turns off or "crowbars" the supply if the output exceeds a given voltage. Most overvoltage circuits are not programmable, however, and are therefore not of much use when the supply is to be operated over a wide range of output voltages. An output of more than 7 V is fatal to a TTL DUT, but a CMOS unit may require 12 V to operate. Overcurrent protection is more commonly programmable. Detection of load currents outside a predetermined limit indicates a short circuit and results in the turning off of the supply. This saves the DUT from further damage and may also save the ATE power supply and the interface from damage.

Although proper supply and regulation of heavy load current are the most common concern, extremely light loads pose problems for some power supplies. An excellent example is the use of a "series pass" power supply to provide "pull-up" or reference voltage through resistors to a 3-state TTL bus (Figure 7.3). The intent of the arrangement is to provide quick centering of the bus voltage when the drivers of the bus are turned off. The uncertainty of value is limited to a short transition time, and the correct operation of the bus output enable lines can be observed by double-threshold receivers in the ATE which will see neither a legitimate 0 nor 1. In the actual application the current through the supply will become nearly zero when the bus turns off, and the power supply may not regulate due to starvation of the pass transistor's base emitter current. The voltage then floats, and the test is invalidated. Addition of a small constant resistive load between the power supply and ground within the adapter assures regulation.

Often forgotten in the power wiring of an adapter are the connector pin ratings of any plugs or connectors in the path. Assigning multiple pins and multiple wiring strands to each power supply circuit avoids both the wire gauge and connector pin

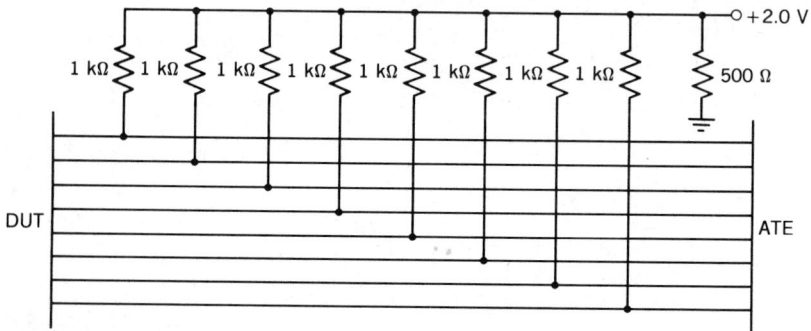

Figure 7.3. Typical "pull-to-center" loading for a TTL 3-state bus. A 500-Ω "keep-alive" resistor has been added to assure power supply regulation.

resistivity problems. Connector pin ratings vary according to geometry and material and should be acquired from the manufacturer, but a rule-of-thumb guideline is to allocate less than 1 A of current to a pc card style connector pin. Each connector in the path contributes to the overall drop, of course, so minimizing the number of connectors traversed is essential to power and ground integrity. Relay and switch contacts in the power circuit must also be accounted for by calculating the drop if they are outside the power supply feedback loop.

The simple use of Ohm's law seems mundane, but poor performance of ATE driven tests is all too often the fault of failure to account for its effects in the power and ground circuits of the interface adapter.

Signal Wiring

The reactive elements of the line model in Figure 7.1 were ignored in power wiring, but become the emphasis in signal wiring. The current involved in signal wiring is insignificant in most cases, so the resistance of the wire is of little consequence unless the wire is extremely long (>100 ft). The step functions that make up digital signals contain very-high-frequency components which must be considered in properly connecting ATE to DUT signals. A common complaint in digital ATE testing is that "noise" in the adapter creates a glitch which causes a false failure. The "noise" is usually generated by one of two effects: crosstalk or reflection.

Crosstalk refers to magnetic or electric field (capacitive) coupling between conductors. In the case of ordinary cylindrical wires used in most ATE adapters, the magnetic field effects are dominant (Figure 7.4), but the capacitive coupling effects become important in the laminated flat conductor geometries of pc cards. The change in current in a signal wire will induce currents in other wires bundled together in an adapter that is proportional to the length of the wires and the rate of change of the current. High-speed logic creates current changes of milliamperes per nanosecond, which can create spikes on neighboring wires great enough to cross the threshold and temporarily alter their logic state.

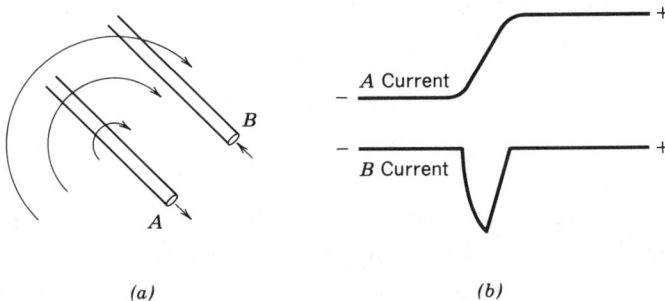

(a)   (b)

Figure 7.4. Changing current in wire *A* of the pictorial produces a changing field which induces an opposite current in wire *B*, as shown in the waveforms.

Crosstalk is best controlled by using conductor pairs in which one conductor carries the signal while the other provides a return path for current from the DUT ground to the ATE ground. In the ideal case a coaxial conductor is used so that the outer ground conductor shields the inner one from external magnetic fields while containing the magnetic field generated by the signal changes in the inner conductor. Twisting a signal wire with a ground return creates an offsetting effect, as the induced current will be common mode. Proper use of either method requires that ground connections at either end be as close as possible to the signal connectors. The ground wires at the DUT connector cannot be simply connected together, they must be bonded to a relatively massive signal ground such as the power supply return lines.

Maintainability of the adapter can be enhanced by using multiconductor cable made up of twisted pairs (either round or ribbon) which are color coded for easy wire tracing. Ribbon cable which has a grounded substrate plane embedded in one side or a shield around the entire cable is suitable, but generally less flexible and more difficult to install.

Reflection is a completely different "noise" effect which stems from the mismatch of impedance at junctions of the adapter and either terminus. Cable characteristic impedances are usually in the 50–100-Ω range. Some technologies, such as ECL, maintain a carefully matched load and driver impedance, and reflections are simply minimized by selecting wires and terminations which match the device impedances.

TTL is unfortunately not a matched impedance technology. Device inputs have a characteristic impedance of several thousand ohms, while outputs have impedances of 100 Ω or so. Additionally, the impedance of both inputs and outputs depends upon the logic state of the node at the time of measurement. A TTL output has an impedance of about 250 Ω in the 1 state and about 30 Ω in the 0 state. Terminating the lines with matching impedances often overloads the TTL drivers unless the terminators are connected to a voltage other than ground or V_{CC}.

Using a circuit whose Thévinin equivalent is a resistance connected to an intermediate voltage is usually handier since the need for a separate power supply is eliminated.

The calculation of resistor values for the upper and lower "totem pole" resistors in the circuit of Figure 7.5 is performed as follows. The specifications for recommended operating output currents for the 74S00 NAND gate are

$$I_{OL} = 20 \text{ mA @ } 0.4 \text{ V}$$

$$I_{OH} = -1.0 \text{ mA @ } 3.4 \text{ V}$$

If the output is a 0, then the voltage across R_1 will be 4.6 V and the voltage across R_2 will be 0.4 V. Neglecting the current in R_2 gives a resistance of R_1 of

$$R_1 = \frac{4.6 \text{ V}}{\text{mA } 20} = 230 \ \Omega$$

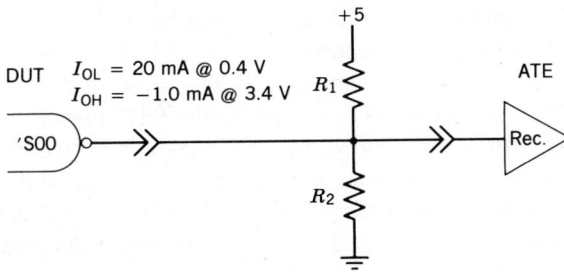

Figure 7.5. Schematic of DUT interface loading for a TTL signal.

Alternatively, when the output is 1 the current through R_1 is given by

$$I_1 = \frac{(5 - 3.4)\,\text{V}}{230\ \Omega} = 6.9 \text{ mA}.$$

Assuming that the ATE receiver input has a high impedance, the current from the top resistor and the current from the output of the 'S00 must be both pass through the bottom resistor R_2. This results in a determination of R_2 as follows.

$$R_2 = \frac{3.4 \text{ V}}{6.9 \text{ mA} + 1.0 \text{ mA}} = 430\ \Omega$$

Of course, the Thévinin equipment of this circuit is a 3.26 V source in series with a 150 Ω resistor. The termination is not perfect, but it is not a bad approximation of the characteristic impedance of a twisted pair wire. The presence of additional fan-outs on the DUT driven by the same gate output changes the calculations and may result in a less desirable termination from a transmission line viewpoint, but the logic drivers should not be overloaded in any case.

In extreme cases where line lengths exceed a few feet (as when connecting to a DUT in an environmental test chamber), buffers must be inserted between the DUT and the ATE within the adapter to amplify the signals. Utilizing buffers with Schmitt trigger inputs or similar hysteresis thresholds reduces the amplification of low-level noise while enhancing the logic signals. The ability to make parametric measurements is lost in such a situation. ATE thresholds which determine a legitimate 0 or 1 are not useful since they only judge the output of the buffers, not the actual circuit. It is desirable to move the ATE receivers to the DUT site, but that is seldom a viable option due to the architecture of most ATE.

The high cost of ATE and the difficulty of changing the architecture of in-place ATE leads to occasional "above-and-beyond" duty for the adapter. A simple case is the use of ATE with fewer I/O channels than the DUT signal lines require. If the circuit can be partitioned during test generation, the interface lines are segregated and the partitions are each mapped to the ATE channel allotment. A set of adapters is then prepared in which each adapter matches one of the partition mappings and ignores the others. The mappings may overlap if constraints are required to keep uninvolved partitions from interfering with the test of the active partition.

If the circuit cannot be partitioned, all inputs to the DUT must be assigned ATE channels, while the outputs are divided into groups which can be monitored with the remaining channels. Separate adapters are prepared for each of the output mappings, or an adapter with multiplexers installed is used to electronically switch the outputs' connections. The test sequence is repeated with each group of outputs connected.

Extensions of the function of ATE are not limited to pin capacity. Signal processing capability can be added on the adapter when necessary. A good example might be a pc board which contains an analog output such as a three-phase synchro motor drive signal. Analysis of the signal to assure correctness would require costly instruments not usually found as part of a digital tester. An inexpensive synchro–digital converter placed on the adapter can convert the signals to ATE readable format (although there will be some slight loss in accuracy). A reasonable test can then be performed to assure functionality of the DUT.

Support circuits such as clock circuits needed for dynamic DUTs are frequently added to the adapter to allow the ATE to pause in the test sequence without loss of DUT internal function. Short pulses which must be detected by the test can be trapped by a flip-flop on the adapter or extended by a one-shot instead of attempting to transmit them to the ATE receiver over lengthy cables. If flip-flops are used, additional ATE drivers must be used to allow reset of the flip-flop during the test. In such cases it may be wise to plan ahead and include the adapter electronics in the circuit model used for test generation. The resets and other controls will be coordinated with the test easily instead of manually, after the fact, with patches to the test.

The final advice in wiring signal channels is an adapter is to keep line lengths as consistent as possible. Uneven wiring creates skew in the signal arrival times and can cause false failures in dynamic tests. In most cases a few inches more or less is not critical, but a few feet can upset the test results. Even with balanced wiring the net effect will need to be taken into account in propagation time measurements if the measurement accuracy is to be within a few nanoseconds.

PHYSICAL CHARACTERISTICS

The reliability, utility, and convenience of an interface adapter depend upon its physical characteristics. Once the electrical specifications have been calculated to assure test accuracy, the physical design must be completed to support the DUT in a convenient manner for troubleshooting, cool it adequately, and provide reliable connections for its inputs and outputs.

The matter of convenience is more important than the word suggests. An interface adapter which requires 10 min to install on the ATE and 10 min more for

DUT attachment wastes time that may not be affordable in a high-volume factory environment. If the same adapter is used in the field for a test of one circuit a day, the installation and loading time will be seen as trivial. The physical characteristics of a portable test adapter are small size, light weight, and simplicity, while a factory adapter is rugged and streamlined in function rather than in form.

Connectors

Functional testing requires the attachment of a few dozen (in the case of IC testing) to a few hundred (pc board test) signal and power contacts to the DUT. The connections must be temporary yet have electrical integrity and reliability. Most of the signals used in a functional test are also used in the operation of the circuit in the final system. Connectors already provided for system interface of the DUT are merely matched with mating connectors on the adapter. Unfortunately, those connectors may be designed to withstand only a few hundred insertions and removals, and the mating connectors of the adapter will wear out rapidly if a more durable equivalent is not used. A typical example of connector technology which has alleviated this problem is the "zero insertion force" or ZIF connector. The differences in operation are shown in the cross-section views of Figure 7.6. A conventional "card edge" connector consists of a plastic trough-like shell with two opposing rows of spring contacts. As the epoxy–glass laminated pc card edge with its copper pads is inserted in the slot, the springs are pushed aside and the wiping action of the springs against the pads cleans surface contaminants and assures a good electrical contact. Several pounds of force may be required to insert the card in the slot. Repeated reinsertions and removals wear the surface of the springs, and occasional misalignment due to warp in the pc boards may permanently bend or break the spring contacts.

The spring contacts of the ZIF connector are held in a spread position by a cam in the base of the connector housing until the card has been inserted into the trough. The cam is then rotated to allow the springs to contact the board pads. The card can be inserted without the spreading force, and the wear to surfaces of the pc card

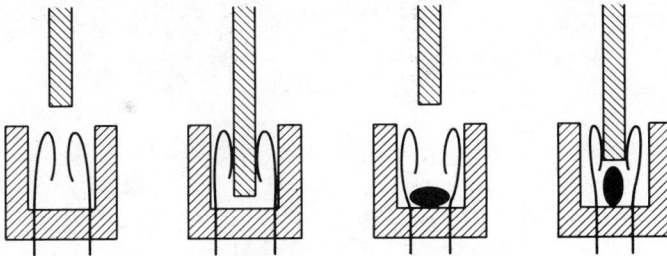

Figure 7.6. Cross-sectional views of edge board connectors. The ZIF connector in the right two drawings has a cam to relieve the spring force of the contacts during insertion.

and springs is lessened. The chance of damage due to warped cards is reduced since resistance to insertion is felt by the operator, and the problem can be remedied. The electrical integrity of the ZIF connection is not as reliable as that of conventional connectors. The ZIF is more susceptible to contamination of the contact surfaces since the wiping action is reduced or eliminated.

ZIF connectors are produced by a number of manufacturers in a variety of shapes and sizes (Figure 7.7) ZIF sockets accommodate most standard IC packages. Unfortunately, the male equivalent of a ZIF connector is not available, so the concept cannot be applied when the DUT has female connectors instead of male.

A more challenging connection problem occurs when the circuit nodes to be accessed for test are not intended for connection in the actual installation of the DUT. ''Test points'' should be collected during the design of a circuit and routed to a suitable auxiliary connector, but the extra connector(s) may not be allocated due to cost considerations, environmental concerns, or oversight. In many cases, test points are strewn about the DUT as unused circuit pads, posts, or plated-through holes. In cases of DFT failure, test points are not explicitly designed into the DUT, and access to critical nodes of the circuit must be made by contacting the leads of components or the pc paths themselves.

Several styles of spring-loaded devices for overcoming these obstacles are marketed. Individual leads can be attached with hooks of various configurations. En-

Figure 7.7. A photograph of ZIF connectors for pc boards and two sizes of IC packages.

tire IC packages can be monitored by "chip clips," which are available in sizes to match most dual in-line package configurations. Other situations may require manual contact using a hand-held probe, although this should be avoided for reliability and convenience reasons.

Testing IC wafers and dies before attachment of the component leads requires a special adapter called a microprobe head (Figure 7.8). Spring contacts with needle points are accurately arranged in a circular or rectangle frame with the needles extending to positions in the center corresponding to pad locations on the die to be tested. A holding fixture with precision alignment guides or an x–y table mechanism capable of microadjustment positions the die under the head, and the two are brought into contact. The outboard ends of the probes are connected to ATE,

Figure 7.8. Wire probes with microscopic tips are mounted in a ring to contact pads on an IC die.

and a functional test can be performed to determine the value of the die before the packaging process is implemented.

Manually manipulated microprobes can be used to augment the fixed micro-probe head for troubleshooting ICs. The manual probing is not useful in a pro-duction environment, but can provide valuable clues in the design verification pro-cess. Several interesting variants which do not require actual contact with the surface of the semiconductor have been developed, but they are not implemented as a major part of the interface between ATE and the DUT so they will be dis-cussed in Chapter 9.

In-circuit testing utilizes contact with circuit nodes without the designation of test points. All circuit nodes are contacted by probes directly applied to pc paths or component leads. A rigid framed bed supports from a few hundred to thousands of spring-loaded nails protruding through holes in an epoxy–glass board. Each nail is wired to a channel of the ATE and may be accessed through a switching matrix or directly by test electronics (Figure 7.9).

The location, placement, and wiring of the nails which are press fit into the bed may seem to be a laborious task. Fortunately, it is easily automated. The monitor points selected are usually a subset of the component lead locations on the pc board to be tested. Only one nail is required for each net, and a useful net on the pc board would at least need two points and probably would have more. A computer program searches the hole location net list which would ordinarily be generated by CAD. The list is intended to operate numerically controlled drilling machines which drill the component holes in the product. The nail locations are selected such that there is at least one nail per net, no two nails are too close together for the larger nail-holding holes to be drilled, and the nails are as evenly distributed as possible. Even distribution assures even force on the pc board under test when it is forced against the bed of nails to make contact.

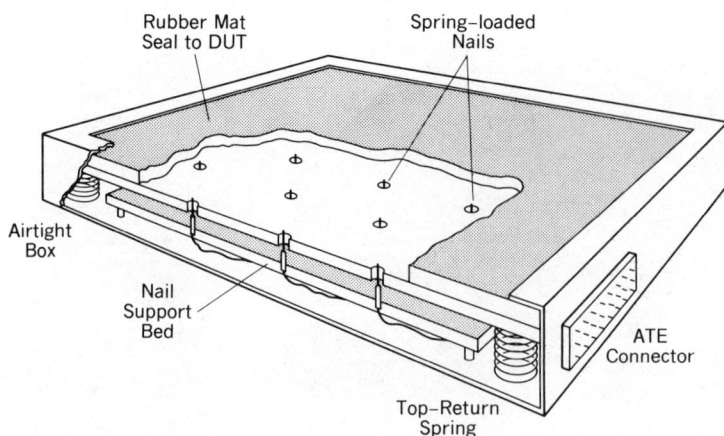

Figure 7.9. The bed-of-nails fixture is an airtight box with protruding spring-loaded contacts (cross section shown).

The subset of hole locations is loaded into the numerically controlled drilling machine which as been fitted with the correct-sized drill for the outer barrel of the nail assembly. After the holes have been drilled, the nails are fitted by hand. A computer program correlates the nail locations with the ATE connector pin on the side of the adapter and generates the necessary instructions to drive an automated wire-wrapping machine. The wire-wrap machine wires the adapter, and the remaining assembly (frame and mat) is finished by hand. Several companies produce kits which include the bed, frame, connectors, and miscellaneous hardware to assemble bed-of-nails adapters matching the leading brands of in-circuit ATE. The entire assembly procedure could take place in a few days, leading to the quick turnaround which has popularized the in-circuit approach.

Reliable electrical contact at low voltages is not easily achieved. The design of nails must assure contact with a potentially uneven surface using only a few ounces of force. Larger forces cannot be used since the total of thousands of nails already reaches hundreds of pounds in some cases. Physical damage to the DUT might occur if forces are not limited. Specialized shapes for the points of nails optimize contact with printed circuit copper pads, plated-through holes, and component leads (Figure 7.10).

The force which holds the DUT against the bed of nails can be supplied in several ways. Mechanical clamps with a frame or a number of posts to distribute the force across the board are used in some operations. The placement of posts is such that components are not crushed in the process. A similar arrangement of posts has been exploited with each post individually weighted and suspended to assure even distribution of force. If the frame which holds the nail bed is a sealed box and a seal is established around the edges of the board with a soft rubber mat, a vacuum can be applied to the interior of the box. Extra holes among the nail holes in the bed admit air from between the board under test and the bed. The vacuum created in this space leaves the air pressure on the outside of the board unbalanced, and the board is pressed against the nails until their combined spring force balances the air pressure.

Connection with the DUT is a necessity for conventional digital testing, and as the old saw goes, "necessity is the mother of invention." Adapters appear with a great variety of clamps, clips, and probes when needed. Of course, many circuits

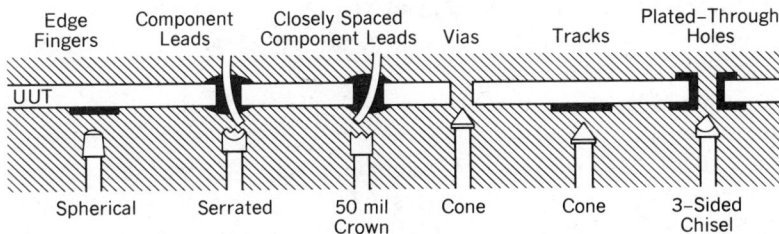

Figure 7.10. Specialized tip shapes for spring-loaded contacts.

are provided with standard connectors, and the mates can be simply procured and incorporated into the adapter design.

Support

Supporting the DUT during the test is a second task of the adapter. If the DUT is an IC, the support may be provided by the ZIF socket which connects the DUT leads to the ATE. Printed circuit boards are usually supported by emulating the support mechanism in their system installation. Very often the DUT is slipped into the slots on insulative posts such that it is held between them. Once the board has slid to the end of the slots, the connectors are engaged and a clamp mechanism holds the board in place. DUTs can be held in place by guide pins, springs, clips, bolts, and other fasteners.

The general guidelines for mechanical design of an adapter suggest that in the case of a pc board, both sides of the assembly should be accessible for the attachment of clips or probes. Support for test point connectors or probes which will be attached as a regular part of the test rather than temporarily held in place for troubleshooting is usually incorporated into the adapter design. Mounting arrangements for the test point connectors or probes and their associated wiring should be included in adapter construction. If the adapter is to be used for several similar types of DUTs with differing test point arrangements, each test point harness should be removable and each should terminate in a common connector which mates to the main adapter. This arrangement prevents extra wires from becoming a nuisance during the test of a particular DUT, but allows easy reconfiguration for other DUT types and eliminates the expense of replicating the entire adapter.

Safety may come into play if the DUT has high voltage associated with it. Shields should be provided to protect the ATE operator from accidental contact in that case. Logic circuits seldom have high voltage, and the open access rule applies generally. Mount the DUT in an orientation which allows easy access to both sides when possible. The in-circuit adapter blocks the noncomponent side of the board, but the contact method makes the situation unavoidable.

The ergonometric placement of the adapter and DUT in a test environment is often restricted by the ATE design or other elements of the test environment such as auxiliary equipment placement. Keeping the DUT within easy reach and direct vision of the test operator is important for maintaining an efficient production environment (Figure 7.11). When a single position does not afford access to both sides of the DUT, a pivoting or flexible adapter design can be implemented. The extra complication of pivots to the mechanical design and the extra flexing of connecting cables are to be avoided otherwise.

Supports should be of sufficent strength to resist flexing the pc board when a single-point probe pressure of several pounds is applied to the board during troubleshooting. If the DUT is to undergo vibration cycling during ATE testing, the support mechanism is critical. Vibration test fixtures are characteristically massive, as seen in Figure 7.12. Thick plates of lightweight metal are used to transmit the

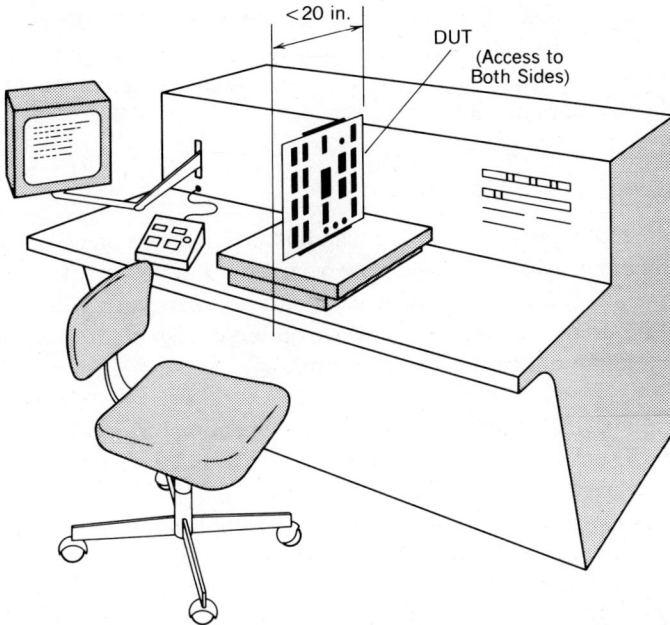

Figure 7.11. ATE with an adaptor and DUT mounted within easy reach of the operator.

vibration without adding resonances [2]. The natural frequency of the holding fix-ture should be outside the range of vibrations to be induced. If the nonresonant condition cannot be met, damping devices such as viscous coatings or laminates can be added to suppress resonance.

Testing during vibration cycling requires attention to the cables which connect the DUT in the holding fixture to the ATE. Ordinary solid or seven-stranded wire

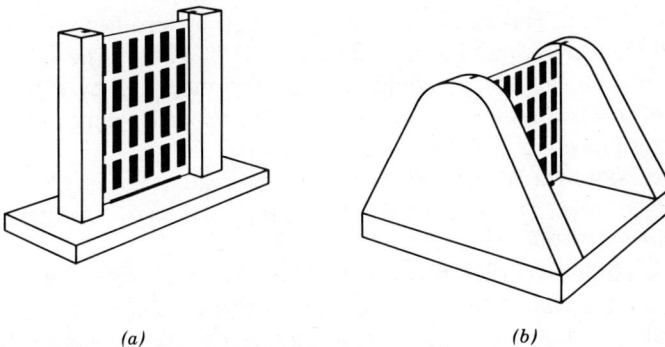

(a)   *(b)*

Figure 7.12. A simple DUT holding fixture and a fixture reinforced for vibration testing. (*a*) Standard duty. (*b*) Reinforced for vibration.

will be fatigued by the flexing. The cables can be supported at the ends by strain relief clamps and positioned such that the excursion of the DUT represents only a small angular motion for the wires. The wire used should be one of the fine stranded or braided flexible varieties if possible.

Thermal Environment Control

Maintaining the DUT at a temperature which is within the desired operating range is another important task of the interface adapter. Although auxiliary equipment such as refrigeration or heating which is not directly associated with the interface adapter may be used, the adapter takes part in the conduction or containment of heat and is therefore critical to proper control.

Ambient Testing. Initial test and debug of an electronic device is ordinarily carried out with the device temperature at or near its nominal operating temperature. Later tests may artificially raise or lower its operating temperature in an effort to test the DUT's temperature margins or in an attempt to accelerate failure mechanisms which would cause faults within the first few weeks or months of operation in the field. With power requirements of ICs reaching the watts range and pc boards often consuming 20–100 W, the simple-sounding task of maintaining the DUT at its nominal temperature for testing is no longer trivial.

Cooling techniques, like support techniques, mimic the final installation of DUT in the system. If the DUT is convection cooled in its final application, convection cooling will probably be adequate in the interface adapter. Assuring the free flow of ambient air around the DUT is often the only concern given to cooling for lower-powered DUTs. The relatively short power-applied time of most test and troubleshooting sequences also reduces the need for extra cooling measures.

High-power or densely populated pc boards depend upon forced air cooling or conduction of heat away from the components by metal heat sinks. Adding a fan to an adapter in an attempt to supply cooling air is unwise for several reasons. The fan is certain to get in the way of the test operator and present an inconvenience if not a direct hazard. The electrical noise generated by small motors is very undesirable in the direct vicinity of the adapter and DUT wiring. The air from blowers mounted several feet away should be ducted to the DUT by a flexible hose instead. Even in this arrangement the hose should be easily removable for short troubleshooting procedures or it will get in the way.

Conduction cooling a DUT involves making intimate contact with the heat distribution surface, grid, or ladder of the DUT and providing a low thermal resistance path to a heat sink. The heat sink might be a finned metal air-cooled heat exchanger or a chiller in which refrigerated liquid is circulated (Figure 7.13).

Detailed thermodynamic treatment of heat transfer is beyond the scope of this text, but a quick analogy may help the test engineer gauge the reasonability of a design. The flow of heat through solid materials is analogous to the flow of current

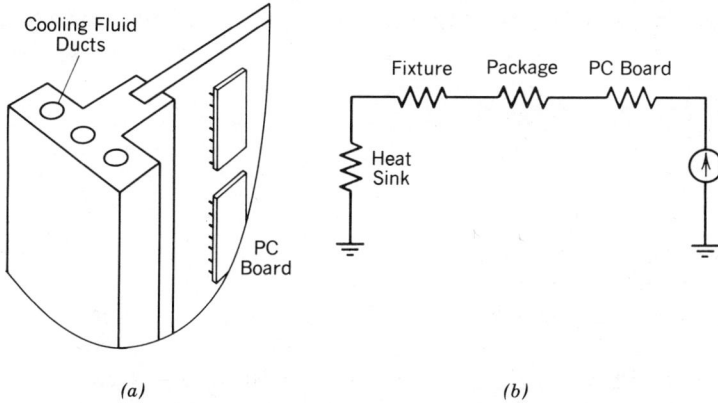

Figure 7.13. Details of a conduction-cooling support post and the equivalent heat flow circuit. (a) Mechanical heat-flow circuit. (b) Electric equivalent.

through a conductor. Thermal conductivity is similar in concept at a macro level to electric conductivity. An electronic model of the heat flow from an IC die to the liquid-cooled heat sink can be made by representing temperature by voltage and heat by current.

Assume that the heat path beginning in the IC consists of metals electrically insulated from one another by thin sheets of insulators which have sufficient thermal conductivity to be ignored. This simplified model will be adequate for estimation. The heat flow through a solid object is given by the equation

$$Q = -kA \frac{dT}{dx}$$

where Q is the rate of heat transfer, A is the cross-sectional area of the material perpendicular to heat flow, x is the distance through the material in the direction of heat flow, T is the temperature (dT is the difference in temperature across the object), and k is the thermal conductivity of the material [3]. Note the similarity to the electrical equation for current:

$$I = -yA \frac{dV}{dx}$$

In the electric equivalent equation the y for electrical conductivity has replaced the k, and the voltage potential dV replaces the temperature of gradient dT. The similarity of the two equations leads to a regrouping of the heat equation to separate the factor $k\,A/dx$ as the thermal admittance of an element of the heat path. The thermal conductivities of a few typical materials used in pc boards and adapter construction are

Aluminum   2.18 W/cm C°
Copper   3.94 W/cm C°
Beryllia   1.17 W/cm C°

If the bottom of the IC package is made of Beryllia and is 1 cm × 0.7 cm or 0.7 cm^2 and 0.05 cm thick, then it represents a heat path with a thermal conductance of about 16 W/°C. Assume the bottom of the package is in intimate contact with a copper strip "heat ladder" which leads to the edge of the board 10 cm away and is 1 cm wide × 0.5 cm thick. In reality, the boundaries of surfaces and the interaction of heat sources on the board complicate the calculations greatly, but for explanatory purposes a single heat source is assumed. The thermal conductance of the heat ladder is about 0.6 W/°C.

The heat sink of the adapter is a block of aluminum which is clamped to the pc board edge in good contact with the copper heat ladder. Between the board edge and the liquid coolant chambers is a block of aluminum 20 cm × 2 cm in cross section. The distance from the board edge to the liquid coolant is 10 cm. The thermal conductance of the adapter block is about 8.7 W/°C. The thermal conductance of the total path is obtained by the same process as the electrical conductance of a circuit is calculated.

$$\frac{1}{Y_T} = \frac{1}{Y_{PK}} + \frac{1}{Y_{HL}} + \frac{1}{Y_{AD}}$$

In this case the thermal conductance of the entire path is 0.54 W/C°. A handier form is the inverse called thermal resistance, which has a value of 1.84 C°/W in this case. The heat ladder is the dominant factor since its dimension is long and narrow. If the part dissipates 1 W, the temperature will be 1.84°C higher than that of the coolant. This number is unrealistic since many objects on the pc board contribute heat. Many thermal modeling programs exist for simultaneously computing the effect of multiple heat sources and sinks.

For a more realistic example, assume that the pc board dissipates 50 W through conduction to the edges. The adapter mounting block will be 5.7°C warmer at the board edge than at the coolant. Actual heat conduction is affected by coatings on the materials and other factors. Accurate determinations should be made for individual situations with the aid of handbooks or computer modeling programs.

Environmental Manipulation. During characterization of a new design, proofing of operating range, or stress screening of electronic devices, DUTs are subjected to temperature and humidity excursions while under test. Maintaining connection and control of an IC or pc board while controlling its environment is a difficult challenge for the adapter designer.

Since electronic devices almost always generate heat, the actual temperature of parts of the DUT will be greater than that of the surroundings. Heat flow is not instantaneous, and therefore a temperature applied to the immediate surroundings of a DUT must be allowed to ''soak''into the device. Adequate time for soaking or stabilization depends on the mass of the DUT and the amount of temperature change from the last stable point.

For small DUTS such as ICs, temperature control can be affected by a stream of conditioned gas. Dry air, carbon dioxide, or nitrogen is heated or cooled to the desired temperature and then piped to the DUT. The gas stream can be merely directed at the DUT if the gas volume is sufficient to create a complete curtain about the DUT, excluding the ambient air by force. Larger DUTs must be enclosed in order to hold a sufficient volume of the conditioned gas. If the enclosure is airtight, the gas may be recirculated through the heater or cooling coils and back over the DUT.

Temperature cycling is a common requirement for stress screening in military applications. A time–temperature profile similar to the one shown in Figure 7.14 specifies the conditions of the test cycle.

The excursions, soak times, power on and off conditions, rate of change of temperature during transitions, and time of test applications may all be specified. Temperature cycling is a poor use of ATE since most of the time is used for temperature stabilization and not testing. A ''hot hands'' test can often be arranged in which a nearby oven is used to establish the DUT at the desired temperature point, and a rapid transfer from the oven to the adapter minimizes the temperature change before test can begin. A short restabilization time in a temperature-controlled adapter may be necessary, but the stabilization time on the ATE is much shorter than the soak time. Many DUTs can be soaked in the oven at once, and the throughput of the system is much better.

Adapter construction for thermal cycling must take two major concerns into account. All wires must be insulated and protected from moisture. When the tem-

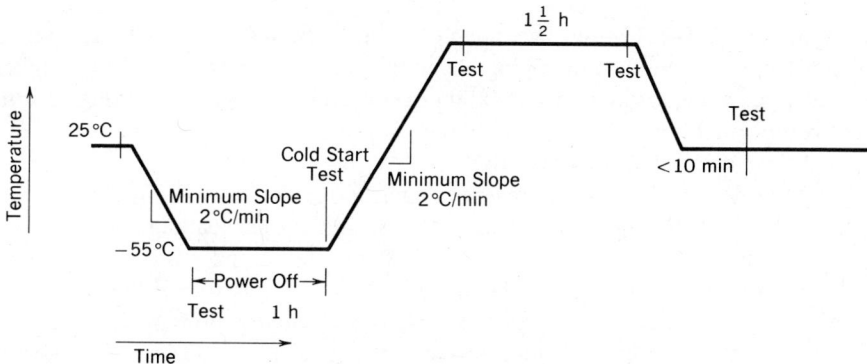

Figure 7.14. Time–temperature diagram defining the environmental conditions for a temperature stress test.

perature is reduced, dew forms on the outside of the adapter container and on the inside if dry gas is not used for cooling. The moisture will change the resistance between signal paths, resulting in the possibility of intercircuit shorts in the adapter. In the extreme, even digital tests will fail erroneously, but parametric tests will be upset first.

Enclosing the DUT in a structure to contain heated or cooled gas makes test probe access for troubleshooting difficult or impossible. Removing the enclosure for troubleshooting causes a temperature change, and the failure may go away in the meantime. Limited access panels or windows in the enclosure are inconvenient, but better than nothing at all. In general, the enclosure or "hood" portion of an adapter used for thermal testing should be constructed of an insulative (thermally and electrically) material and should be as small as possible to reduce the amount of gas that must be changed to transition from one temperature to another.

In an attempt to reduce infant failures of a product in the field a burn-in test is sometimes used. Burn-in tests often last for a week or more during which the units to be tested are held in an oven at an elevated temperature. Power may be applied to the units, and they may even be exercised to some extent, but constant test application is usually not performed. The DUTs are tested after burn-in, but the hot hands test is not necessary since the effect sought by the elevated temperatures is aging not direct stress.

Occasional function testing during the burn-in cycle may be desired to replace early failures immediately instead of waiting for the complete cycle to finish. The adapter necessary for this operation is quite complex. Burn-in ovens (or chambers) contain racks upon which the DUTs are held. Connecting all the DUTs to a power source and power monitor satisfies the power-on requirement without a great deal of complexity, but providing a switched mux path from each pin on each DUT to an ATE channel rapidly becomes cumbersome and costly.

SUMMARY

Although the DUT–ATE interface adapter is often the last item of the test set to be considered, it is vital to obtaining accurate, repeatable results from any digital test. Adapters are also useful in extending the capability of ATE by adding custom circuitry for glitch processing, pin muxing, or conversion of nondigital signals to a digital format for ATE interpretation.

Carefully designed adapters preserve signal integrity. The wiring must be adequate and shielded from interference and crosstalk. Proper loading should be designed with the DUT specification in mind. Power leads and ground nets should be robust and well filtered within the adapter as near as possible to the DUT.

Physical device requirements such as cooling and support along with ergonomic design for convenient troubleshooting guide the mechanical design. The overall design for an interface adapter is a particularly eclectic undertaking, and interdisciplinary advice should be sought to assure compliance with DUT specifications during test and troubleshooting.

REFERENCES

1. W. H. Hayt, Jr., *Engineering Electromagnetics*, McGraw-Hill, 1967, p. 367.
2. W. Tustin, *Environmental Vibration and Shock*, Tustin Inst. Technol., Santa Barbara, CA, 1977, ch. 21.
3. J. B. Jones and G. A. Hawkins, *Engineering Thermodynamics*, Wiley, 1960, p. 661.

CHAPTER EIGHT

ATE LANGUAGES

ATE is automated because its actions are controlled by a computer. Programming the computer provides a blueprint for the actions desired. It is inconvenient to program the controlling computer at the machine language level for each test to be performed. A higher-level language and an operating system are installed and kept resident on the computer to simplify programming and standardize the test approach. A particular piece of ATE seldom provides for support of more than one language.

Beyond the restrictions placed by the target ATE, a uniform approach to test programming aids in the efficiency of the test engineer and the operator by reducing the training required for each new DUT encountered. Organization and documentation are just as important to quality control and maintainability of test programs as they are to production software. In some cases the software is written in a standard language in the hope that it will be portable from one type of ATE to the next.

Although the idea of a standard language is appealing to test managers and engineers, none of the available programming languages used for other than test applications contains the appropriate commands and constructs for test specification. ATE manufacturers have adapted BASIC, Pascal, C, and other languages to testing by adding or substituting instructions to deal with test data and timing. In some cases, completely new languages are derived for execution on a particular manufacturers line of ATE. The operating system which supports the test language may even be specialized. In single-user ATE the operating system and the test language may become merged, but a multiuser system which supports both test program development and test application needs a distinct operating system.

TEST PROGRAMMING

In general, programming is a somewhat creative endeavor. A good programmer is one who rigorously adheres to the rules and grinds out a well-documented result, but a great programmer is the one who derives new algorithms, invents new forms of a solution, or elegantly streamlines the process. Test programming is bounded on one side by the restrictions of the ATE and its language, while the other bound is the definitiveness of the test goal. Therefore, test programming requires good programmers, but seldom great ones. The areas of ATG and fault simulation are better arenas for great programmers. Moreover, the structure and content of a good test program are so guided by the hardware of the ATE and the hardware of the DUT that an extensive knowledge of engineering is required rather than a proficiency in computer science.

Most ATE languages are nevertheless intricate enough to require several days to several weeks of specific training for the engineer/programmer. Most of the training consists of learning the several dozen to several hundred commands, acronyms, and their associated functions. I/O and file storage procedures appropriate to the operating system must be mastered. The operating characteristics of hardware subsystems such as pin electronics and power supplies must be understood, and the corresponding software warnings (or lack of them) must be digested. After several different ATE systems have been mastered, the sea of details gives way to an understanding of the underlying essentials.

The essentials of test programming are easily expressed in terms of the objectives of the program and the typical program flow. Variations in either the goals or the structure of the program may occur as a consequence of the type of DUT to be tested or the ATE to be used, but the sections which are not applicable to a particular situation are usually obvious. A full-featured example will therefore be used.

Objectives

It is far too simple to suggest that the objective of a test program is to perform a test. ATE can contain a bewildering array of subsystems, each with numerous options and operating instructions. Attached to the ATE can be accessories and communication channels which extend its influence toward the environment or the operator. Effective use of ATE systems involves sequencing and coordinating events with timing precision ranging from picoseconds to minutes. Orchestrating such a performance requires careful consideration of a series of objectives, not just a single goal of performing a test.

Environmental Control. Although the term "environmental control" usually connotes only temperature and humidity regulation, it is used here in the broader sense of adjusting or controlling all nondigital aspects of the DUT surroundings. Aspects of the environment such as temperature or vibration are usually controlled by external subsystems which may be manually controlled. The use of microcontrollers in such subsystems provides an excellent opportunity for incorporating their control into the overall test program through I/O ports of the ATE. Whether or not temperature and vibration should be included in a test sequence is a matter of test specifications and product flow economics. Such issues are discussed in Chapter 12.

The most prevalent environmental factors included in test programs are controlled by internal subsystems of the ATE. Voltage levels and current limits provided to the DUT are programmed, and the power supplies are switched on and off in sequence if necessary under program control. Programmable power supplies smooth the setup from one type of DUT to the next. Automatic control of the levels also reduces the change of manual adjustment errors which might damage a DUT or invalidate a test result. Voltage margin testing would be very tedious without programmable power supplies to vary voltage parameters for each successive pass of the test.

ATE languages usually include a "SET" operation which defines the voltage and current to be provided on a particular power bus. A second command is usually used to turn on or connect the power supply so that all supplies can be programmed and then sequenced on without gaps in the turn-on sequence. Separate connect commands allow the matrices of testers which have switchable DUT interfaces to be routed "cold" before current is applied. Relays can carry more current than they can switch, and hot switching reduces their life. Also, the changes of inadvertently applying momentary power to an erroneous DUT pin during switching are eliminated. Examples of the particular syntax of these commands in several tester languages will be reviewed later.

The interface adapter is an important part of the DUT environment. Ordinarily, the functions of the interface device are static—limited to fixed resistor loading and interconnection. It is conceivable that multiplexers, variable or switchable loads, and synchronization devices might be included on the interface adapter. Direct test language commands cannot exist since the ATE designer has no idea what capabilities will be added later via the interface adapter. Controllable signals can usually be obtained through spare tester I/O pins or special ports such as synchronization triggers.

Control of Stimulus Application. For analog applications the control of stimulus may take on a wide range of meanings. In digital applications (with which this text is concerned) a stimulus is almost always a square-edged, accurately timed signal which moves quickly between two predefined voltage levels. Several stimulus sources of differing voltage excursion or timing may be needed for a test, and the test program must define each and then coordinate their application.

Definition of a stimulus is the process of supplying the values which specify it to the hardware which produces the stimulus. If the stimulus for a given DUT input were to be supplied by an instrument such as a pulse generator, the definition arguments would include the waveform excursion limits (amplitude) in the form of two voltages. Further specifications would be the pulsewidth, pulse direction, and repetition rate. A single ATE language statement could name the signal, define its type (which would select the source equipment), and pass the specifications as arguments to a routine which communicates with the instrument and configures it properly.

Most digital applications use either stored input vectors or algorithmic patterns as signal sources for the majority of DUT inputs. Instruments could be used for communicating with the DUT through a bus such as the RS-232 port or a Manchester serial bus, but few instruments are capable of the flexibility and multiline capability needed to generate test vectors. Those capable turn out to be stored pattern or algorithmic units themselves, and therefore ATE is their own right.

The definition needed for stored pattern stimulus sources is similar to the instrument setup in that it contains voltage levels and timing information. Since the stored pattern stimulus source is usually a broad bus of signals, specifying each with a separate statement would be unproductive. These signals are usually grouped into families by a simple statement which lists their connector pin after the family name. The specifications for all signals in the family are then provided by a single statement.

```
INPUT:A 12,13,16,27,43,56,57,58,59,62
INPUT:B 23,24,25,26,33,34,35,36
OUTPUT:A 5,6,28,29,30
OUTPUT:B 31,32,45,48
DRIVEA1 = 4.0 V
DRIVEA0 = 0.5 V
DRIVEB1 = 10.0 V
DRIVEB0 = -10.0 V
RECEIVEA1 = 3.4 V
RECEIVEA0 = 0.8 V
RECEIVEB1 = 8.0 V
RECEIVEB0 = -8.0 V
```

Timing definition for stored pattern stimuli depends on the flexibility of the ATE. Simple ATE sources assume timing to be fixed in relationship to an internal ATE clock. Intermediate systems allow specification of stimulus mode and application time after some internal start of cycle time of the main clock. The mode specification defines the signal as a "return to zero"(RZ) or a "nonreturn to zero" (NRZ) signal. The botton two waveforms in Figure 8.1 show the difference.

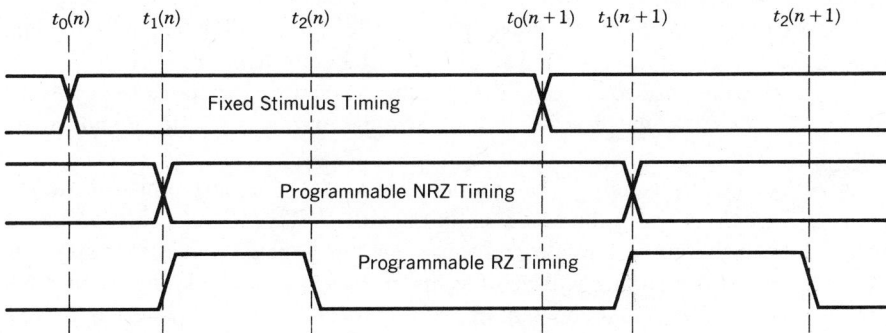

Figure 8.1. Timing diagram of test stimulus application in three modes.

The RZ waveform needs to be defined in terms of both rising and falling edges, while only the rising edge of the NRZ waveform must be included. The rest state of the RZ is also normally programmable. That is, it can be defined as either an RZ or a "return to one" signal.

Some advanced ATE also provides the capability to synchronize the stimulus to a signal external to the ATE. In this manner the DUT's own clock can be used to slave the ATE stimulus application. Real-time testing of this type is quite helpful when the DUT is dynamic. Examples of ATE language signal specification will be covered later in the chapter.

Response Analysis.   Stimulus application results in a DUT response, and the ATE language must specify how to analyze that response to complete a test. Response analysis mechanisms parallel stimulus generation mechanisms, and the language constructs to control analysis parallel those which specify stimuli. The voltages and timings are to be interpreted differently, but the form is quite similar.

As in the case of stimulus generation, analysis can be performed en masse by a broad comparator bank made up of the pin electronics and stored expected response. Alternatively, there are instruments which perform complex waveform analysis on a single signal or a small group of signals. Analysis definition for the later can range from simple transition couting to statistical analysis.

Program control of the pin electronics comparators is similar in form to the driver control in stimulus application. Individual DUT signal pins are assigned to families of receivers by a simple listing of pin numbers. The definition statements which specify comparator operation may take several forms, depending on the sophistication of the receiver–comparator electronics. The simplest form is a fixed timing single threshold comparator. Timing for the DUT output sample is slaved to a point near the end of the tester master clock cycle, leaving only the definition of logic threshold to the program. Each family of pins can be examined with respect to a threshold voltage, and any pin potential above that voltage at the time of the sample strobe will be reported as a 1, while any pin potential below that level will be reported as a 0.

Dual threshold comparators and programmable strobe timing represent the full-featured pin electronics case. One threshold is used as the minimum level that will result in a 1 report, and the other in the maximum level that will give a 0. Anything between the levels will be reported as a Z or an indeterminate in most cases. For TTL logic a signal between a logic 0 and a logic 1 can be generated by a 3-state driver in the off state (thus the Z). For other logic drivers the result is indicative of an error.

Strobe timing can usually be programmed in terms of integral units of the internal ATE clock. As with the stimulus timing some ATE provides external clock tracking. Since the entire stimulus response cycle is lashed to the external clock in those cases, strobe timing will be in terms of the external clock or a frequency-multiplied slave. The actual compare operation against the stored pattern expected response is a hardware function which uses the result of the comparators and sets a software readable pass or fail flag.

Instrument style analysis commonly used for digital testing includes voltage, pulsewidth, and frequency measurement. A subroutine sets us the instrument and connects it to the proper DUT pin via a specialized bus, and trigger instruction causes the measurement to be made and reported. Software analysis such as comparison to limits is then performed to decide upon the pass or fail before continuing the test.

Operator Interface. The essence of ATE is the automation of a complex task. Perhaps the visible difference between a manually operated instrument test bench and an automated tester is the narrowing of the operator interface. A skilled technician or engineer is required to learn the front panel controls of several instruments and then configure and connect them to perform each step in a manual test. The test program is written by someone who has learned the capabilities and operating procedures for the test equipment, and that knowledge is encapsulated in the program. A simplified operating menu is presented to a less-skilled operator.

Typical operator interface tasks include option selection and the presentation of setup instructions and troubleshooting information. Operating system functions such as the method of inputting test program selection are usually decided by the ATE manufacturer when the operating system is written. A typical scenario of operator interaction might be as follows:

```
        ***ATE-OP-SYS LEVEL 4.06***
>load 2058392 {operator's input}
>READY {operating system response}
>run
Functional Test for CPU-Module 2058392
Revision J
Select test mode
1)Stop at End Only
2)Stop on Fail
3)Step Mode
4)Loop Test
>1                  {operator's input}
Install DUT on tester using adapter A-304
Press Return to test, Space for additional
       instructions, Escape to cancel
><cr>               {operator presses return key}
FAIL
Press Return to repeat test, Escape to menu
><esc>              {operator presses escape}
```

```
Select test mode
1)Stop at End Only
2)Stop on Fail
3)Step Mode
4)Loop Test
>2                      {operator input}
Install DUT on tester using adapter A-304
Press Return to test, Space for additional
      instructions, Escape to cancel
><cr>                   {operator presses return key}
FAIL step 1032 pins 52,86,87,101
```

The example scenario shows the typical terseness of messages and responses which typify a production environment. The additional help information is detailed, but hidden unless called upon to avoid operator boredom. Essential troubleshooting information is presented if the test is run in a troubleshooting mode, but left out if product grading is the test purpose. Graphic displays and printout facilities can be added in many systems to improve operator comprehension or reduce failure logging tasks. The objective of operator interface is to present unambiguous options and instructions without requiring lengthy operator response. Response error checking should be built in where possible, and default values should be avoided unless there effect can be assured to be harmless to the ATE and the DUT.

Process Flow. As ATE becomes a part of the manufacturing process its output information can be used to provide management product yield statistics. Test programs log the part type and sometimes individual identifiers such as the serial number of the DUT along with its test results. Additional information from the operator can be included to allow rework notes such as ''damaged component at location R65.'' If troubleshooting results are entered, fault classification can be recorded, and a later statistical analysis provides management with a distribution chart which shows unusual tendencies.

Direct action to control the process flow of the DUTs can be controlled by the program if automated binning equipment or robots are involved in product handling. Failure of a test could result in a subroutine call which would instruct the robot to place the DUT on an appropriate conveyor to the rework station, while the pass condition would place the DUT on the conveyor to packaging.

Since the computers and/or operating software of ATE are usually altered to support the test function, most statistical analysis of management data is left to other computers. Data logging is sometimes replaced by a communication task in which the management information is transmitted to a data collection and analysis computer via a bus rather than having it stored for later collection.

Program Flow

As with all other aspects of test program construction, freedom of program flow or statement order is variable across ATE systems. Restrictions may be total, as in the earlier "fill in the blank" ATE programs, or they may be minimal and only concerned with commonsense order, such as "power supplies must be defined before they can be turned on." Most test engineers work from a shell program which has been debugged rather than reinventing the program format each time a new DUT is to be tested. Not only is the programming time greatly reduced by this practice, but the results are more consistent, leading to less operator training and better documentation readability.

Careful structuring of a program maximizes test efficiency and reduces program debug time. Maximizing test efficiency and reducing the risk of damaging the ATE or DUT are goals which can be easily accomplished by following the same general procedures used by a manual test sequence. Test technicians have developed a cautious procedure to save their time and avoid smoke. Far from being accidental or arbitrary, the procedure selects test sequences in a hierarcy based on the potential hazard of detected faults to later tests, the probability of occurrence of faults detected by a particular sequence, and the effectiveness of a sequence in covering numerous fault classes. The overall aim is to reduce average test time by encouraging most failures early in the test sequence. Later tests will be performed only if earlier tests are passed.

DUT ID. If the DUT has an electronic key such as a resistor value for identification, it should be tested first. Why waste time testing the wrong DUT!

Isolation/Continuity. Each of the DUT interface pins is measured for resistance to ground. This test uses only low-current sources and precedes power turn on. Shorted power nets, open ground or power pins, and shorted signal pins can be detected without secondary damage to the board or ATE.

Power-in Specification. Power is applied and, after a short pause for stabilization, checked against voltage margin or current limits. This procedure limits the damage caused by breakdown type shorts and assures that the remaining tests are being conducted in a proper environment.

Static Functional. This is where the test vectors from the digital test generation system are first applied. For troubleshooting ease the vectors should first be applied in a pseudostatic mode if the nature of the DUT allows. In this mode the test can be stopped at the first failing test step, and the DUT can be examined with a simple logic probe or a voltmeter.

Dynamic Functional. When the test specification requires a full clock speed functional test, it should be run after the static version. Even though the test vectors may be the same, the speed increase may bring out timing faults which are not detected in the static mode. Troubleshooting requires an oscilloscope or a logic analyzer, and this extra complication is unnecessary for most stuck-at faults.

Memory Tests. ROM, RAM, and PLA tests could reasonably be performed before or after the dynamic functional tests. The speed sensitivity tests which are part of many memory algorithms require the same troubleshooting style as the dynamic functional, and the failure probabilities depend on the proportion of memory circuitry in the DUT.

Parametric. Some measurements of a parametric nature such as oscillator frequencies, may be better placed immediately after the power-on test. This applies in the case where the functionality of the device requires a strictly controlled signal and the oscillator cannot be segregated from the function and replaced by the ATE clock source. Otherwise, measurements such as frequencies, voltages, currents, and resistances are measured last. The tests often call upon instrumentation which is slow to set up even under program control. In some cases the DUT must be adjusted or calibrated by the operator on the basis of parametric tests. If the functional test can be performed on an uncalibrated DUT, the operator frustration in attempting to calibrate a malfunctioning DUT can be avoided by placing this test last.

Since the capabilities of ATE vary considerably, sections of the program will need to be omitted or modified for a particular application. In many cases several distinct ATE systems with particular attributes will be used in succession, and the appropriate sections of the overall test program will actually be written as separate test programs in different ATE languages. The sections and order given are intended only as a guide. Program structures must be added to provide for operator interface, failure-handling subroutine calls or branching, and auxiliary data collection and logging.

Branching

The implication of linear order contained in the previous description of test program elements is not necessarily correct. Although digital test sequences tend to be linear from the overview perspective, branching is a well-used function at several lower levels. Obviously, each test step in a stored pattern digital test contains a potential branch as the outcome of the actual to expected DUT output comparison results in a pass (continue) or a fail (halt). Several other opportunities for branching occur occasionally.

Fail Subroutines. Due to the frequency of need for program code to deal with failure of a conditional branch which corresponds to a DUT test, a designated ON__ERROR__GO__TO is usually constructed. Its form may be a subroutine in some instances, but in other implementations a simple jump destination label is used without the capability to return to the main program.

A subroutine which handles failures should contain several optional paths for action chosen on the basis of a logical flag which reflects the operator's mode choice from the entry menu. The example subroutine assumes that the test step which resulted in the call left its sequence number in the TEST__COUNT register and that the difference between the expected output and the actual output is readable from a 512-bit register called DIFFERENCE.

```
PROCEDURE Fail_Test
VAR
  I : integer;
BEGIN
  CASE mode OF
    'S' : GOTO Test_Halt {Stop on fail
    return control to operator}
    'P' : BEGIN {Print out failures but
    do not stop}
      Writeln ('Failed Test',
      TEST_COUNT);
      Write ('Pins In Error:');
      FOR I = 1 to 512 DO
        BEGIN
          IF TEST_BIT
          (I, DIFFERENCE)
          {TEST_BIT is a function which is true
          iff the bit in I position of the register
          named is a '1'}
          THEN Write I;
        END; {of FOR loop}
      END; {of Case Clause}
    'N' : SET FAIL_INDICATOR {System function to
    light indicator on operator's control panel}
  END; {of CASE}
END; {of procedure returns to main prog and
continues test}
```

The example subroutine is not rigorous or particularly applicable in any ATE system, but is meant to show the form. Use of a fail subroutine shortens the in-line code of a test program greatly since each test step needs only a test and jump statement for the fail condition. Some ATE operating systems contain an implied fail jump statement that does not need to be included in digital test vector sets. A single statement enters the digital test stream mode in such a case, and a hardware interrupt starts the fail-handling routine by jumping to a previously arranged location.

Determining Sequence Branching. In Chapter 3 the idea of initializing a circuit by a determining/homing sequence was advanced. A determining sequence is a series of input vectors which allow the state of a circuit to be determined. Once the determining sequence is successful it can be followed by a selected homing sequence to force a desired state. The entire routine would be placed in front of the stored pattern functional test section of the test.

The simplest case and the most commonly used is a loop which performs a repetitious operation such toggling a DUT input pin and monitors a group of outputs for a given state. The state sought is the beginning state for a functional test, so no further homing sequence is required.

An important aspect of the example loop in Figure 8.2 is the specification of a maximum number of trials to be made before failure. Omitting this condition test will result in an infinite loop if the DUT fails to respond as expected. The DUT outputs compared to the expected state need only be a small subset of the total number of outputs. If necessary, a multistatement sequence can replace the simple clock issuance. The test engineer must be careful that the input sequence will always cause a good circuit to pass through the state tested for before the maximum try limit has been reached or a false fail may occasionally result.

Although the simple clock and test loop is the most prevalent, more complex structures can be useful. The problem encountered all too often is that the routines become too complex to write and comprehend easily. The example flowchart of Figure 8.3 tests a status word output of the DUT and issues one of several commands to the DUT accordingly.

The opportunity to use an initialization routine of the latter form is rare compared to that of the earlier type. Circuits which have been designed with testability in mind should not require either routine.

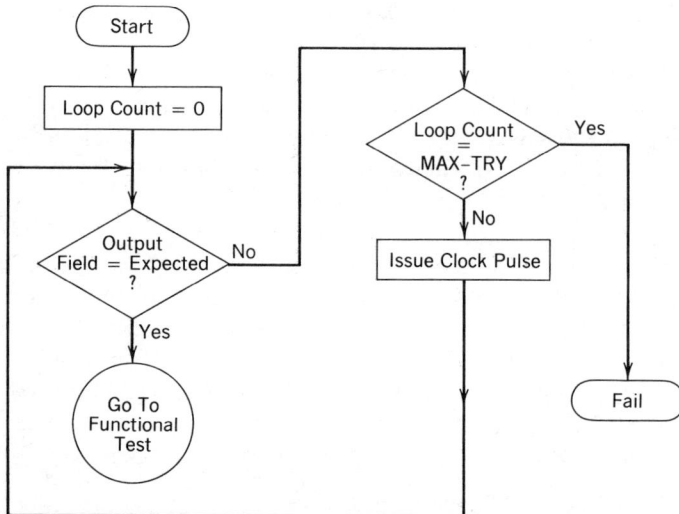

Figure 8.2. Conditional initialization flowchart.

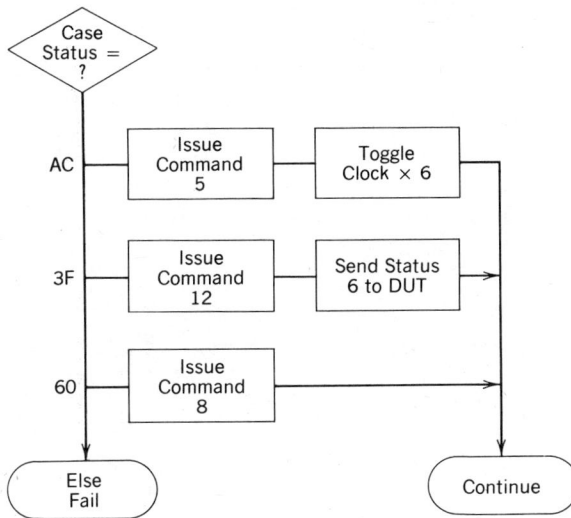

Figure 8.3. Multiway branching conditional initialization flowchart.

Diagnostic Branching. A theoretical use for branching that finds more application in analog testing than in digital is diagnostic branching. Manually generated tests may take the form of a chain of "go–nogo" tests, each of which branches on failure to a more thorough diagnostic test aimed at a narrow scope of circuitry (Figure 8.4). ATG programs seldom work in this fashion, although the increased use of techniques such as AI may revive the idea.

The underlying concept hinges upon the existence of key tests which cover many faults with detection, but have poor diagnostic resolution. Built-in self-test (BIST) features in a circuit create this situation when viewed from an ATE standpoint. A test sequence from the ATE may initiate the circuit's self-test features and branch to one or more secondary diagnostic routines only upon failure.

Diagnostic branching shortens the test time for good circuits while providing in depth fault resolution for faulty ones.

Operator Intervention. Even the slowest ATE is much faster than the human operator. Thus, when a response is required from the operator, a pause must be inserted in the test program. The ATE operating system often provides for this eventuality by implementing the keyboard read function as a wait for carriage return loop. But the need for operator intervention itself is often indicative of a choice to be made, which in turn means a program branch.

Escape branches are often neglected when the operator response is perceived as an "OK, I'm ready" end to a pause. For example, the operating system may take preliminary data from the operator and prepare a test program which prompts for the installation of the DUT on the adapter and prints "press any key when ready." The lack of contingency planning in such a statement is poor practice.

```
                    ┌─────────┐
                    │  Start  │
                    └─────────┘
                         │
                         ▼
                 ┌──────────────┐              ◇ Case
                 │ Initialization│              Status =
                 └──────────────┘                 ?
                         │
                         ▼
                 ┌──────────────┐              ┌──────────────┐
                 │   Kernel     │              │    Bus       │
                 │  Self–Test   │              │  Isolation   │
                 └──────────────┘              └──────────────┘
                         │
                         ▼       No
                      ◇ Pass ──────────►       ┌──────────────┐
                         │                      │ Instruction  │
                         │ Yes                  │  Set Test    │
                         ▼                      └──────────────┘
                 ┌──────────────┐
                 │     CAM      │              ┌──────────────┐
                 │  Self–Test   │              │  Microcode   │
                 └──────────────┘              │   Check      │
                         │                     └──────────────┘
                         ▼       No
                      ◇ Pass ──────────►       ┌──────────────┐
                         │                      │  Address     │
                         │ Yes                  │  Hit Test    │
                         ▼                      └──────────────┘
                 ┌──────────────┐
                 │  "Random"    │              ┌──────────────┐
                 │  Logic Test  │              │ Walking Ones │
                 └──────────────┘              │  Miss Test   │
                         │                     └──────────────┘
                         ▼
                    ┌─────────┐
                    │  Done   │
                    └─────────┘                   ╭──────────╮
                                                  │  Return  │
                                                  │ Control to│
                                                  │ Operator │
                                                  ╰──────────╯
```

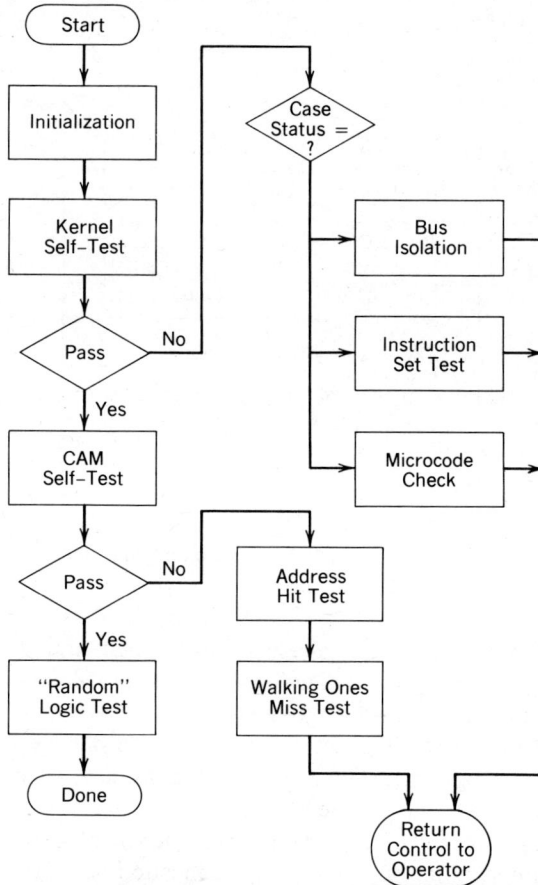

Figure 8.4. Branching in the flowchart directs the test to diagnostic routines only if a go–nogo test fails.

Unless the ATE designer has been very careful, the chance of accidentally pressing a key (particularly the space bar) by merely leaning on the keyboard while shuffling documentation or handling the DUT is great enough to cause an occasional false start. Premature power on may damage the DUT being inserted. A preferable operator intervention loop requires a specific key, as in "Press the Carriage Return when ready." Suppose, upon examining the DUT, the operator finds that the wrong test program has been called up. The addition of an "or ESCape to cancel" to the options leaves a graceful unconfusing exit.

A more complex loop structure for operator interface is required for calibration or option-setting operations. If the test reads a frequency and finds it out of range, the operator may be prompted to adjust a potentiometer and respond. The frequency may then be remeasured and the process repeated until a reading in range or a "give-up" point is reached.

The example routing of Figure 8.5 leaves considerable judgment to the operator, including how much to turn the potentiometer and when to give up and declare the DUT faulty. If the interface were with a robot rather than a human operator, the exact details would have to be included.

Implementation of particular branch structures varies among ATE systems. Simpler systems have only the built-in implied branch of a hardware interrupt on fail. Elaborate systems support a rich set of branch and subroutine call statements similar to Pascal or other popular programming languages. Precise description of any particular ATE language syntax would be lengthy and of limited value and is

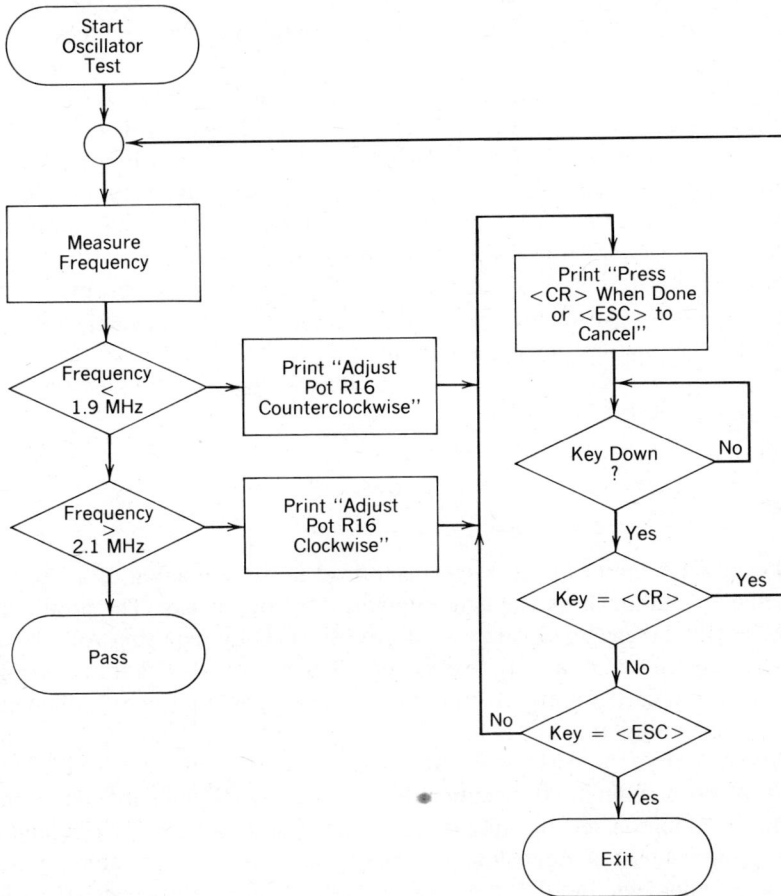

Figure 8.5. Operator interface looping in the flowchart allows repeated attempts at manual adjustment.

therefore left to specific sources such as manufacturers' training classes or technical manuals.

Documentation

Before pursuing specific examples of ATE languages and programs, the issue of documentation should be introduced. Although the development of ATE programs is far more restricted and predefined than ordinary programming, documentation is still a necessary part of the package. Documentation in the case of ATE programming is aimed at two distinct readers: the test operator and the test engineer who must maintain the test program.

Test program operation requires some form of documentation to assure correct DUT and interface installation, to prompt operator intervention at appropriate times, and to supply troubleshooting aids and information. Messages from the program usually can be displayed or printed out to the operator, and these may be sufficient in simple cases to guide operation. The program should be structured to allow all but the most essential messages to be eliminated from the second and subsequent passes through the test. Troubleshooting aids usually include the DUT schematic (in paper form), a listing of ROM or PLA data used in the DUT, and any conversion charts which may be needed to relate ATE connector pins to DUT pins, locate parts on the DUT, and so on. Any or all of these aids can in some cases be automated and stored with or called by the test program.

Maintenance of the test program requires embedded comments or, in extremely complex programs, a flowchart. Schematics, mechanical drawings, and a parts list pertaining to the interface device will assure proper understanding of the relationship of the program to the DUT and help physical repair operations should the interface device become damaged.

ATLAS

Several economic and technical forces have lead to the formation of a "standard test language," although its use is not universal by any means. The proliferation of ATE-specific languages frustrates test professionals by requiring ATE specific training before programming any new system. Translation of tests from a test generation facility to each variety of ATE requires a separate translator, although the results may look similar.

In 1966 an arm of the Airlines Electronic Engineering Committee (AEEC) with the express sponsorship of Aeronautical Radio Inc. (ARINC) identified a need for and began working on specification of a standard test language [1]. The language was based on a subset of English and primarily served the programming needs of analog test equipment. In 1976 the result called ATLAS (Abbreviated Test Language for Avionic Systems) was reformatted and adopted by the IEEE as Abbreviated Test Language for All Systems. By then it had become a well-structured programming language with hundreds of key words related specifically to test and measurement tasks.

Specific details of the language are contained in ANSI/IEEE Std. 416-1984 (the date is the last version as of this writing and is periodically updated). A companion standard (IEEE Std. 771-1984) is a guide to application of the ATLAS language. Rather than give a full explanation of the language, this text will try to explain the structure and give a flavor of the typical use in mixed analog–digital applications. Each of the aforementioned standards is a book in its own right.

Since ATLAS is a tester-independent language, all actions are explicit rather than implied, as the branch statements in some ATE specific languages may be. ATLAS is not in broad use in the digital test arena, and none of the major brands of digital ATE implements ATLAS directly. Translators are available in some instances. Military and avionic applications are most frequent users of the language.

ATLAS Structure

Structure in ATLAS is a two tiered matter in which many alternate forms of branch and loop control exist within an overall block program structure reminiscent of Pascal. Features of control within the blocks are of mixed derivation, in some cases appearing Pascal-like while in others looking like BASIC. The committee origin of the language is reflected in the overly rich selection of control mechanisms.

Block Structure. ATLAS test procedures consist of a main program and a group of referenced procedures. The main program and all procedures may contain a "preamble" section and a "procedure" section. The preamble section defines data structures and procedures much the same as the "declaration" part of a Pascal program does with its TYPE, VAR, CONSTANT, LABEL, FUNCTION, and PROCEDURE statements. All of the listed Pascal declarations are handled in ATLAS by the two statements DEFINE and DECLARE.

The DECLARE statement allows labels, constants, and variables to be named for later use in the program or procedure. Storage requirements and format are specified as part of the DECLARE statement. DECLARE statements are effective for the block level in which they appear and in procedures defined within that block unless the procedure specifically alters the declaration. All DECLARES in the outer program block are global.

Procedures are described in a DEFINE statement prior to their being invoked. Data are passed to and from procedures via formal parameters specified in the DEFINE statement along with the name of the procedure and its code. Each procedure may have its own preamble section for defining local variables, constants, and labels.

An interesting and useful anomaly among the reference statements of the preamble section is the FILL statement. Test programs often use large data structures such as stimulus and response arrays, and the define constant statements become tedious after only a few numbers are included. The fill statement is essentially an I/O statement which transfers a block of data into a defined structure for later use in the program.

The body of any procedure resembles the body of the main program which is called the ''procedure'' block. The procedure section of an ATLAS program is a series of statements to be executed in the order listed. Branching and loop control constructs allow flexibility in the execution order using simple or conditional transfer of program execution to labeled or numbered statements.

Blocks are delimited by pairs of statements which serve only to show structure. BEGIN, ATLAS PROGRAM and TERMINATE, ATLAS PROGRAM are used to bracket the entire program; BEGIN, BLOCK and END, BLOCK segment the statements to improve readability; and DEFINE, 'name' PROCEDURE and END, 'name' surround the statements which make up a procedure description.

```
000010 BEGIN, ATLAS PROGRAM 'MicroBoard'$
000020 DECLARE, DECIMAL, STORE,
'PwrVoltage'$
000030 DECLARE, CONN, STORE, ...
|
|
000100 DEFINE, 'PwrCheck', PROCEDURE
(Supply No, Uplimit, Lolimit)$
|
000150 END, 'PwrCheck'$
|
End of Preamble for main program ...
001000 BEGIN, BLOCK, 'Set-Up'$ |
001070 END, 'Set-Up'$
002000 BEGIN, BLOCK, 'Ram-Test'$
|
002450 END, 'Ram-Test'$
|
|
100000 TERMINATE, ATLAS PROGRAM,
'MicroBoard'$
```

Flow Control. Within the procedure section of the main program or any PROCEDURE definition, control of the sequence of instruction execution depends on a number of branch constructs. They include the unstructured GOTO, the simple branch IF, and loop controls FOR and WHILE. The statements contain some overhead to simplify compiler design.

The GOTO statement directs program flow to a line number without regard to labels. It has both an unconditional and a conditional form as follow.

```
350  GO TO,  STEP 520$
480  GO TO,  STEP 2463 IF NOGO$
```

The NOGO in the statement represents a negative condition of the condition flag from the last COMPARE operation. For compiler convenience a location in the program pointed to by one or more GOTO statements in marked by a BRANCH FROM STEP # statement. Complex branching decisions make use of a DECISION TABLE which operates to collect the results of several calculate, measure, and compare operations and then to branch on the results like a Pascal CASE statement.

Adherents to structured programming will find the IF statement more satisfying. The IF ... THEN ... ELSE ... END, IF structure separates a section of code which is to be executed conditionally with the same appearance as the Pascal equivalent. Longer code segments are included in PROCEDURE definitions and called by PERFORM statements.

Indexed loops are controlled by FOR ... THEN constructs of the form

```
FOR 'IndexName" = lower THRU upper BY step THEN$
```

followed by the body of statements and terminated by END, FOR$, A WHILE ... THEN ... END, WHILE$ construct operates similarly, using a Boolean condition instead of a calculated one.

Although the loop control mechanisms of ATLAS offer fewer options than general-purpose programming languages, they suffice for writing structured, readable procedures without undue compiler complexity. Some additional test environment program controls are added which pertain to the task of tester control and operator interface. Pauses, synchronization marks, and wait for manual intervention statements simplify ATE programming and help to standardize handling of software–hardware interfaces.

ATLAS Action Statements

Statements in ATLAS are made up of a verb followed by one or more nouns and/or descriptive phrases concluded by the ATLAS statement terminator "$." Each line is labeled with a statement number to define its position in the sequence. A flag is prefixed in some cases to indicate to the compiler when a certain type of statement such as a comment or a branch statement has been encountered. Great variety exists in the vocabulary of verbs and their associated phrases. User-defined variables and names can be included through the preamble definition if a particularly suitable one cannot be found in the language definition. Since the possibilities are so numerous and since the applications are generally analog in nature, only a few examples will be given as explanation.

```
023080 APPLY, DC SIGNAL, VOLTAGE 5V ERRLMT +-
   .25V, CURRENT MAX 2.0A, CNX HI A42 LO A01$
045000 VERIFY, (VOLTAGE), AC SIGNAL, FREQ
   2.5MHZ, UL 4.5V LL 0.5V, CNX HI J2-04 LO COMMON$
045320 MEASURE, (VOLTAGE INTO 'RegOut'), DC
   SIGNAL, VOLTAGE RANGE +4V to +2V, CNX HI B54 LO
   B55$
```

The first statement applies power to the designated connections (CNX) according to the specification and is accompanied by limit detection for both voltage and current. The form can be used to apply ac stimuli and pulse stimuli as well. VERIFY is a comparator form of the response analysis command which in this case is checking the amplitude of a DUT oscillator. The result of VERIFY is a Boolean condition and not a number. When a value is desired, the MEASURE function can be used as in the third example, which assigns the result of the measurement to the variable 'RegOut'.

Digital Testing with ATLAS. Digital testing is a newcomer to ATLAS, and accommodation has been afforded through additonal "macrostatements." Digital test requires the application and comparison of large blocks of data ordinarily stored and handled by the ATE hardware directly via a test pattern buffer. A compiled software handler would be too slow in many cases. The DO statement defines the mode of digital test and initiates the action. Parameters such as the stimulus application timing, response strobe timing, and voltage limits for the signals must be defined prior to execution of the DO command. Connection tables for the stimulus and response signal must also be previously entered.

The proper sequence for using the DO, DIGITAL TEST command involves several steps. Storage locations must be associated with the labels 'stim', 'stim-interval', 'resp', 'resp-delay', 'ref', 'mask', 'error', 'error-index', 'fault-count', 'step', and 'conn-set' by means by the DECLARE statement. The FILL statement directs the ATE to load the pattern set into the arrays 'stim', 'mask', and 'resp'. Analog instructions such as SETUP and APPLY prepare the drivers, receivers, and load conditions. Finally, a series of DO and FILL statements trigger the tester to cycle through buffer after buffer of digital data until the digital test is completed.

Performance of the digital test is a hardware function, and the mode of test is specified by the DO macroinstruction but carried out in real time while the software waits for the completion flag. Modes of STIM-ONLY, RESP-COMP, STIM-RESP-COMP, STIM-RESP-MATCH, and STIM-RESP-SAVE are used for most digital tests. The STIM-RESP-MATCH differs from the COMP function in that it halts the operation if the DUT data field defined matches the specified conditions. This form is useful for initialization of the DUT. The other modes store or compare the DUT outputs and interrupt if the outputs do not match the expected or 'ref' file.

The following example program segments taken with permission from the IEEE

Std. 416-1984 show the alternate methods of explicit bit manipulation and macrostatement construction of digital tests in ATLAS.

Simple Digital ATLAS Program [2]:

```
000001 BEGIN, ATLAS PROGRAM $
    10 DECLARE, BOOLEAN, STORE, 'INPUT-FAULT' $
    15 DECLARE, DIGITAL, STORE, 'PCB-RESP', 34 BITS,
       '50' BNR 1/'51' BNR 1/'52' BNR 1/'53' BNR 1/'54' BNR 1/
       '55' BNR 1/'56' BNR 1/'57' BNR 1/'58' BNR 1/'59' BNR 1/
       '60' BNR 1/'61' BNR 1/'62' BNR 1/'63' BNR 1/'64' BNR 1/
       '65' BNR 1/'66' BNR 1/'67' BNR 1/'68' BNR 1/'69' BNR 1/
       '70' BNR 1/'71' BNR 1/'72' BNR 1/'73' BNR 1/'74' BNR 1/
       '75' BNR 1/'76' BNR 1/'77' BNR 1/'78' BNR 1/'79' BNR 1/
       '80' BNR 1/'81' BNR 1/'82' BNR 1/'83' BNR 1 BITS $
C      *** FIELD NAMES COULD ALSO BE ALPHA-NUMERIC, E.G. 'PL-1' ***
    20 DEFINE, 'INPUT-FAIL', PROCEDURE $
       OUTPUT, INPUT FAULT ON UUT $
       FINISH $
    30 END, 'INPUT-FAIL' $
    40 DEFINE, 'DIGITAL-FAIL', PROCEDURE $
       OUTPUT, UUT HAS FAILED DIGITALLY $
       FINISH $
    50 END, 'DIGITAL-FAIL' $
000100 DEFINE, 'TTL-INTERFACE', DIGITAL CONFIGURATION $
    10    DEFINE, 'PCB-IN', DIGITAL SOURCE, DC SIGNAL,
          LEVEL LOGIC-ZERO VOLTAGE 0.4V EERLMT +-0.1V,
          LEVEL LOGIC-ONE VOLTAGE 3.2V EERLMT +-0.1V,
          ILLEGAL STATE INDICATOR 'INPUT-FAULT',
          CNX HI 1 2 3 4 5 6 7 8 9 10 11 12 13 14 15 16 17
                 18 19 20 21 22        LO JO $
    20    DEFINE, 'PCB-OUT', DIGITAL SENSOR, (VOLTAGE), DC SIGNAL,
          LEVEL LOGIC-ZERO, UL 0.5V LL 0V,
          LEVEL LOGIC-ONE, UL 5.0V LL 2.4V,
          CNX HI 50 51 52 53 54 55 56 57 58 59 60 61 62 63
                 64 65 66 67 68 69 70 71 72 73 74 75 76 77
                 78 79 80 81 82 83        LO JO $
    30 END, 'TTL-INTERFACE' $
C      *********** MAIN PROCEDURE ************* $
000200 ENABLE, DIGITAL CONFIGURATION 'TTL-INTERFACE' $
       CALCULATE, 'INPUT-FAULT' = FALSE $
    10 ENABLE, ESCAPE TO PROCEDURE 'INPUT-FAIL'
                 IF 'INPUT-FAULT' TRUE $
    15 ENABLE, ESCAPE TO PROCEDURE 'DIGITAL-FAIL'
                 IF NOGO $
C      *** SAMPLE BIT MANIPULATION STATEMENTS *** $
C      *** N.B. ALL UUT CONNECTIONS ARE SENSED ***** $
000300 FILL, 'PCB-RESP', B'1111111111111111111111111111111111' $
       FILL, '58', '59', '60', '70',
             B'0', B'0', B'0', B'0' $
    05 STIMULATE, ONE, ON CNX HI 3 5 6 7 10 13 20 22 $
    06 PROVE, (VALUE) REF 'PCB-RESP', ON 'PCB-OUT' $
    10 FILL, '58', '59', B'0', B'0' $
    15 STIMULATE, ZERO, ON CNX HI 5, 6,
                  ONE, ON CNX HI 1 $
    16 PROVE, (VALUE) REF 'PCB-RESP', ON 'PCB-OUT' $
    20 FILL, '50', '51', '61', '79', '83',
             B'0', B'0', B'0', B'0', B'0' $
    21 STIMULATE, ZERO, ON CNX HI 3 7 10 13 20,
                  ONE, ON CNX HI 5 $
    22 PROVE, (VALUE) REF 'PCB-RESP', ON 'PCB-OUT' $
C      ****** ETC..... ********* $
 800000 REMOVE ALL $
 900000 FINISH $
 999999 TERMINATE, ATLAS PROGRAM $
```

The following portions of a program (not for the same DUT as previously) show the SETUP and DO statements for a digital test using macros [3].

```
123462 DEFINE, 'DIG-IN', SOURCE, TYPE PARALLEL,
          VOLTAGE-ONE 5.0 V EERLMT +-0.5V,
          VOLTAGE-ZERO 0.5 V EERLMT +-0.5V $

123464 DEFINE, 'CONN-IN', SOURCE, CNX-STIM HI J1-1 J1-2 J1-3
          J1-4 J1-5 J1-6 LO J1-0 $

123466 DEFINE, 'DIG-OUT' SENSOR, TYPE PARALLEL,
          VOLTAGE-ONE 5.0 V EERLMT +-0.5 V,
          VOLTAGE-ZERO 0.5 V EERLMT +-0.5 V $

123468 DEFINE, 'CONN-OUT', SENSOR, CNX-RESP HI J1-11 J1-12
          J1-13 J1-14 J1-15 J1-16 LO J1-10 $
            .
            .
            .

123470 SETUP, DIGITAL TEST, 'DIG-IN', 'CONN-IN' $

123472 SETUP, (VALUE), DIGITAL TEST, 'DIG-OUT', 'CONN-OUT' $

123474 DO, DIGITAL TEST, STIM-RESP-COMP, STIM 'INPUT' (1 THRU 10)
          WORD-RATE 500 KWORDS/SEC, RESP-DELAY 0.5 USEC,
          RESP 'OUTPUT' (1 THRU 10), REF 'ANS' (1 THRU 10),
          WORD-RATE 500 KWORDS/SEC, MASK-ONE 'MASK 1'
          ERROR 'E-LIST' (1 THRU 10) 'CONN-IN', 'CONN-OUT' $
```

The instruction set of ATLAS is a very rich one with modifiers which further define the nouns and give an English appearance to the programs. The coverage here is not intended to be complete enough to tutor the reader in ATLAS programming. The brief discussion is only meant to give a flavor of the language in digital applications for comparison with dedicated ATE languages.

ATLAS handling of digital test is not as tester independent as its other aspects. A lack of easily applied memory algorithms contributes to its lack of popularity in the digital arena. As the only "standard" test language available, however, ATLAS has become an exchange medium for test program expression and documentation. The U.S. Department of Defense and the United Kingdom Ministry of Defense support ATLAS, and the adoption of ATLAS by the ANSI/IEEE standards group has virtually guaranteed its continued development. In applications where mixed analog–digital DUTs are prevalent, the use of ATLAS has the distinct advantage of organizing and documenting the entire test sequence as one test program rather than requiring separate ATE language descriptions of each segment of the test.

AUTOMATED PROGRAM GENERATION

Regardless of the language, the test programs for a family of circuits (i.e., pc boards from the same system) are likely to resemble one another closely. Continuity of programming style is desirable since it presents the ATE operator with a familiar interface and training requirements are lessened. A common approach to test programming by a team of test engineers is to create a prototype program, critique and refine it, and then copy its form for the remaining DUTs. The prototype program may be stripped of its DUT dependent data and used as a shell. To create a new test, the ATG output data and the connector configuration data for the target DUT are filled into the blanks of the shell.

When program generation volume is high, an automatic program generator (APG) can be constructed. The APG consists of templates for the various sections of the test program and interfaces to the test engineer and other databases. The test engineer supplies requirements and direction, while the APG translates data and compiles a completed program from the templates. Such conveniences save time while resulting in consistent test programs and the proper documentation.

SUMMARY

Test programming in the digital realm bears only a superficial resemblance to computer science. The requirements of the product, customer, and ATE restrict the form of a test program severely. After an initial setup of the ATE involving connector definitions and parameter specification, the typical digital test program consists of a lengthy series of stimulus application and response evaluation statements. Many ATE systems reduce the body of the program to data tables which are automatically loaded and executed when encountered.

Applications which call for testing analog–digital hybrid circuits or program controlled troubleshooting sequences begin to resemble conventional software with a variety of apply, measure, and branch statements. The ATLAS language originated from such a background and has become quite robust. Standard ATLAS provides a method of writing non-ATE-dependent test programs which can be hosted on several types of systems. Its adoption by the military has maintained its presence in spite of a slow adaptation to digital test.

REFERENCES

1. *IEEE Guide to the Use of ATLAS*, IEEE, New York, 1980, pp. 20–22.
2. *IEEE Standard ATLAS Test Language* (IEEE Std. 416-1984), IEEE, New York, 1984, pp. 11–50.
3. *IEEE Standard ATLAS Test Language* (IEEE Std. 416-1984), IEEE, New York, 1984, pp. 13–30.

CHAPTER NINE

DIAGNOSTICS AND TROUBLESHOOTING AIDS

If the preceding chapters have been taken in order, the reader has seen a discussion of how to design and apply a test to a digital circuit by means of ATE. Although fault location has been mentioned, the primary emphasis has been on the detection of faults. In some circumstances, detection is sufficient because the entire unit is to be scrapped rather than repaired, but in general, the natural follow-up to detecting the presence of a fault is to find it and fix it.

Diagnostic analysis can be applied manually or automatically. In computer systems the diagnosis is often begun by internal software or built in test circuitry and completed by external testing or manual troubleshooting. Diagnosis typically proceeds from the functional description down to the gate or logic description level. This chapter will attempt to describe the manual method of troubleshooting to provide a foundation, followed by descriptions of several automated and instrument-augmented methods. Before getting to details, however, the goals of diagnosis and its theoretical background should be made clear.

FAULT LOCATION

Test generation methods are focused almost solely on detection of faults. An efficient test sequence is seen as one in which the maximum number of faults is detected in a test vector set of given length. The goal is to provide a test sequence which can be applied quickly and which will at the same time assure DUT functionality with a high degree of confidence. The ultimate test sequence might be described as a single "magic vector" which results in a pass or fail for 100% of the faults possible. A completely self-testing circuit may even appear to provide this very scenario when viewed from the outside. The self-test is activated, and the result is a green light indicating a pass condition. But what if the circuit fails the test?

If a personal computer fails its power up confidence test, the owner is faced with the problem of deciding what to fix. The next step is usually to apply a set of diagnostic tests aimed at fault resolution. A memory test may be applied, then a disk drive test procedure, then various interface tests, and so on until the problem is narrowed down to a replaceable unit. If the unit is a subassembly, the diagnosis may be moved to a different level. For example, if the personal computer problem turns out to be a disk drive failure, the owner is likely to want the drive repaired rather than replaced as a memory chip would be replaced. The disk drive will probably be taken to a shop where external test equipment will be used to pinpoint the problem to the speed control PC board and finally to an IC, which will be replaced.

Deterministic Detection Shortcomings

Information about the potential cause of failure is inherent in the test generation method to some extent, as long as deterministic methods are used to generate test vectors. An ATG program produces each series of test vectors toward a goal of detecting a particular fault selected from the list of possible faults in a given class. It would seem on the surface that a simple trace of the goals of the ATG program or the test engineer performing manual test generation for each test vector or subset of test vectors would provide the fault location insight needed.

The focus of the test generation goals is too narrow to be used as a fault location guide. Three phenomena occur regularly to thwart this approach: accidental fault detection, fault detection ambiguity, and detection of faults outside the target fault class. The combination of these possibilities leads to ambiguity in test results, but the term ''fault detection ambiguity'' will be defined later in a more restricted sense.

Accidental Fault Detection.   Accidental fault detection can be understood by considering the random test vector selection method. Random manipulation of the circuit inputs has a probability of detecting faults. The intention of random test generation is to stumble across an input vector or a set of vectors that activates some function of the circuit and propagates the results to an output. A deterministic test generator activates functions of the circuit precisely. Along the way, some parts of the circuit are triggered, although the fault target is not contained in that section. In a complex circuit, activating a single event in a focused locality of the circuit usually creates side effects for the rest of the circuit that are at least as efficient at fault detection as random patterns.

Manual test generation which is unsupported by fault simulation is doomed to suffer the accidental detection phenomena, but ATG and manual generation for complex circuits are normally supported by some variety of fault simulation. Since fault simulation has a broader view of the situation, the accidental detections are discovered. The parallel, deductive, and concurrent methods all take the approach of trying a particular set of test vectors against all possible faulty circuits in a given class. Although the test engineer may have been focusing on an s-a-1 fault at a particular flip-flop output, the fault simulator is likely to produce a long list of detected faults the first time the master clear is activated.

Recording a list of faults which the fault simulator tallies as detected on each test vector is an excellent answer to the problem of accidental fault detection. Furthermore, if the differences between the good and faulty circuit outputs are paired with the assumed fault which produced them in the simulation, a reference for later troubleshooting linking the assumed cause and effect can be created. The reference is not completely unambiguous for several reasons that will be explored.

Fault Detection Ambiguity. During fault simulation and test generation, fault equivalence appears as a bonus toward meeting goals. A test which covers a fault covers its equivalent faults and therefore has a higher yield of fault detection. When the fault simulator reports the faults detected and their corresponding faulty output signatures, fault equivalence causes overlap in the lists. Also, fan-out from a circuit node may cause overlap in the lists, as the conditions for detection of a fault may be a subset of the conditions necessary for other fault detection. The overlap causes ambiguity in the list.

As an example the circuit of Figure 9.1 is under test and the previous test step left the counter in a state of all 1s. As this is the first time the counter has definitely reached this state (it powered up as unknown and the first clear applied did not achieve any detection), the test engineer selects this pattern to force a clear and test for pin 1 of chip G42 s-a-1. In addition to providing the expected outputs for a good circuit (primary output pins 34, 36, 38, and 40 s/b 0 and pin 103 s/b 1). A diagnostic message is provided in the form of a table or list. The list of possible

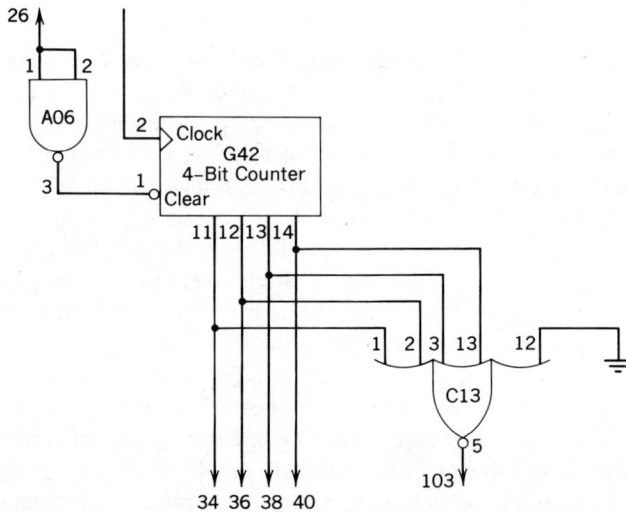

Figure 9.1. A sequential circuit diagram. Several faults are detected by the same test.

faults and fault signatures returned from the fault simulator for this particular test step will appear as shown in Table 9.1.

The first entry of the table includes the fault which the test engineer intended to detect. It also includes four equivalent faults which cannot be separately detected by examination of the primary circuit outputs. Any test for G42-01 s-a-1 will also be a test for A06-03 s-a-1, A06-01 s-a-0, A06-02 s-a-0, and primary input pin 26 s-a-0. These five faults are only implicated if the set of incorrect output pins includes all five of the pins as shown in the left-hand column of the first entry.

The next four entries have two-pin fault signatures, which eliminates the possibility of the presence of the faults in the first entry. The elimination is based on an assumption of single stuck-at faults as the sole basis of analysis. The same assumption reduces the possible cause list to a single fault for each entry. Only an open circuit or an output driver failure at G42 pin 11 (either results in an s-a-1) will cause both pins 34 and 103 to be incorrect. An open in the connector pin P1-34 will not, on the other hand, cause 103 to fail; only its own signature will be in error.

A long list of possible faults is associated with the symptom of pin 103 in error because those faults are equivalent. The final entries are simple one-symptom–one fault relationships. A tester which has stopped on this step and simply indicates a failure leaves a considerable fault location problem, although the test engineer's notes (if no fault simulation were used) might simply suggest that the test detects G42 pin 1 s-a-1. With the help of a fault simulator the extra detections can be uncovered, but some of the faulty circuits simulated create the same outputs even though each contains a distinct single fault.

Extra-Class Faults. Anticipating the fault classes which might occur in a circuit is difficult, and the facilities to simulate and test for all fault classes are never

TABLE 9.1 Diagnostic Table

Faulty Output Pins	Possible Causes
34, 36, 38, 40, 103	A06-01, s-a-0, A06-02, s-a-0, IN-26 s-a-0, A06-03, s-a-0, G42-01 s-a-1
34, 103	G42-11 s-a-1
36, 103	G42-12 s-a-1
38, 103	G42-13 s-a-1
40, 103	G42-14 s-a-1
103	C13-05 s-a-0, OUT-103 s-a-0, C13-01 s-a-0, C13-02 s-a-0, C13-03 s-a-0, C13-13 s-a-0, C13-12 s-a-0
34	OUT-34 s-a-1
36	OUT-36 s-a-1
38	OUT-38 s-a-1
40	OUT-40 s-a-1

available. Some compromise must be made which simplifies the problem. The reduced fault set chosen in the example is the "single stuck-at" fault set. Fault location can be thwarted by the occurrence of faults from outside the target fault class.

Suppose that a gouge in the circuit board causes an open in all four of the paths from the outputs of G42 to the primary pins, but the gouge interrupts the circuit after the connections to gate C13 are made. The test will fail, and the fault signature will include 34, 36, 38, and 40, but not pin 103. None of the table entries from Table 9.1 covers the condition. In fact, if the "closest match" is chosen, the first entry will send the test technician on a wild goose chase. An experienced technician might expect a multiple fault or at least discover the error in diagnosis after the first measurements are made. Common instances of extra-class faults include power and ground pin opens on ICs and intercircuit shorts.

The overriding message in the above example and discussion is that the fault location ability or "diagnostic resolution" of a test is never as good as it appears due to the assumptions made in the test generation process. There are methods of improving the diagnostic resolution of a test, however. In addition, there are tools and techniques for using the test information and direct measurements to improve fault location capabilities in either the factory or field testing arena.

Test Structuring for Fault Location

Careful examination of the example described in Figure 9.1 and Table 9.1 raises some questions about the assumed test sequence and affords an opportunity to suggest ways of improving the fault location capability of the test at the same time. The first question which arises concerns the method of getting the counter to the all-1s state described as the previous step's state without having detected some of the s-a-1 failures previously. The assumption which satisfies the outcome is that the counter was parallel loaded with all 1s during some previous test step and has never previously had any 0s on any of its outputs.

It would seem just as likely (even more so in view of the lack of parallel load mechanisms in the schematic shown) that the counter was cleared from an unknown state to all 0s and then counted to a state of all 1s. If that were the case, the first master clear would have resulted in detection of all the faults listed in the table except those in the first entry. The reasoning is that the initial state of the counter is unknown, and therefore the master clear may or may not have been responsible for the state of all 0s on the counter (it may already have been in that state). The state of the good circuit can nevertheless be determined after the master clear, and any fault which would cause a differing output can then be detected. The second master clear in the test step of the example would only result in the listing of the first entry since the others have already been detected and removed from consideration.

An improved test sequence for fault location proceeds as follows. Parallel load the counter with a (0101), then count it to a (1111) or load a (1010) and (1111) in succession. In this sequence fewer faults are detected in each step, but the counter outputs will have been tested for both s-a-1 and s-a-0 before the OR gate or the master clear is tested. Spreading the fault detection over more test steps produces clearer, more direct fault location messages from a fault simulator.

This philosophy runs contrary to the efficiency goals for minimizing the number of test vectors needed to detect a given level of faults. It also assumes manual test generation since there are no ATG programs at present which offer this type of subgoal control. As a result, other test structure techniques and the use of post-test-generation tools have become dominant in fault location and diagnosis.

TROUBLESHOOTING

Manual diagnosis at the point of test application is called "troubleshooting." An examination of successful troubleshooting techniques for digital circuits will place a realistic foundation on automated techniques and reflect on the test generation output information which is of most value.

When an experienced test technician troubleshoots a circuit several levels of knowledge are employed. General knowledge about electronics and testing must be related to specific knowledge about the DUT, the ATE being used, and the test program controlling the test. The resulting analysis spans both spatial and temporal dimensions as the technician traces the schematic and steps through the test program.

Hierarchical Representations

Circuit representations on both the functional and gate levels are likely to be used by a test technician during troubleshooting if the information is available. The laws of electricity, logic, and arithmetic must be applied to the appropriate representation as in Figure 9.2.

In digital testing most technicians are supported by logic-level circuit representations of the specific DUT. The logic drawings make reference to standard parts or custom ICs as a parts list or by a special symbol or label in the drawing. Preferably, a block diagram or register-level description of the circuit is provided to supply higher-level information. The lower-level representation is retrieved from separate custom IC schematics, microcode listings, or vendor specifications for standard parts. Troubleshooting can be successfully performed with only the lowest-level information, but the analysis is inhumanly tedious.

Starting Point Selection

In field troubleshooting problems the cause is often a single fault. In manufacturing the faults often occur in multiples, as parts are missing or misoriented. Either case can exhibit a bewildering mess of symptoms as fault effects propagate and fan out

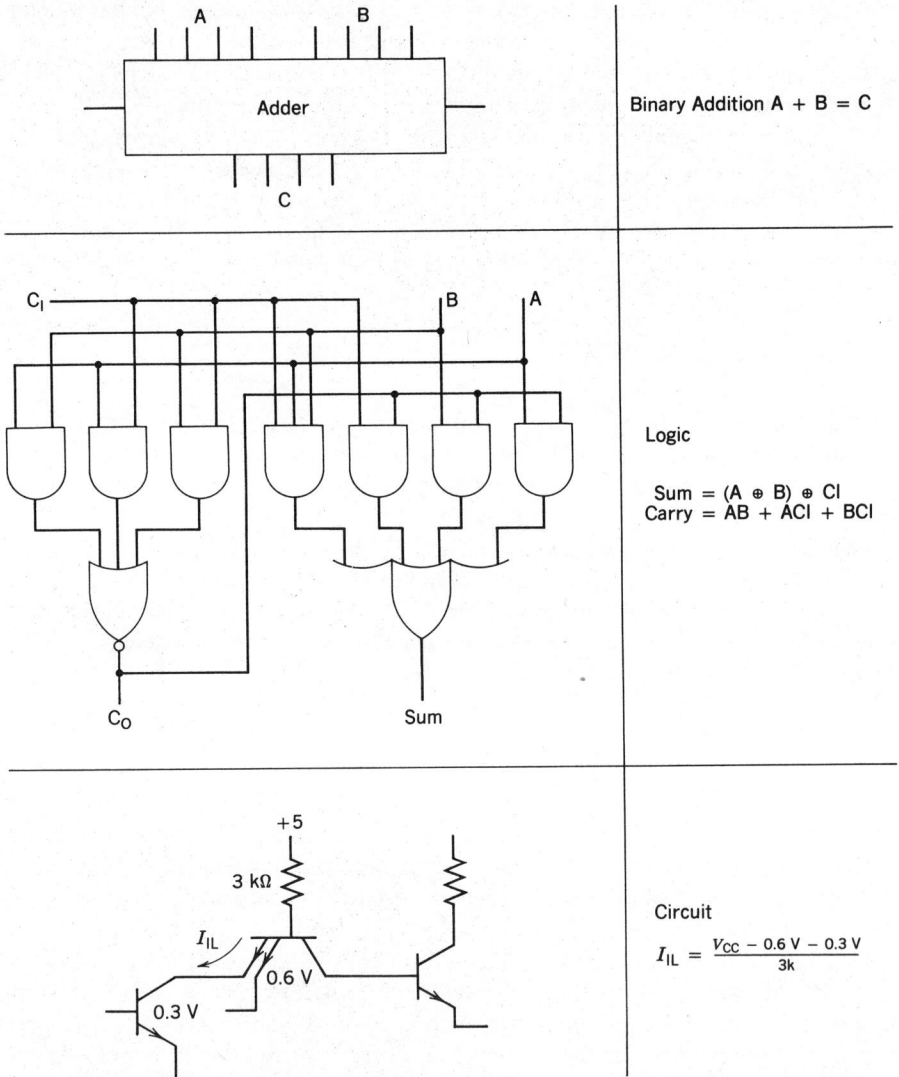

Figure 9.2. Circuit descriptions appear in several levels of abstraction. Each level contains information of use to the troubleshooter.

through the circuit. It is just this predicament which best makes use of the upper levels of the circuit representation. It is the ability to correlate the symptoms and select a probable starting point for circuit tracing that is a head start for a skilled technician and difficult to automate.

Not only must the failing outputs be identified as primarily from a single bus, for example, but the symptom of being in a voltage range between a valid logic 0 and a valid logic 1 must be a related to the function of the bus enable. The leapfrog reasoning leads to a single point where the bus enable fans out to supply each of several interface chips and the trace slows to a logic-gate-level step-by-step process. Work is underway to understand and replicate this sort of reasoning using AI techniques, but since automation of this part of the process is not widespread, the detailed troubleshooting explanation will proceed from a simpler beginning.

Diagnostic Backtracking

Let the technician operate a bit more mechanically by selecting as the starting point for investigation into the faulty circuit the first failing pin. If a DUT has been installed in an ATE system and the test has been started under the "STOP ON FIRST FAIL" mode, the condition now facing the technician is the first appearance of a fault effect at a primary output. Further testing would be confusing in most cases since the fault effects may enter feedback loops in the circuit and multiply in number.

If multiple outputs have incorrect values on them at the first failing test step, any one of them can serve as a starting point for backtrace. The starting point selection process often saves time by reducing the number of steps in a backtrace, but it is not strictly necessary. Any faulty output must be the result of the fault (or faults) since good circuits do not create faulty outputs. Backtracking from the lowest-numbered failing pin will eventually reach the fault and so will a backtrace from the highest-numbered failing pin (or any other failing pin).

Spatial Tracing. Logic backtracing in combinational circuits is spatial in nature. Since no flip-flops exist, all faulty outputs must be caused by a fault which is active in the same test step as the failing output. The explanation of sequential circuit analysis will benefit from this restricted treatment since both spatial and temporal tracing are used in that case. Initially considering only spatial tracing and combinational circuits saves some confusion.

The analysis technique used is akin to the backtracing that accompanies ATG algorithms such as the D algorithm, except that in this case the algorithm is divergent if not contained by actual circuit values. In essence, each step looks backward through the circuit and searches for possible causes for the effect seen at this point. The assumption at each step must be that the inputs to the gate or device being examined may be faulty and that the observed output effect is a consequence of the faulty inputs.

The process of investigating the assumption follows a basic routine except when ambiguous structures such as buses or feedback loops are encountered. These steps require the use of a complete nodal table of good circuit values for the failing test step. A good circuit tested under the same circumstances can also be used for a reference.

1. Select any failing primary output of the circuit to be the circuit point for examination.
2. Trace the schematic back from the circuit point under examination to de-

termine the driving gate. If the output is driven by a bus or wired logic structure, see the special case rules that follow.

3. Compare the output of the driving gate with the known good value. If it is correct, the intervening circuit path is defective. If it is different, proceed.

4. Check all the logic values of the inputs to the driving gate. If they are correct, the chip containing the driving gate is suspect. If one or more of the inputs is found to be incorrect, select one of the incorrect nodes to be the next circuit point to be examined and go to step 2.

5. Suspect components should be checked for possible fault causes other than an internal malfunction before removal and replacement operations begin.

   a. Check all static logic inputs which do not appear in the nodal state table (unused inputs and grounded or pulled-up enables).

   b. Verify the power and ground connections to the component.

   c. Inspect the component and its immediate vicinity for visual defects such as incorrect part installed, misalignment of part, solder shorts, unsoldered leads, or circuit breaks.

Bus structures provide additional problems in that the gate involved is implied and bidirectional in action. An IC AND gate allows inputs to determine the output but generally prohibits output values from affecting its inputs. A wired-AND gate formed by connecting the outputs of several open collector TTL gates consists only of wire, and thus the value of the result or output is imposed upon all the inputs as well. If a short to ground occurs anywhere on the net, all points of the net will be at a logic low or 0.

Physical troubleshooting of buses will be discussed later. The only logic-level analysis that can be made of the circuit is to examine each of the gates which drive the bus in an attempt to skip a level of tracing and find a faulty path before its effects become tangled in the bus.

In the absence of sequential circuit elements or feedback loops which make the circuit sequential, the simple procedure described will locate the faulty component. Once the procedure has identified a faulty component or path, a forward trace from that point to all outputs will verify coverage of all fault effects by the defect. Any remaining unexplained erroneous outputs should be used as starting points for an additional search by the same method to find multiple faults.

The technique described is neither efficient nor intelligent. It is given as a basis for improvement rather than a means to an end. Application of functional analysis at each examination point (usually the output of a gate) can improve efficiency by reducing the number of measurements or "probes" to be made. When a driving gate or a complex chip is encountered, determining the value of a few controlling inputs can eliminate from consideration a portion of the other inputs. If a multiplexer's select inputs are correct and selecting input 3, it is of no use to check inputs 0, 1, and 2. Even if they are incorrect, a correctly functioning multiplexer should ignore their influence.

Adding intelligence will be discussed later, in addition to the possibilities of adding perception of nonlogic information such as parameters (circuit voltage, currents, etc.). For the sake of completeness the manual troubleshooting method will be pursued through sequential circuits before diverging into improvements.

Temporal Tracing. Encountering a sequential circuit element during backtrace brings the dimension of time into play in the investigation. Treatment of a simple latch or flip-flop element is a matter of determining the conditions which would sensitize the faulty output to one or more of the inputs and tracing back in time to find the last existence of such conditions before resuming the spatial search at that point in time. In real circuits, sequential elements may have inputs which act combinationally on the outputs as well as those which require a sequence of input events to change the results.

A successful tracing operation must first examine combinational inputs, then examine asynchronously acting ones, and finally backtrack to the last period of sequential sensitivity. For example, if one of the outputs of an octal register with an output enable line and a clear input line as well as a clock input has been found to be incorrect in value, the preferred sequence of probing for logic values is as shown in Figure 9.3.

In the case of sequential logic elements the simple technique used before will not work. If the logic signal which is causing the problem is the one driving the output enable or the clear signal, then it will be detected by the usual walk around the chip with the probe. But if none of the inputs is found defective by this routine, it does not necessarily mean that the chip is defective. The fault effect may have been stored at the last rising edge of the clock. Lack of a sensitive path to an output could have kept the fault from being detected at the output until the present step. In this instance the fault effect could have been blocked by the output enable signal until the present step.

In general, the added rules to contend with sequential circuit elements are added between steps 4 and 5 of the previous list. They are again stated in such a way as to require minimal functional knowledge of the circuit, although more is required than for strictly combinational circuits.

4.1. If the driving gate or component is combinational, proceed to step 5.

4.2. Determine which control inputs could trigger the storage of faults from data inputs and search backward in the expected states node table until one of the controlling inputs produces a sensitivity to corresponding data

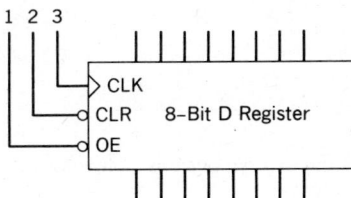

Figure 9.3. Octal D register. Inputs have been numbered in the preferred order of investigation.

inputs. Sensitivity is a level for latch-type clocks and a transition for edge-triggered devices.

4.3. Using the values of the expected state table for the time frame of the test step found above, examine the sensitized data input values and proceed as in step 4.

The idea of spatial backtracing was to arrive at the site of the fault by beginning at the point at which its effects became visible and working toward the circuit inputs. The circuit inputs represent, in terms of propagation, the earliest point in the circuit at which a fault could possibly happen. Temporal backtracing is similar. The tracing begins at the earliest time at which the fault effects can be observed and proceeds backward in test steps to the fault activation time. The earliest time possible is the first test step.

The diagram of Figure 9.4 shows the direction of progress toward finding the fault. There may be more than one path and more than one starting point for a single stuck-at failure, but the paths will always converge on the fault unless a feedback loop is present.

Feedback loops present a difficult form of sequential circuitry. The distributed nature of a feedback flip-flop is not recognizable during application of the preceding procedure. A technician often recognizes the loop by simply having remembered examining the same gate previously. Automating this memory involves a formal technique. Recording each gate as it is examined in step 2 of the procedure after comparing it to the existing list of gates covered will emulate the memory.

Once a loop is encountered, as evidenced by encountering a gate output twice, the problem becomes one of how to determine when and where the loop was entered. The fault may exist within the loop, or it may be outside the loop with the fault effects trapped within the loop. "Breaking the loop" requires an in-depth analysis of the logic structure and the event timing which pertains to the loop. Often, the technician separates the schematic of the loop from the surrounding logic by sketching a subset schematic.

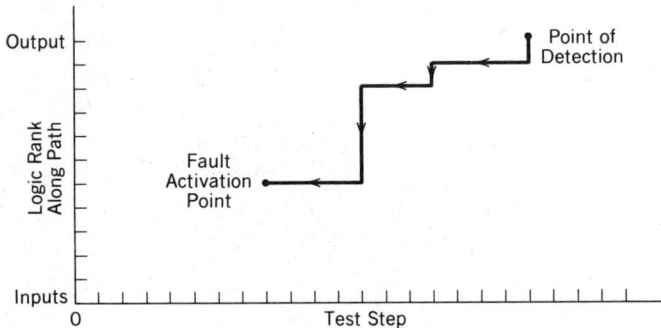

Figure 9.4. A graph of troubleshooting progress in the two dimensions of testing.

As the elements that make up the loop are identified, each is examined as a potential original data source for the loop. In Figure 9.5 the loop, which has been separated from the test circuit, was found to contain an error when the left input of the OR gate (A) became a 0 and sensitized a path to an output.

The loop in Figure 9.5 has numerous opportunities for data entry. In fact, only the inverter gate (D) can be ruled out as a candidate. This elimination does not eliminate D from suspicion of having a fault, only from the list of entry points for effects from outside faults. Each of the candidate entry points is considered as a possibility, and a search backward through the expected state table is made to find the latest previous condition which broke the loop at any of the points. The inputs to the gate which was responsible for closing the loop the last time are then examined for the test step just prior to the closure.

In the example the loop is identified during a test step in which the states shown exist for each of its nodes (loop values are circled). The temporal backtrace shows that the last time the loop was intentionally broken, the far-left-hand input of gate C was intended to be a 0. Checking that value against the actual circuit may reveal that the CUT displayed a 1 for that node during that time frame, in which case the fault effect was entered into the loop at the time and can be traced back from the point thus located. Alternatively, the inputs can be discovered to agree, leaving the question of a faulty loop or a later corruption.

A simple traversal of the loop at the test step which has been found to be the latest open loop condition will suffice to discover any loop fault. If the state of each node of the loop agrees with the expected value, the loop must have been corrupted between the entry of data and its later observation. Any point in the loop can now be observed via a probe while stepping the test forward in time until a change in value indicates the time of corruption.

The loop must then be circled once more, checking all inputs to each gate in

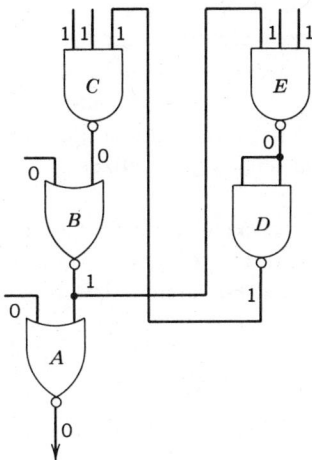

Figure 9.5. Backtracing example circuit with a feedback loop.

the loop to discover the fault effect that resulted in corruption. Backtracing by the algorithm can then resume from this newly discovered fault effect. Loop breaking is an arduous task not easily automated.

AUTOMATED DIAGNOSTIC TECHNIQUES

The examples and procedures given were reduced to the simplest working level. Sophistication aimed at shortening the troubleshooting process or generalizing it to include dynamic- or mixed-level-logic structures was ignored to present as clear a picture as possible of the tasks involved. Nevertheless, the routine can be seen to be lengthy and involved, particularly when loops or sequential elements become involved. The cost of troubleshooting complex boards is sufficient to attract many attempts to automate part of all of the process.

Basic logic tracing is only a part of the problem. After the fault is localized, automated techniques are needed to examine the potential physical faults and evaluate their effects in light of the observations. Only when functional and physical data are brought together can the problem be identified. Tools and techniques for achieving all or part of the objectives will be discussed.

Tracing Automation

Automating the simple algorithm first presented would seem to be easy. The computer which drives a particular ATE could certainly be given a description of the logic circuit in terms of gates and connections from which to work. The information concerning the present state of the CUT internal nodes must be obtained, and a file of corresponding expected states for each test step must be stored and available for comparison.

The technician performs several subtle tasks using commonsense rules along the way, and these must be included in the automation to successfully troubleshoot without intervention. If the technician makes all the measurements using a voltmeter, a commonsense judgment is made at each measurement of a node voltage as to whether the voltage represents a logic 0, a logic 1, or some other value such as a floating 3-state level, ground, or supply voltage. If the measurement does not "make sense" because the probe has slipped off, the technician realizes it and reprobes the point.

Front-end "automation" of the logic probing task has been available for years. Logic probes and logic clips sense the voltage on a node or the voltages on all pins of an IC individually and compare the potentials to fixed thresholds. The result can be read from a light or set of lights. A typical probe lights a green light if the value is less than the logic 0 threshold and a red light if the value is more than a 1 threshold. If neither light is lit, the voltage is between the thresholds, indicating an open circuit due to bus logic in the high-impedance state or a misprobe failing to make contact.

Advanced logic probes can detect non-steady-state signals. If both lights light, for example, the signal is oscillating, and if either light blinks, the signal is a pulse. Pulse-latching circuits can be installed to capture fast pulses. The most sophisticated systems are logic analyzers, which allow the sampling of multiple logic probes to be synchronized to an external system of clocks from the circuit. Logic analyzers sense and compare the signals to thresholds as do logic probes, but the results are recorded in memory for each sample time and displayed as digits on a CRT screen.

Logic probes and logic analyzers seldom have sufficient resolution or enough output states to detect and report the difference between proper logic levels and a ground or power state. At the end of the backtrace a measurement must still be taken to determine if the logic signal is in the incorrect state because of an internal chip fault (usually resulting in 0.1–0.6 V) or because it is shorted to ground (less than 0.1 V). The speed and ease that logic probes lend to backtracing has led to their nearly universal use for most of the process.

Guided Probes. Adding the adjective ''guided'' to the probe device denotes automation. The actual probe is usually a logic-sensing device similar to one of the logic probes already mentioned. Guidance is provided by a computer working with two databases as shown in Figure 9.6. One of the databases is the circuit topology which identifies components and their interconnection to support spatial backtracing. The other database contains the state of each node which might be probed for each test step. Since the primary inputs and outputs might just as well be included in this database, it may be shared with the ATE, which would extract the input and output states to use as test vectors.

In fact, the guided probe computer and the ATE control computer may be

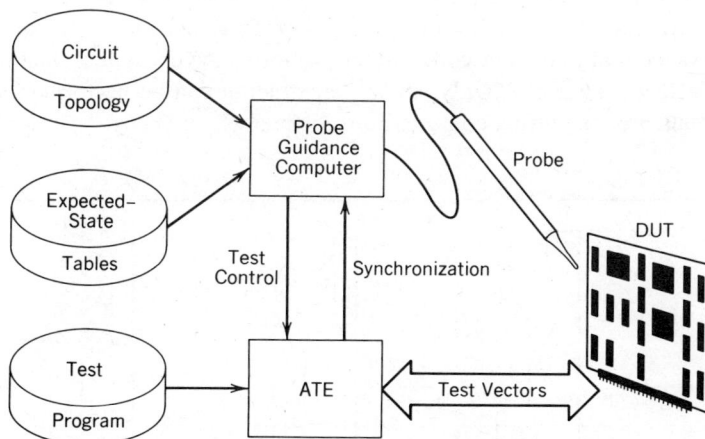

Figure 9.6. Guided probe system diagram includes several sources of information and the ATE.

merged. If they are not one and the same, a communications link must be established between them to synchronize the guided probe state sampling with the test vector application time and to allow the guided probe to request reapplication of the entire test set. The latter situation is necessary since the test cannot back up in time as would be desired by temporal backtracing and since guided probes only record a node or a few nodes for each pass.

When the ATE and guided probe are intended to be portable, the size of the databases becomes important. The topology database is not usually large and not very reducible in any case. Data reduction is quite viable on the state table database, however, A simple technique records only the nodal states that change at any given step. A further reduction uses the LFSR techniques.

Each data strobe clocks all the flip-flops in the register. Actual registers have more flip-flops in the line than shown in Figure 9.7, but the principles hold. The logic probe input is on the left-hand side and is clocked into the first flip-flop either inverted or noninverted, depending on the state of the last flip-flop in the line. The exclusive-OR feedback connections are configured by completing the feedback or ignoring it to make a unique polynomial generator. Each stage of the register is an increasing power of 2, and the connections correspond to a binary coefficient of the term (either 0 for no connection or 1 for a connection). The resultant device divides the incoming serial stream of data by the fixed polynomial, and the answer is read in parallel from the flip-flops when the stream is finished.

In most applications the data for the LFSR guided probe state tables are merely learned by probing each node of a good circuit and running the test end to end. Each node creates a unique answer from the LFSR, which is recorded and represents an entire column from the usually bulky state tables. During backtracing, the node of a suspect circuit is probed and tested in the same manner, and the LFSR output is compared with the recorded value for that node to determine correctness.

Although the LFSR technique (introduced as "signature analysis" by Hewlett-Packard Corporation) requires only a small storage space for "signatures" (four or five hexadecimal digits per node) which comprise the entire test state database, there are a few drawbacks. Only spatial backtracing can be performed since the temporal data are lost during compression. Signals which change at different times

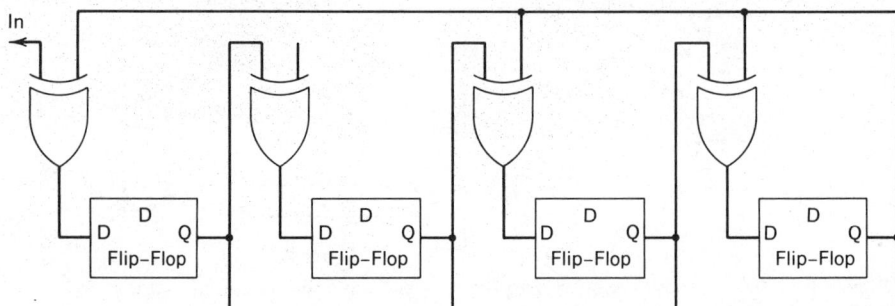

Figure 9.7. LFSR configured to calculate a "signature" of the incoming serial bit stream.

create different signatures, but since only the end result is stored, there is no information as to when the signal first became different than the expected signal. The value of the LFSR is also lost if the good circuit node being recorded has not been initialized or becomes unknown during the test.

Careful construction of the test vectors and the ability to disable the probe recording if the node state is suspected to be uninitialized can give reasonable results with this technique, but applications which allow greater data storage can benefit from the more precise information.

A few minor changes are needed in the algorithm given earlier to fully automate it. The changes install a consistency check operation, which is a branch based on agreement between the value sensed at the input to a device and the value sensed at the driver for that particular node [1]. If the driver of a node is good, but an erroneous value is sensed at a gate input which is driven by the node, either the wire is broken or one of the readings is in error. The branch fails if the two values are different, and the input pin is reprobed. Reaffirmation of the situation results in a message asserting that the wire is broken. Failure to reaffirm the input pin value as faulty causes the algorithm to fall back another step and recheck the remainder of the chip's inputs and outputs. If the fallback procedure results in returning to an output pin, an inconsistent data error has occurred.

There are many slightly different implementations of the guiding algorithm with augmentation designed to overcome one or more of the more difficult troubleshooting situations. Typical shortcomings of basic algorithm guided probes are as follows:

1. Lack of Loop-Breaking Capability. If a loop is detected by the encounter of a node twice in the sequence, manual intervention is required. Test sequences must be rearranged during generation to avoid activating loops if basic algorithm-guided probes are to be used.

2. Inability to Resolve Bus Structure Faults. Manual intervention includes modification of the CUT by temporarily disconnecting drivers (unsoldering an IC lead) or by the use of secondary diagnostic tools such as current-measuring device.

3. Static logic limitations. Pulses which occur within a single test step may not be detected or may not be properly interpreted by the algorithm.

4. Two-State Logic Limitations. Differentiation between ground level and a logic 0 or the inability to sense or interpret the off state of a bus forces manual intervention during the final stages of fault tracking and contributes to the bus resolution problem.

The requirements for a particular application may change the emphasis on the need for advanced features, but this list should serve as a guide in evaluating the capability of "guided probes."

Effect–Cause Analysis. Although the basic guided probe algorithm and its minor variants have been used extensively, major improvements to its efficiency can be made. Effect–cause analysis [2, 3] is a powerful method which uses deduction of the internal states of the logic circuit to replace the storage of predetermined expected state tables or signatures. Test vectors are still required for the ATE to excite the CUT and primary outputs must be supplied for the go–nogo portion of the test, but the derivation of internal states is unnecessary.

The algorithm analyzes the values obtained by probing the CUT, comparing them only to the truth tables of the circuit components in search of inconsistencies. Failure of a primary output begins the analysis. The primary inputs and the circuit topology information are used to deduce the state of internal nodes at this particular test step using an extension of the deductive algorithm for fault simulation. As the internal states are deduced they are compared with the actual results of the test for consistency. At the end of the process, only those fault effects which match the primary output failure conditions remain, and their fault causes are read from a list kept along the way.

No *a priori* expected state tables are required, and the calculation time, although greater by far than that required by the basic guided probe, is considerably less than full fault simulation for the test step since only fault conditions which match the actual results are simulated. This scheme has been successfully applied to both combinational and sequential circuits with the addition of a few backtracing rules.

Expert Systems

Algorithmic improvements such as the effect–cause analysis address part of the problems in simple guided probes, but many of the problems are not embodied in the algorithm per se. That is, lack of information about the circuit and inability to incorporate commonsense knowledge is not within the realm of algorithmic improvement. The rise of computer programs which operate in unconventional ways to allow the relating of facts and rules in an approximation of the human thought process has provided new tools. AI and, in particular, the subset called "expert systems" can be adapted to the diagnostic problem.

The concept of an expert system involves understanding and codifying the method used by a human expert to solve a problem. Many levels of information are assembled in a relational database called a knowledge base, and a set of rules or routines for extracting facts from this knowledge base in response to external questions or stimuli is built in a computer. Usually, a unique architecture computer called a symbolic processor is used for an expert system since the facilities for knowledge base searching and structure have been enhanced in these machines.

The knowledge base for a diagnostic expert system contains specific information about the circuit topology and general information about the function of devices and the nature of possible faults. The ability to cope with diverse facts and the flexibility to improve on its knowledge by adding relationships or facts in the same way a human technician learns is key to solving complex troubleshooting problems [4].

A network of links, as seen in Figure 9.8, connects items representing individual instantiations of ICs in the knowledge base in the same way the circuit paths on the PC connect them. A second network of the characteristics of each type of IC (flip-flop, NAD, etc.) is linked temporarily at run time with the first. The specific knowledge about which device is the input to which other device is thus connected to the deeper knowledge about what happens when the input of a flip-flop (or any device) undergoes a logic transition. Further knowledge bases about fault types and part histories can provide information to a much greater depth than that available to most guided probe algorithms. Each item block can contain its own tree of details such as the stock number and the equivalent vendor part numbers for a component referred to by a model label.

Operations are performed according to routines, sometimes called methods, or by rules which invoke a search of the knowledge base in an attempt to satisfy a goal. Although the format of the rules is dependent on the language and the expert system used, most can be expressed in the IF. . .THEN. . . format.

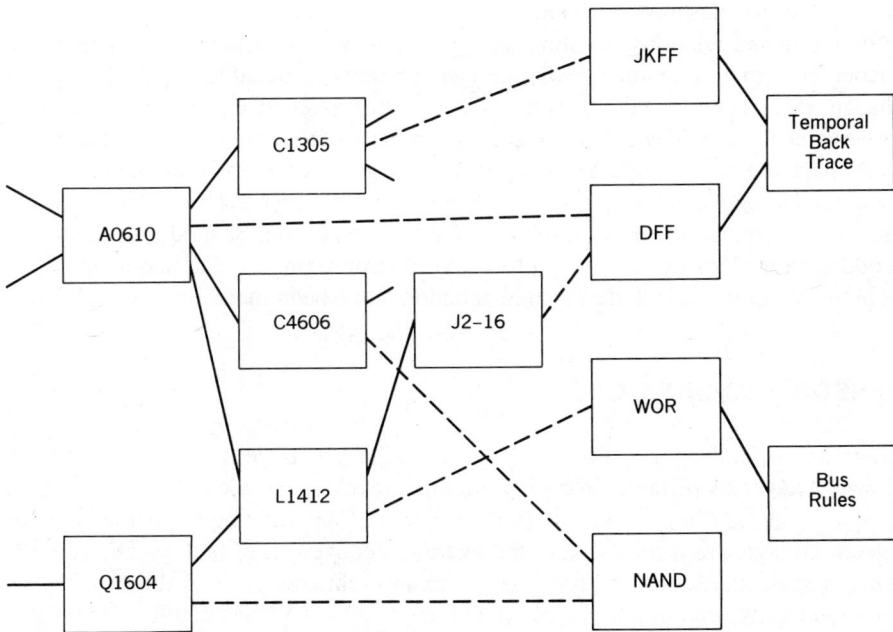

Figure 9.8. Knowledge base structure for an expert system probe. Dotted lines are run-time linkages which relate general knowledge to the CUT.

```
IF (failed_output = '0' AND drive_device =
'D-FF') THEN
     Check (drive_device.CLR),
     Timeback
     (drive_device.CLK,Up_transition),
     Check (drive_device.DATA)
ELSE
     IF (Measure (failed_out) < 0.1 v)
     THEN
          Output (''shorted net'',
          failed_output)
     ELSE Output (''failed chip'',
     drive_device)
     END;
END.
```

The rules are derived from the procedure followed by a test engineer or a technician in the process of troubleshooting actual circuits. Achieving a thorough yet compact set of rules is the heart of the process (called knowledge engineering). Careful structuring of the knowledge base is imperative for efficient operation since solving the troubleshooting problem involves repeated access and traversal of the information and relationships stored within. As new logic structures or devices are encountered, new rules or knowledge base entries can be added to deal with the change. This single aspect is the most powerful asset to expert system guided probes since it provides an easy way to keep the system current with technology.

Perhaps the next natural automation step for guided probes is the adaptation of robot technology to physically manipulate the probe. Since most circuit structures are planar, a linear or cartesian robot would be appropriate. Several manufacturers produce manually directed servo probes for microprobing of ICs, and adaptations of automatically controlled x–y table actuators have been made in prototype [6].

SENSOR TECHNOLOGY

There are several options for impoving the logic backtrace process, as has been shown. Diagnosis of faults in digital circuits includes a physical aspect as well as a logical one, and improvements to the process of determining the actual state of the circuit node are just as vital in the overall effectiveness of fault location. Problems in logic state sensing arise due to circuit architecture, as in the case of bus fault ambiguity, and due to the physical dimensions of VLSI circuitry. Hand-held voltmeter leads are of little use to a VLSI chip test technician. Economic pressure also forces a search for automated replacements for the technician's eyes and sense of touch.

Current Probes

Use of bus-oriented architectures elevates the importance of solving the bus structure ambiguity problem encountered during logic backtrace. Consider as an example the circuit of Figure 9.9, which shows one of the data lines on a multidriver bus which fans out to several gates. The output of gate A01 is s-a-0 due to an internal fault which has enabled the lower transistor of the output circuit.

The test sequence for the circuit containing this bus has stopped with E2 = 0 and all other enables at a logic 1. The data on the bus should be a 1 and have been enabled to an output through gate Q06, and the logic backtracing algorithm has led to the bus through Q06. Since the bus is merely a wire, the faulty value of 0 appears everywhere, and it is impossible to tell which output is causing the error with only a logic-sensing probe.

A test of the circuit which should be driving the bus according to the expected state table shows that all of the inputs to A02 are correct, but its output is in error. Unless a secondary procedure is invoked, the component A02 will be replaced, only to reveal that the problem has not gone away. The failure assumed incorrectly is an internal s-a-0 at the output of A02. For TTL logic this would suggest that A02 is sinking current when it should be sourcing current. A useful diagnostic technique measures the current in each portion of the bus circuit rather than the voltage.

Schematically, the output drivers and input transistors of the circuit appear as in Figure 9.10. The currents which should be flowing are I_{OH} from gate A02 and the three I_{IH} input currents from Q06, F08, and N16, which by Kirchhoff's law must sum to the negative of the A02 current. Since all the inputs should be reverse biased, the currents are on the order of a few hundred microamperes.

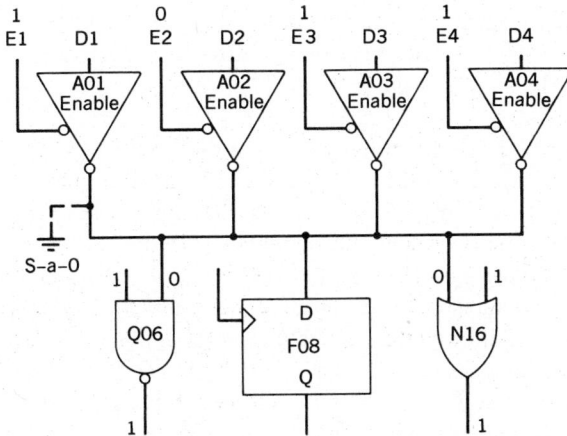

Figure 9.9. Schematic of one bit from a data bus. A short at the output of A01 creates an s-a-0 fault.

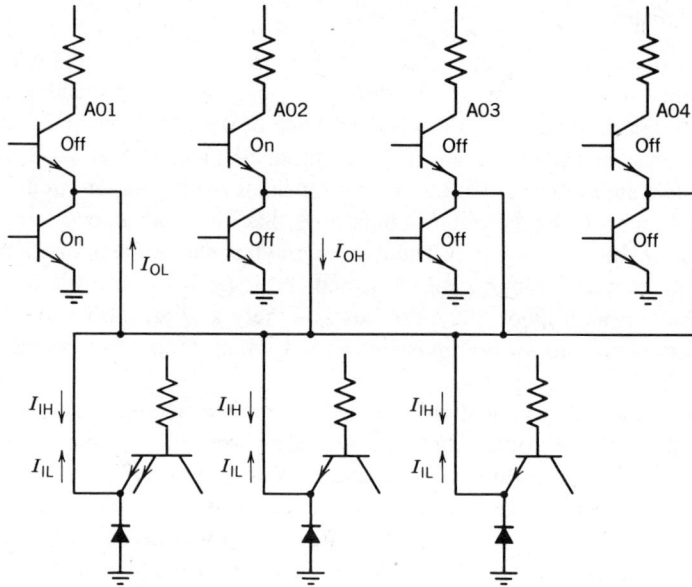

Figure 9.10. Transistor-level schematic of the circuit in Figure 9.9. Several current sources contribute to a current sink.

Due to the stuck transistor in A01 the actual currents flow out of the three driven gate inputs with a total I_{IL} of a few milliamperes. The current from A02 is still flowing in the same direction, although it is considerably higher since the bus voltage is at about 0.6 V rather than at the intended 4.2 V. The erroneously turned on gate A01 sink the total current from all of these sources. A simple diagnostic procedure of measuring the current in each component lead associated with the bus will reveal that A01 has a current flow of several milliamperes when it should be off.

Hall Effect Sensors. The remaining problem is the measurement of currents without destroying the pc board to insert ammeters. Two techniques have been developed for printed circuit boards. The first uses a Hall effect magnetic sensor which senses the magnetic field around the conductors of the pc board or around the device pins themselves. The Hall effect sensor (Figure 9.11) is a semiconductor slab to which four contacts have been orthogonally attached around the edges [5].

A current is passed through the slab between two opposed contacts, establishing a flow of carriers within the slab. If a magnetic field is present, the velocity of the electrons in the magnetic field will create a force which will turn them toward the side. The net result will be the accumulation of a voltage across the slab which is proportional to the current and the magnetic field. With a constant excitation current the magnetic field is the only independent variable, and the voltage can therefore measure the magnetic field and thus the current that created the field.

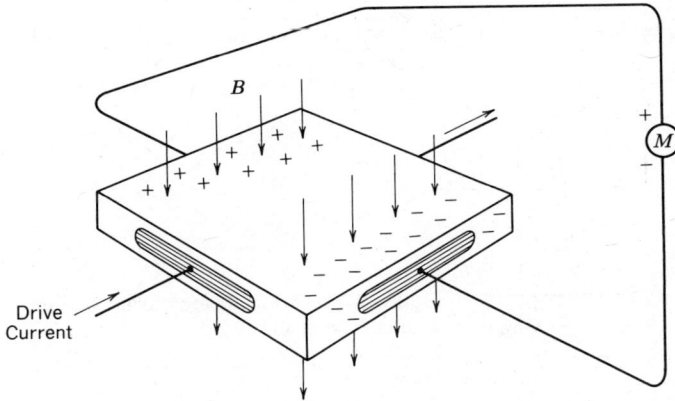

Figure 9.11. Hall effect sensor formed by passing a current through a slab of semiconductor. Electron drift curvature creates a voltage proportional to the magnetic field.

The slab of silicon can actually be made quite small and can be placed in a hand-held probe tip. Although the Hall effect probe is convenient and in some cases quite useful, it is not precise since the magnetic fields that must be sensed vary inversely to the distance from the conductor to the sensor. Placing the probe the same distance from each conductor to be measured can be difficult in some cases. Crowded board layouts also make it difficult to find a conductor in the desired path which is accessible but sufficiently isolated from other currents' magnetic fields to avoid confusion.

Current Bucking. The second popular current sensing system uses two physical probes and relies on the electrical principle of superposition. In accordance with Ohm's law there is a small voltage drop in any portion of any current path. The current bucking probes use that voltage drop indirectly to measure the current.

The probe mechanism is a current source controlled by a sensitive voltmeter. As the probes make contact with a current-carrying circuit path as shown in Figure 9.12, the voltmeter (actually a differential operational amplifier) detects the voltage drop and turns on the current source. The voltage sensed is

$$V_{\text{LOGIC}} = I_{\text{LOGIC}} R_{\text{WIRE}}$$

and the current from the probe will create another voltage drop

$$V_{\text{PROBE}} = I_{\text{PROBE}} R_{\text{WIRE}}.$$

By superposition the net voltage drop for the section of wire will be

$$V_{\text{NET}} = (I_{\text{PROBE}} + I_{\text{LOGIC}}) R_{\text{WIRE}}.$$

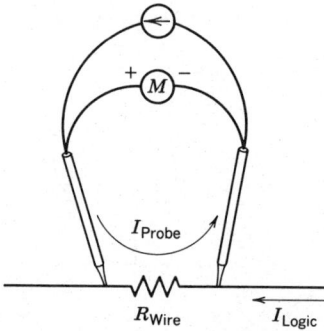

Figure 9.12. A voltage-controlled current source used as a current probe.

The operational amplifier (voltmeter) in the current probe actually senses the net voltage, of course. Feedback from the voltmeter to the current source is arranged to seek a zero voltage reading, which implies that the current injected by the current source will quickly settle to a value equal in magnitude to and opposite in sign from that of the logic signal on the path. Note that the result is completely insensitive to the actual resistance of the path, except that the resistance must create a drop that can be sensed by the operational amplifier in the probe mechanism.

Using the current-bucking probe requires placing two hand-held probes on the current path and observing a readout or a meter needle which shows the relative strength of the current. Either current probe method provides a tool for resolving faults on bus structures.

Microprobes

In the early 1960s the traditional voltmeter probe became obsolete in some applications as IC technology introduced 100-μm geometries for electronic circuits. Resourceful designers produced the "microprobe," which consisted of a fine wire needle with a tip a few micrometers wide. Now the move to submicrometer geometries has obsoleted needles. The replacements do not actually make physical contact with the target on a chip. Noncontact probing avoids the damage which might be inflicted by what could be described as the collision of the microprobe and the IC.

Electron Beam Probing. Scanning electron microscopy (SEM) is a technology that has been used to examine the structure of ICs for some time. Its use has been limited to static imaging for the purpose of failure analysis and scientific reseach until recently. Two modes of measurement that make use of the energy differences in secondary electrons freed from the surface by the impact of the primary beam have become useful tools for voltage measurement in a real-time operating mode of ICs [7].

The electron beam prober is essentially an electron microscope, consisting of an evacuated column with an electron gun and beam control system (Figure 9.13).

Electron
Gun

Electronic
Shutter

Deflection
Electrodes

Focus

Detector

Target

Stage

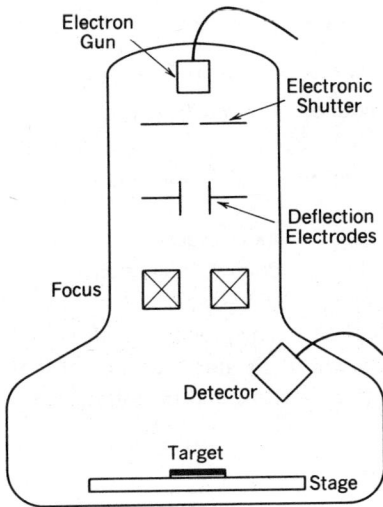

Figure 9.13. Electron microscope diagram with beam scanning and secondary electron detector.

Beam control includes focusing, deflection, and for one of the modes of use, a fast electrostatic shutter. In the single-point probe mode, the beam deflection is adjusted to illuminate the node of interest, and the shutter timing is used to select the test step of interest. The test is run, and the beam strikes the spot during the correct step. Secondary electrons freed by the beam have a potential which is proportional to the beam voltage, the material struck, and the voltage above system ground of the target. Since the first two are constants, the secondary electron energy is a measure of the voltage potential of the node. A detector measures the energy of the secondary electrons, and the control computer plots the energy as a function of time to give an oscilloscope-like waveform image of the potential at the node.

Using the shutter as a stroboscope in the second mode of operation can produce a "stop action picture" of the entire IC surface. As the test sequence is repeated continuously the shutter is opened only during the proper sampling time in the same test step of each pass. A 100-ps strobe can be achieved. After each test pass the beam is moved ahead in a raster scan fashion. The secondary electron energies from each strobe pulse are recorded by the computer and, after the scan is complete, used to control brightness on a CRT display which mimics the scan pattern of the beam. A gray-scale image of the voltages across the IC for the instant of time selected in the result.

Although the electron beam probing technique is fascinating and very useful in IC analysis, it is expensive and cumbersome as a production troubleshooting tool. The necessity of working in a vacuum makes preparation slow.

Laser Probing. An alternative beam probing technique uses a laser beam to illuminate a 1.8-μm spot on the IC's surface [8]. The beam power is low enough to avoid damage to the surface, but high enough to excite a measurable photocurrent in the area. A monitor of the power line for the chip can detect the current by comparing the power supply current to an unilluminated pass through the same test conditions. Furthermore, the photocurrent induced is proportional to the surface voltage potential, and the variations caused by that potential can be sorted out of the power line waveform. The result is a reading of the potential of the point probed at any time in the test.

Since the laser light cannot be easily deflected for spot selection, the IC is mounted on a high resolution x–y table controlled by a computer. Interference from the atmosphere is negligible at such a short range, so that laser prober can be operated without the vacuum chamber of the electron beam prober. Setup is fast enough with the aid of the attached camera and monitor for initial alignment that the laser probe can be used in a production test mode for test and troubleshooting ICs.

ADAPTIVE INSPECTION

When the logic tracing and probing techniques have located the improper voltage on a node, there remains the task of associating the phenomenon with a physical defect. Assuming that a component is always responsible for the defect will create a considerable amount of false rework. The technician uses the optical and tactile senses to check for trouble in the vicinity of the node showing improper voltage. A hot component is a reasonable sign of an internal short, and a reversed component or a solder splash can be found by comparing the appearance of the circuit to a known good circuit or a picture of a known good circuit.

Optical Inspection

Optical technology is still a long way from emulating the eye–brain system of the human technician, but dozens of manufacturers have successfully designed and produced optical inspections systems which are nonetheless very useful. Resolutions of less than a mil allow adequate inspection of circuit board traces, and the speed with which simple good circuit comparisons can be made is sufficient to replace some of the in-line inspection during the manufacturing process.

Pattern comparison and design-rule-checking optical inspection systems usually consist of a camera (either conventional vidicon or charge coupled device arrays), an illumination system, and an x–y table for positioning or scanning the circuit board beneath the camera. The comparison is usually performed by a specialized signal processor against a stored, simplified image of a good circuit.

Inspection of populated pc boards or hybrid chips requires a higher level of processing. The exact image of one populated board seldom matches that of the next due to the irregular nature of component markings and tolerance in component placement. A pattern recognition and decision-making scheme in the control computer must be called upon to make the pass/fail determination [9]. This higher level of processing is becoming available with the aid of research under the AI–machine-vision title.

Linking the optical inspection system to the guided probe controller could overcome some of the processing time restrictions by narrowing the range to be inspected. Close range inspection of solder joints and circuit paths in addition to component placement checks promises to reduce the human error and speed the process of pinpoint troubleshooting. Adding depth perception [10] will help deal with problems such as components that are only partially inserted in sockets. Research is underway to develop fast techniques of full color, stereoscopic vision with object recognition that could rival that of the human inspector (at least on a Monday morning).

Thermographic Inspection

Just as the technician can find a shorted component by feeling for the hot spot on the board, the thermograph can detect many faults by their temperature signature. Thermography is an area of sensing that has surpassed the human capability, as temperature differences of a few tenths of a degree centigrade can be resolved over distances of a millimeter [11]. A thermograph of a CUT not only shows what component is hot, but it shows which corner of the component is the hottest!

Thermographic equipment has become plentiful and capable [12]. Sensitivity is sufficient to detect and precisely locate a circuit trace short to ground without requiring the application of damaging high current. High-resistance pc paths can also be detected by comparing the board signature to a KGB. Difficult-to-inspect areas such as circuit paths under components can be detected as the infrared radiation passes through layers of pc board and packaging material.

Adaptive control of thermosensing and thermographic equipment can allow the guided probe computer to locate cracked solder joints and other difficult-to-pinpoint faults. The thermal imaging process is somewhat slow due to the necessity of thermally stabilizing the CUT test before a reliable image can be obtained, but its help in locating the most difficult of faults is valuable enough to suggest that thermography will become a commonplace diagnostic tool.

SUMMARY

Performing a digital go–nogo test is only the beginning for most products. Analyzing the failure symptoms and troubleshooting the fault is essential to pinpoint the repair needed. Troubleshooting has traditionally been a skilled human technician's task, but recent developments are pushing toward the total automation of locating the fault as well as detecting its presence.

Test generation methods impact the fault isolation methods by providing information about the internal expected states of the DUT logic. Conversely, the goal of being able to locate faults to an acceptable diagnosic resolution influences not only the test generation method but the structure and built-in test attributes of the circuit design. Circuit partitioning and nodal observability are critical to the ability of ATE tests to identify the failing node.

Using the information provided by fault simulation or KGB recording requires semiintelligent algorithms known as guided probe controllers. Advanced guided probe mechanisms provide step-by-step probing instructions to guide a technician or a robot in measuring the logic states along the fault propagation path in reverse, from the failing output to the fault site. This backtracing involves solving ambiguous structures such as loops and buses to determine the source of the fault effects.

Upon finding the fault site the task turns to recognizing the physical cause of the fault. Automation in this realm includes visual and thermal inspection. Complete automation of the diagnostic task is now theoretically possible, but the integration of all the diagnostic tools such as current probes, waveform analyzers, laser probes, and thermographic imagers requires a capable computer system and complex programming. Eventual reduction of the human involvement in troubleshooting is inevitable due to the cost and inconsistency factors. Expert systems show great promise for fulfilling this function due to their abilities to emulate the expert technician and to learn.

REFERENCES

1. M. Namgostar, *Digital Equipment Troubleshooting*, Reston, Reston, VA, 1977, p. 222.

2. M. Abramovici and M. A. Breuer, "Multiple Fault Diagnosis in Combinational Circuits Based on Effect–Cause Analysis," *IEEE Trans. Comput.*, C-29(6), 451–460 (June 1980).

3. M. Abramovici and M. A. Breuer, "Fault Diagnosis in Sequential Circuits Based on Effect–Cause Analysis," *IEEE Trans. Comput.*, C-31(12), 1165–1172 (Dec. 1982).

4. L. Apfelbaum, "An Expert System for In-Circuit Fault Diagnosis," in *Proc. 1985 Int. Test Conf.*, IEEE Comput. Soc., pp. 868–874.

5. J. P. McKelvey, *Solid State and Semiconductor Physics*, Harper and Row, New York, 1966, p. 199.

6. T. Moore and S. Garner, "Auto-Probing on the L200 Functional Tester," in *Proc. 1985 Int. Test Conf.*, IEEE Comput. Soc., pp. 618–628.

7. Y. Goto, K. Ozaki, T. Ishizuka, A. Ito, Y. Furakawa, and T. Inagaki, "Electron Beam Prober for LSI Testing with 100 ps Time Resolution," in *Proc. 1984 Int. Test Conf.*, IEEE Comput. Soc., pp. 543–549.

8. F. J. Henley, "An Automated Laser Prober to Determine VLSI Internal Node Logic States," in *Proc. 1984 Int. Test Conf.*, IEEE Comput. Soc., pp. 536–542.

9. S. J. Jones, "Flexible Inspection System (FIS) for Printed Circuit Board Production," in *Proc. 1985 Int. Test Conf.*, IEEE Comput. Soc., pp. 403–412.

10. S. F. Scheiber, "Machine Vision Moves into Three Dimensions," *Test Measurement World*, 6(4) 60–71 (Apr. 1986).

11. H. Boulton, "New Concepts of Applying Thermographic Testing to Printed Circuit Boards and Finished Products," in *Proc. 1985 Int. Test Conf.*, IEEE Comput. Soc., pp. 413–418.

12. H. Boulton, "Infrared/Temperature Tester Survey," *Test Measurement World*, 6(4), 81–95 (Apr. 1986).

CHAPTER TEN

MEMORY TESTING

Deriving an adequate test for a block of memory is a task which appears deceptively simple. A black box description of a memory block only specifies an array of storage for digital data. The array may be permanent as in the case of ROMs or temporary as in RAMs. The misnomer of RAM for a memory which may be both written into and read from is continued here due to common practice. Actually, both RAMs and ROMs are random access memories since any piece of data can be accessed with the same delay or access time. A black-box-level test might simply write both 1s and 0s into each storage location and read them out again to verify operation.

Such a straightforward approach turns out to be ineffective for many classes of faults which commonly occur in memories and inefficient for much of the fault class that it detects. Effectiveness and efficiency are both critical issues in memory testing due to the geometrical nature of some of the faults which occur. An analysis of the structure of memories and the faults which are important in memory applications will reveal a set of tests which may be applied in accordance with testing goals. No single test suffices in the manner with which logic elements may be characterized.

When selecting test schemes and algorithms for memory testing, two goals guide the results. For any given fault class it is desirable to find the shortest test sequence which will detect the faults of the class. Growth in the size of memory structures continues to apply pressure to this goal. If an algorithm requires 10 test vectors per address, testing a 1K memory will require 10,000 steps. At a speed of 1 MHz the test will take only 10 ms to complete. As larger memories are used the same algorithm can be applied, but a 64K memory array requires 0.64 s to test. A 256K memory will need 2.56 s, and a 1M memory needs 10 s to be tested.

An algorithm which uses only 10 steps per address is a short one, as will be seen, and yet the test time is measured in seconds. Production line tests must test hundreds of thousands or millions of memories a day, and a test time of 10 s limits the throughput of the tester to about 7500 memories per day, allowing only a few seconds to change from one DUT to the next. Shortening the test time means less test equipment and labor are required and the product can be produced more cheaply.

Since there are several fault types of interest in memories, the second goal is to detect as many faults of the different classes as possible with each algorithm. Writing all 1s into a RAM, reading them, and then repeating the process with 0s detects simple stuck-at faults for the memory cells, but it does not cover addressing faults or adjacent data lines shorts. Using a checkerboard pattern for data solves the adjacent data line short problem, but does not cover the address faults.

STRUCTURES AND FAULT MODELS

Memory blocks may consist of a single ICs, many ICs interconnected on a pc board or substrate, or subcircuits of a complex IC such as a microcomputer. In order to clarify purely memory-related issues, memory block testing will be approached generically as though the arbitrary memory were a stand-alone device before the implications of embedded memories are discussed.

Block Diagram Representations

Closer definition of the memory model reveals new possibilities for faults and better definition of what fault effects may be expected. Zooming in from a black box functional model to a logic equivalent model gives a block diagram as shown in Figure 10.1. Each bit of data in the memory array is considered to be stored in a separate latch arranged in rows and columns. All the latches in a single row can be accessed simultaneously, resulting in the appearance of a data word on the column lines. The row selected is determined by the decode of an address imposed on $\log_2 n$ input lines where n is the number of rows.

Configuration of the matrix in the model is not critical. There can be one or more columns (usually a power of 2). The number of rows is a power of 2 leading to a binary encoded address. Early memory devices had only a few addresses or a few hundred addresses, resulting in a simple combinational address decoder. The stuck-at faults associated with these decoders are the same as those for any combinational decoder. A stuck input line to the decoder (address line input to the memory) will disable access to half the memory.

The effect at the output of the memory is not a simple stuck value. Consider the case of the least significant address bit s-a-0. When address (0) is accessed, the correct data will be found since the least bit is a 0 in the normal memory. If address (1) is accessed, however, the stuck bit will override the least significant 1 and the data in address (0) will be accessed again, instead of that in (1). Address (2) will be correct, (3) will actually access (2), and so on. The apparent failure will be a mirroring of half of the memory onto the other half.

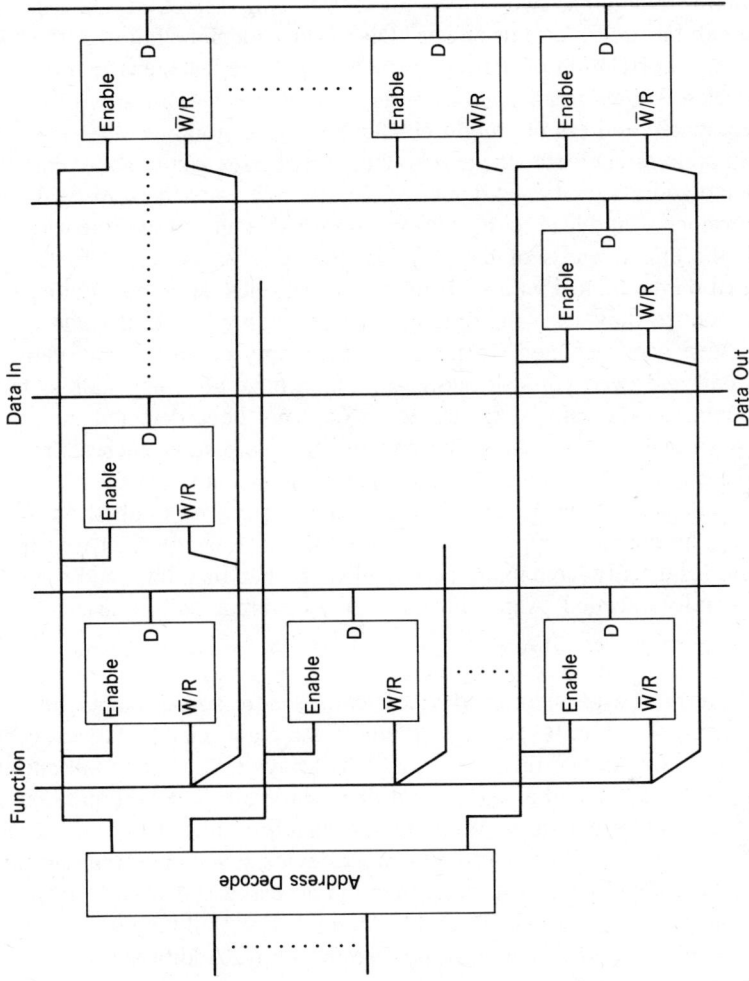

Figure 10.1. Memory block diagram with cells represented by single-bit latches.

235

The simple example has two points. First, even single stuck-at faults in memories often exhibit multiple fault effects. Diagnosing memory failures, therefore, usually requires correlation of the results of two or more test steps. If the data in address (1) were known in advance as in the case of a ROM, a test which simply read and compared address (1) might conclude that the storage cell(s) associated with data in address (1) were stuck. A few more accesses would show that every odd-numbered address contained defective storage cells, and the statistical oddity of this occurrence would lead to the better diagnosis that an address line was stuck.

Second, the effectiveness of memory tests depends on the data stored in the cells, not just on the logical values given the control lines. Had the memory in the example contained only 1s at the time of the test reading, the faulty address line would not have been detected. In fact, if the memory contained any pattern in which the odd and even pairs of addresses which have all other address bits in common contained the same data, the test would not have detected the fault in question. A simple data pattern which contains unique data in each address will suffice if the memory has more data lines than address lines, but the pattern will contain redundancies if the reverse is true. Such a pattern is called an address pattern since the common approach is to write the address of each row into its cells. Memories usually have many more addresses than data bits, and a different routine is therefore needed. Algorithms and their coverage will be discussed later in the chapter.

Address Decode. Large memory devices require more address input lines than are convenient to fit on an IC package. Memory structures changed twice as array sizes grew to accommodate this limitation. The arrays first became tall and thin, using only one data line and a few control lines while all remaining package pins were devoted to address signals. When the package pin limit was again reached, partial addresses were latched into the memory device at separate times, reassembled within, and then decoded to determine the accessed cells. The two parts of the address are called "row select" and "column select," although they may not represent the actual internal layout of the memory array. Additional opportunities for faults exist in the control lines which sequence the address acquisition process and in the latches which hold the partial address segments.

In the memory implementation depicted in Figure 10.2, the "row address select" or RAS strobe latches 8 bits of address and the "column address select" or CAS latches 8 more bits. The 16 bits of address are then rearranged internally to control access to a 4096-row, 16-column array. A single data line is input to the array through a demultiplexer and output from the array through a multiplexer.

Possible faults on the RAS and CAS lines are obviously different for multiplexed address memories in that the lines do not exist in direct address versions. The latch structures themselves are transparent from a stuck-at viewpoint since stuck-at faults on latch inputs and outputs are equivalent to address bit, row, or column stuck-at failures. Timing faults are added in this configuration since the relationship of address information and the RAS–CAS signals is critical to the proper function of the memory.

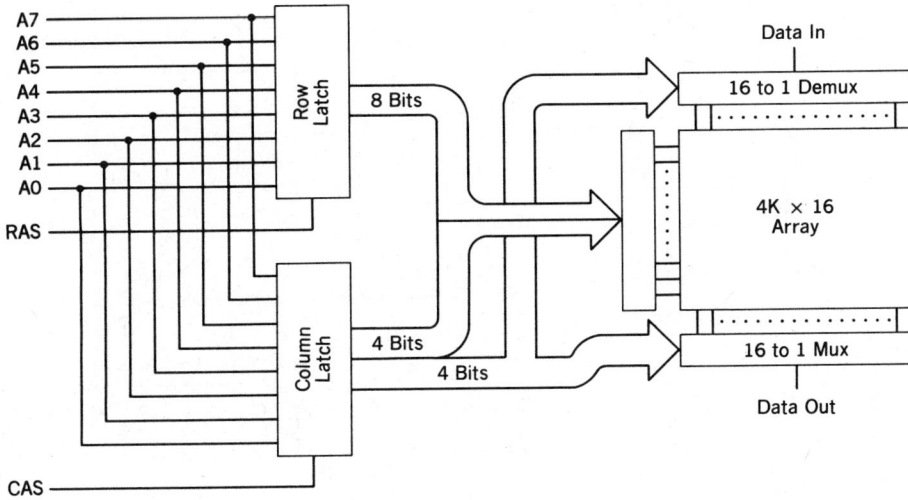

Figure 10.2. Block diagram of a multiplexed address memory.

As ICs have become denser and the cost per gate has fallen a structure of memory called "content addressable memory" or CAM has moved from an item of theoretical interest to a frequently used form of memory. Memory management subsystems use a high-speed static memory configured as a CAM to keep track of which logical pages currently reside in physical memory.

A CAM (Figure 10.3) is not designed to output its stored data, but only to signal a "hit" when the data word presented on its inputs matches one of its stored words. Addressing the memory is actually a dual function in which the address for writing is a conventionally decoded binary number which points to one of the physical locations (words) within the memory regardless of its content. During a read operation the data to be matched are presented in a word which is simultaneously matched against all the stored words to see if there are one or more hits. Each cell has an associated EXCLUSIVE-OR circuit which contributes through a wired-OR to the word match signal. If all bits in a word match, an OR gate which connects all the word match signals conveys the hit to the outside.

The faults which present challenges in a CAM are not the storage-associated ones, but the matching circuit faults. Activating the EXCLUSIVE-ORs and manipulating the dual addressing scheme to perform a meaningful test is a more significant problem that detecting conventional stuck-at memory cell faults. Algorithms to accomplish some of these objectives will be suggested later.

Cell Models. For the stuck-at modeling level the cells appear as flip-flops or latches. A ROM cell is fixed in value at the time of test and can therefore be modeled as a wire to ground or to the power supply. The simple representations of Figure 10.4 model functions, but do not show the actual implementations.

Both the ROM and RAM models are static representations, although RAMs are

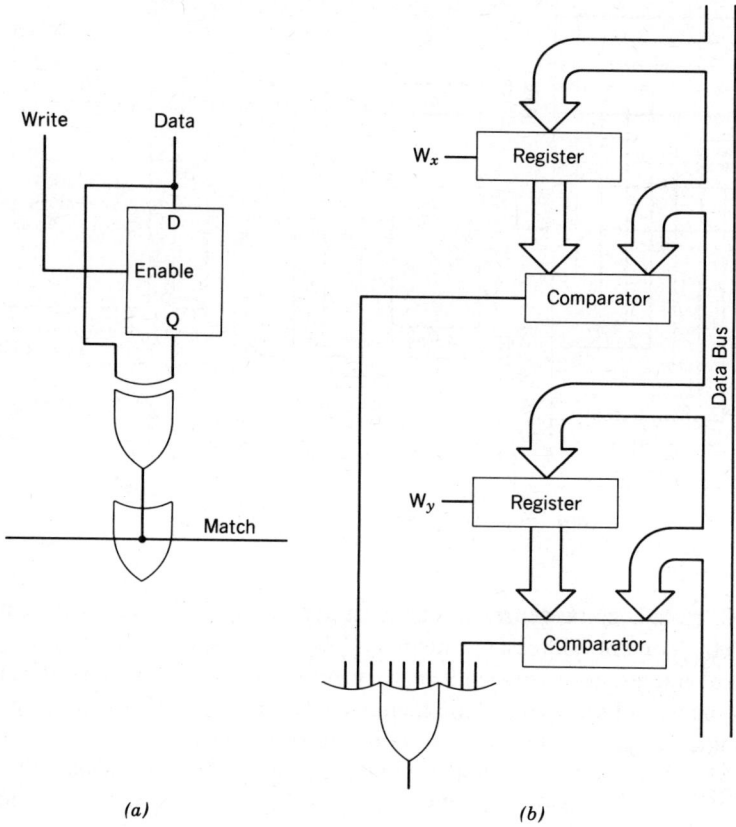

Figure 10.3. Content addressable memory. Individual word registers are paired with matching comparators.

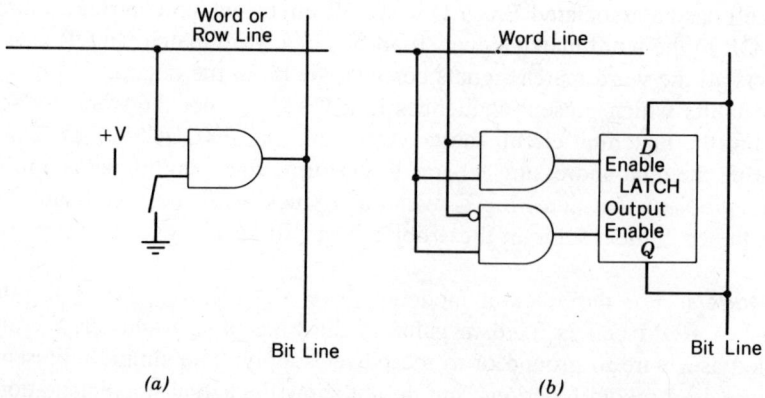

Figure 10.4. Gate-level representations of memory cells. (a) ROM. (b) RAM.

often dynamic. The stuck-at fault level assumes static fault characteristics, and therefore the models are static. Note that the AND gate of the ROM cell must be an open collector type (in the TTL convention) for proper operation of the bit line. Prior to the test the jumpers of each cell have been installed (or removed) to create a data storage array. When a row or "word" line is activated by the address decoder, a path is sensitized from the data jumper to the bit line, which is read by an amplifier and presented to the memory outputs. The RAM latch shown has a bidirectional bit line, and the write enable (WE) determines whether data will be latched from the bit line or enabled onto it. In either case, only the cells enabled by the word line pointed to by the address bits will respond.

The storage cells of a memory array exhibit stuck-at faults which appear in test results as single stuck bits. The modeling of the cells depends upon whether the memory is a ROM or a RAM, but generally does not represent the more subtle neighborhood interactions or time dependences that affect actual memories. Physical realizations must be examined to determine which of the more subtle faults is likely and how they might be detected.

Cell Structure

A quick review of the typical physical structures used in semiconductor memories will serve to illuminate some common failure modes. The additional structures of pc boards which make up higher levels of memory organization are subject to the same faults as the logic sections of pc boards and will not be explicitly discussed here. Examples are drawn from MOS technology since the highest volume of memory devices produced is built with this technology. The structures can be implemented in TTL or ECL technology and appear generally parallel in form.

ROM Cells. ROMs are constructed in several ways to accommodate the economies of mass production on one hand and ease of programming on the other. Minor variations in technology and programming mode create numerous choices of performance and cost, but three basic styles will exemplify most of the potential fault modes.

For high-volume applications in which the data contained in the ROM are determined well in advance of use and do not vary from unit to unit, the mask programmed ROM is usually the least costly. Referred to simply as a ROM, the array of transistors is laid out on the substrate without regard to data content until the gate metallization layer is formed. Each cell consists of a single transistor which bridges between a bit line and ground (Figure 10.5). The metallization mask which lithographically determines which areas of the chip are to be connected to the word lines (address row lines) is altered to reflect the data to be stored.

Each bit position which is to store a 0 is given a MOS transistor gate connected to the appropriate word line. If the actual gate is formed in a lower silicon layer, the metal added connects the gate contact to the word line; otherwise, the gate is formed with the metal. Bits which are to be 1s are left unconnected. When a word

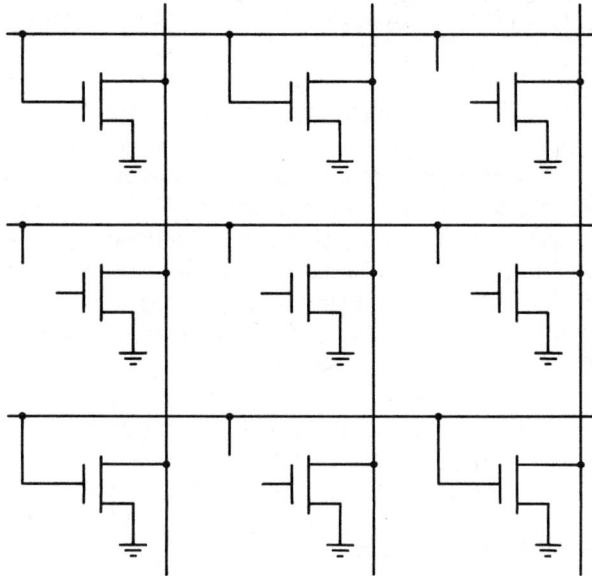

Figure 10.5. Data are recorded in mask programmable ROMs by the presence or absence of a connection to the MOS transistor gate.

line is activated by the address decoder, connected gates turn on the respective transistors, and the bit lines which they drive are shorted to ground.

Faults in this structure are generally stuck-at in nature. Extra metal may short a transistor gate to a word line, creating an s-a-0 cell. Absence of a connection where one is intended will cause an s-a-1 cell. Other shorts may cause an entire column or row to become stuck in the same way faults in the address or data lines leading to the array can be affected.

A simple cell which best exemplifies the removable jumper aspect of the gate-level model shown previously uses links or jumpers of an easily melted material (Figure 10.6). The links may be directly placed between the word line and the bit line or they may be used to connect transistor gates as in the previous example, but the mechanism is essentially the same. The diodes prevent bit lines from becoming shorted together through word lines.

At the time of manufacture all links are in place, and if the cells are configured as shown, they will all contain 0s. That is, when any word line is activated by reducing its potential to near ground, all bit lines will also be pulled to ground. During programming (from whence the ''P'' in PROM arises), selected link fuses will be melted or ''blown'' by the application of a high-current pulse. The absence of a fuse will be detected and interpreted as a 1 during subsequent reading since activating the word line will not short out the corresponding bit line.

Again, the faults are primarily stuck-at in nature. In testing PROMs there is a higher incidence of single bit failures since the data control for the melting or

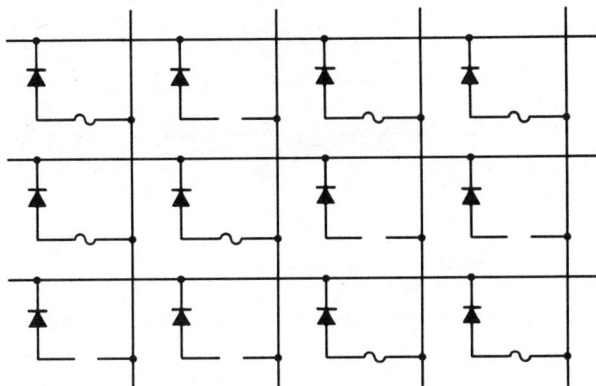

Figure 10.6. All the links in a field programmable ROM are installed during manufacture. Data are recorded later by melting some of the links.

"burning" process is not as good as the mask programming. A proven mask seldom develops errors, while each programming session gives opportunity for a failure to properly blow a link. The current pulse may be insufficient in time or amplitude, or the material from the melting may fall back too close to the link gap, allowing metal migration to regrow the link at a later time. Content verification is an essential part of PROM testing.

Erasable ROMs (EPROMs) exploit a completely different phenomenon to provide a semipermanent form of storage. The example shown in a floating gate EPROM which can be made with minor variations to be erasable by ultraviolet light or by the application of an electrical potential in excess of its operating potential. The former configuration is termed a UV-EPROM, while the latter is called an EEPROM, for electrically erasable PROM.

A conductive gate is completely surrounded in the insulative silicon dioxide layer above the channel of a p-channel enhancement MOS transistor as shown in Figure 10.7. During programming a high reverse voltage is applied across the p–n junction of the substrate and the drain. The pulse voltage is sufficient to cause avalanche breakdown, and high-energy electrons are injected into the oxide from the junction. Electrons gather on the gate and charge it as though it were connected to an external voltage source. After the programming pulse is removed the charge on the gate is insufficient to cause avalanche or significant tunneling, and the lack of direct connection leaves the charge trapped.

The semipermanent charge creates a conductive channel from drain to source in the same way a connected gate would. Gates programmed with a charge conduct, and the others do not. The entire array can be erased by exposing the surface of the IC to ultraviolet light. Photocurrents induced in the dielectric discharge the gates. The programming can be performed over and over. It is possible to degrade the oxide dielectric by repeated programmings, and data retention time is known to suffer slightly. Achieving an adequate charge on the floating gate is critical. Too

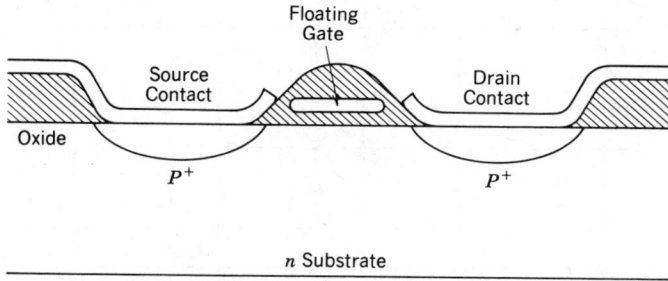

Figure 10.7. Cross-sectional view of a floating gate EPROM transistor.

great a pulse will damage the transistor junctions, but too small a pulse will create a high "on" resistance transistor which may not output the correct state under voltage or temperature extremes.

Faults caused by incorrect data are accompanied in this case by faults caused by weakly programmed cells. Tests for these cells are best performed under voltage margin or tolerance conditions. The capability of most ATE to control the application of voltage or current pulses encourages many users of PROMS on pc boards to merge the test and programming functions. When large PROMs are to be programmed, a reduction in the programming pulsewidth for each cell adds up to a significant overall programming time reduction. Shorter pulses run a higher risk of inadequate charge transfer to the gate, creating a higher resistance cell. The programming process can be optimized by testing each cell after attempting to program it. If the programming fails, a longer pulse is applied and the test cycle is repeated.

Static RAM Cells. The block diagram model of a memory array given earlier in the chapter best fits the case of a static RAM. Individual latches at each cell location are a reasonable behavioral-level description since they can be written and read, but they hold data between writings without external action. A comparison of the actual structure of a static RAM cell described at both the gate and transistor (MOS) levels reveals the classic "cross-coupled" latch form (Figure 10.8).

Static cells not only parallel the cross-coupled latch in construction, they share the same fault set characteristics. Stuck-at faults cover the common faults well. Timing faults tend to be minimal and related to familiar concepts such as access time, which is made up of propagation delays and setup and hold times on the input signals. Intercell faults are a function of electromagnetic disturbance, and the contribution from the digit and word lines is as significant as the cells themselves. The tests required for adequate testing of static memories are a subset of those required for dynamic RAMs.

Dynamic RAM Cells. The simplest structure of a RAM cell is the MOS dynamic one-transistor cell. Simple construction has led to extremely compact layout, re-

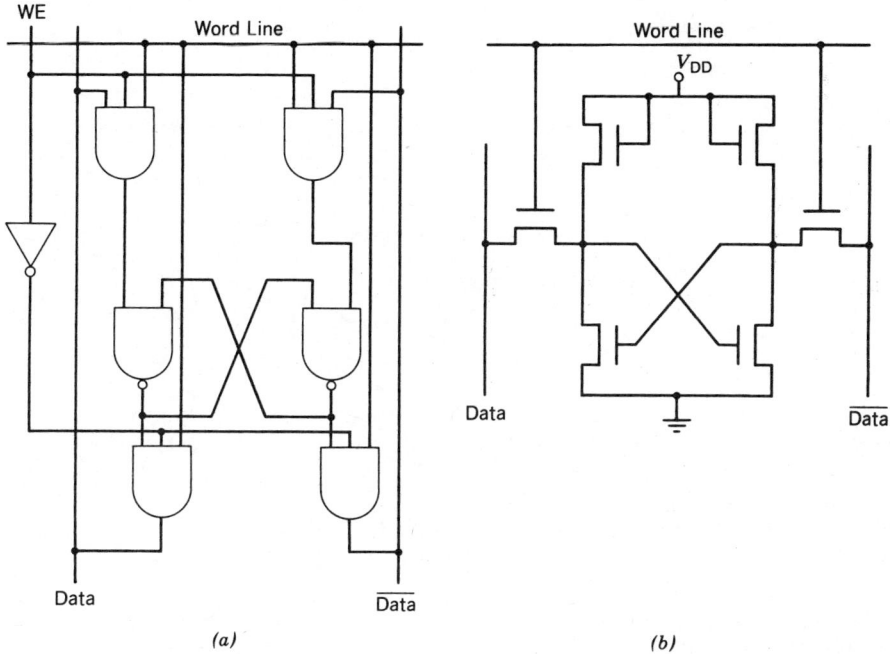

Figure 10.8. Gate- and MOS-transistor-level diagrams of a static RAM memory cell. (a) Gate level. (b) MOS transistor level.

sulting in 256K cells and more on a single IC. Simplicity of operation was traded off, as a refresh mechanism must be installed and properly used to maintain correct data in the storage array.

Storage in the dynamic cell is accomplished in the form of charge on a capacitor. The capacitor is formed by shovel-shaped extensions of the bit or data lines seen in the diagram of Figure 10.9. The upper layer of conductive silicon is separated from the substrate by an insulative oxide layer. The narrow stub that connects the shovel head to the bit line has the word line metal passing over it at a point of thin oxide covering which forms the gate of a MOS transistor. The elegant design permits extremely dense packing of the memory array.

Use of charge storage in a capacitor results in a dynamic memory. Charge leaks through the insulator, and after a few milliseconds or a few hundred milliseconds an insufficient amount remains to give a reliable signal when the cell is read. All the cells in the array must be periodically read and rewritten to refresh the memory. Refresh failures refer to memory cells which, through defects in the insulator or cell structure, fail to retain a charge for the specified maximum allowed time between refresh operations.

Downscaling the array means reducing all the dimensions to allow larger memories on a given die size. The dimensional reduction creates or enhances several fault types. Capacitive and inductive coupling between signal lines increase with

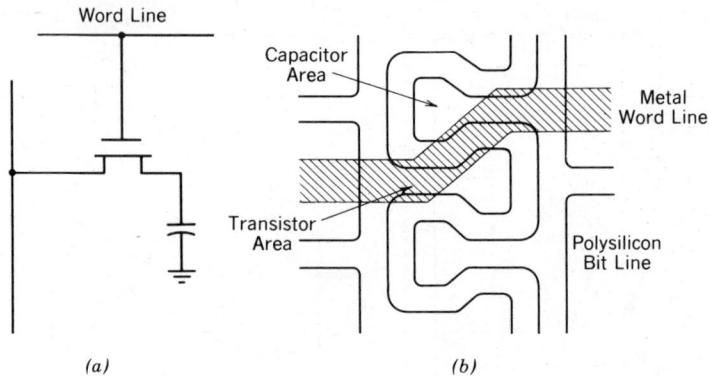

Figure 10.9. Shovel-shaped extensions of the bit lines in the mask correspond to both the transistor and the capacitor in its schematic equivalent. (a) Schematic. (b) Mask layout.

corresponding increases in crosstalk. Shrinking the cell capacitor lessens the charge that can be stored and the resulting signal available when a cell is read. Signal-to-noise ratio as seen by the sense amplifier is reduced. Preventing leakage of the small charge in the cell requires better control of processes which form the semiconductor capacitor.

These faults cannot be modeled as single stuck-at faults. They involve the interaction of two or more signals or cells, or they occur as a function of timing and parameters such as voltage. Failures caused by interaction and timing conditions require relatively elaborate and carefully constructed test sequences for effective detection.

Sense Amplifiers

Timing and parametric sensitivity carry through from the cell to the sense amplifiers as a technological consequence of the cell output. Static memory cells and ROM cells have a voltage or current output. Interfacing to static cells is a simple matter of selecting a path for the signal. The output of a dynamic cell is a charge. The draining of the charge during the read operation results in a small pulse of current which must be detected or "sensed" and used to establish a stable digital signal for presentation at the memory output. A brief description of the structure and operation of a typical sense amplifier will aid in identifying potential failures in its application.

In order to detect a small charge a balanced or differential amplifier is used. The opposite transistors of the amplifier (Figure 10.10) also create a flip-flop for latching the pulse, thereby creating a stable signal. The balanced nature of the circuit is assured by the inclusion of a dummy bit line and cells within the array, which provides a reference signal against which the memory cell output can be weighed. Word line selection is performed in the dummy cells but no data are written into them [1].

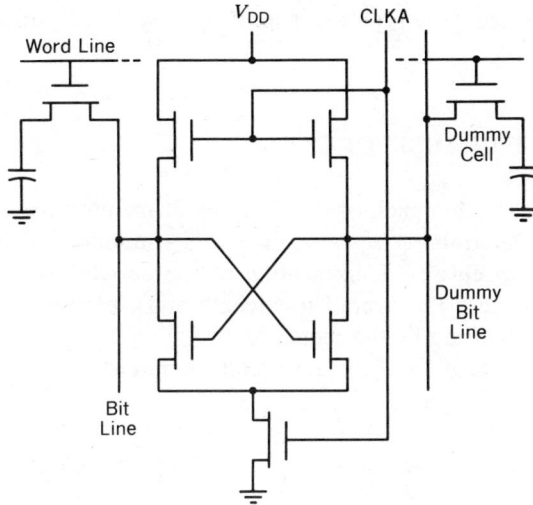

Figure 10.10. Sense amplifier schematic. Note the resemblance of the core circuit to a flip-flop.

Prior to switching the data cell output onto the bit line, both the true and dummy bit lines are precharged to the same level. The load and source transistors above and below the sense amplifier flip-flop are turned off, isolating the cross-coupled transistors. Enabling the cells onto the bit lines changes the amount of charge on each line and results in one of the gates of the cross-coupled pair having more charge than the other. In this sensitized state the flip-flop can be influenced by very small charges.

When power is restored to the cross-coupled transistors, the transistor with a higher charge on its gate is driven by the feedback to an ''on'' state, while the other is cut off. Sense amplifiers are subject to several faults as a result of their environment and the manufacturing process. Bit line noise caused by capacitive coupling may upset the sensing process. Insufficient charge in the cell may not establish the correct state during the strobe time. If the transistor gain or parasitic resistance of the sense amplifier transistors is not properly controlled in manufacturing, higher-charge inputs than normal may be required to reliably set the flip-flop, and the memory may fail under voltage or temperature extremes.

Sense amplifiers, are by design, sensitive to small differences in inputs from the bit lines. During write operations the bit lines are driven by the input data buffers in order to charge the addressed cell to its storage value. The lines themselves have an associated capacitance and are charged likewise. At the beginning of the read operation, bit line recharge circuits are switched on to restore the bit lines to a neutral position. When the lines and the sense amplifiers are not adequately neutralized before the bit selection for reading takes place, the correct operating point of the amplifier may not have been reached and the small signal from the selected bit may not be detected. The problem is termed a "sense amplifier recovery failure."

MEMORY FAULT CLASSIFICATION

Each of the sections within each of the varieties of memory has a set of distinct prominent faults. Determining a test approach is simplified by categorizing the fault types into a few classes. Identification of appropriate test sequences which cover each class reduces the overall test length and clearly shows which faults might expect to be detected by the sequence.

A theoretical classification of memory faults commonly used [2, 3] for RAMs divides faults into four groups.

1. Static stuck-at faults cause one or more memory cells to appear to be in one state regardless of any read or write operations performed on the address(es). The use of "appear" refers to the tendency for stuck signal lines in the decode or output buffers to give equivalent results to those which occur when an actual cell or group of cells is stuck.

2. Transition faults refer to the failure of a cell to assume a new value when the complement of its present value is written into it. Failures in the write control circuitry map into this class as well as certain cell faults.

3. Coupling faults are interactions between two or more cells or lines which allow actions on a cell to affect the contents of another cell. Capacitive coupling or leakage between cells creates such faults.

4. Multiple access faults result in cells other than the addressed cell being read or written along with the addressed cell. An address decode failure or a short between two bit lines may cause multiple access failure.

Although these classes map the static functional faults well, several common fault types occur in the dynamic domain in which the timing of events is critical to testing. Some of the timing faults apply only to dynamic RAMs, but others apply to all memories.

5. Pattern insensitive timing faults include failure to meet timing specifications such as access time, minimum write pulsewidths, setup and hold times, and maximum refresh time. Excess capacitance in the output drivers or manufacturing process variations of other kinds may result in these faults.

6. Pattern sensitive timing faults are characterized by the sense amplifier and write recovery problems in which the preceding action conditions a part of the memory system such that a later attempted operation fails.

Parameters such as voltage and temperature contribute to the susceptibility of memories to failures of the above classes, but no faults other than the obvious chance of destruction due to overvoltage or overtemperature are solely related to these factors. Parametric conditions are applied in the presence of one of the test algorithms to enhance its ability to detect certain types of faults.

MEMORY TEST ALGORITHMS

Regularity in the structure of memories leads to the statement of memory test sequences as algorithms rather than as a list of test vectors to be applied. In addition, the correct outputs of the memories can be predicted as a result of the algorithms, alleviating the need for simulation. Outputs can also be recorded from a known good device as the algorithm is performed by an automated tester. Recording or ''learning'' of outputs is very useful for ROMs since the contents of memory are not predictable from the test algorithm directly, but require device-specific content listings from the programming operation.

Algorithmic ATE

Performing a test on a memory block requires proper application of power, manipulation of the address and data inputs in an orderly fashion, and comparison of the outputs to a calculated or previously recorded result. The opportunity to use calculated inputs and outputs changes the structure of ATE which is efficient at memory testing.

The hypothetical memory tester shown in Figure 10.11 consists of several registers, a block of memory, and the multiplexers and latches to route register contents to and from the DUT interface. The four registers (two address and two data) have simple logical operations built in. Each register can be incremented or decremented, and the entire contents can be complemented bitwise. The control mechanisms have been deleted for clarity. Having provided alternate address and data registers simplifies several algorithms where a foreground operation must be carried out on a target cell, while a distinct background operation must be applied to other cells. Uses for the register functions and data paths will become evident as individual algorithms are explained.

RAM Algorithms

Four basic algorithms will be presented which detect the fault types previously discussed [4]. There are many variations [2, 3, 5]. IC layout and processing variables alter the effectiveness of some patterns. New algorithms have been derived in some cases to combine the steps of other algorithms and detect more fault classes with fewer overall steps. Algorithm complexity is given as the number of steps required per cell. Test time for each algorithm can be calculated by filling in N in the complexity expression and multiplying by the time required per test step. A single write operation has a complexity of N, so the time required to write all cells

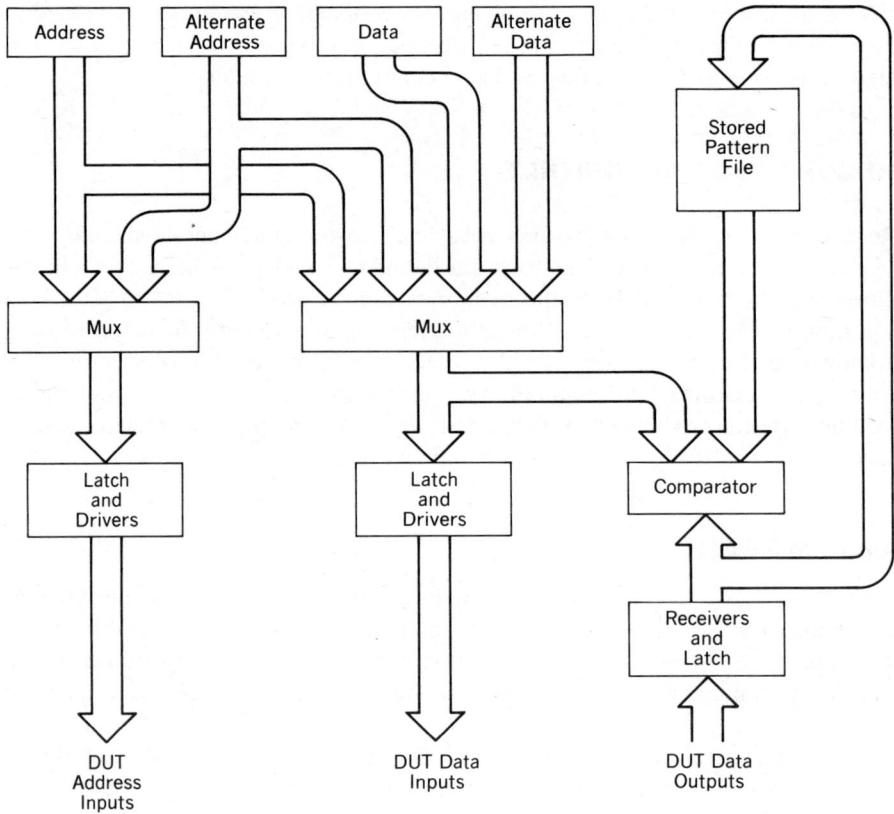

Figure 10.11. Block diagram of an automatic tester optimized for memory testing.

in a 150-n 64 K memory is 9.83 ms. During application of these patterns the refresh requirements of dynamic RAMS must be fulfilled when the pattern does not accomplish refresh by changing all data at least once within the allotted refresh interval.

Address Pattern. The address pattern algorithm writes a checkerboard of data into the array (a column of alternating 1s and 0s if the memory is an $N \times 1$ array), and then reads the data in such a way as to detect a change of data with each address change. In memories which have multiple columns the traditional address pattern writes the address of a row into that row and then reads and verifies the contents later. The address size of memories now being produced has caused a change in the procedure to the one illustrated here.

If the algorithm were to be performed on the ATE previously described, it would make use of both address registers and one of the data registers. After the initial write, one address register would be set to the maximum address of the memory array while the other would be set to zero. The decrement function would be used for the high register, and the increment function would be used for the low register. Complementation is the only function needed for the data register once it has been originally loaded.

Two procedures are needed: one to write the background pattern and one to read and test the cells in an order which assures a different data output for each address change. The entire procedure is then repeated with background data complemented as illustrated in the flowchart of Figure 10.12.

Once the background pattern of alternating 1s and 0s has been written, the test consists of reading the first address, the last address, and the first address again. Since memories always have an even number of addresses, the data for the first and the last addresses will be opposite. Therefore, there is no confusion as to whether the address actually transitioned. The second reading of the first address ascertains that the address transition can be made in both directions successfully. The routine is repeated with the second address and the next to the last address and so on until all addresses have been accessed in both directions.

The procedure detects stuck-at cells since all cells are written to both states and verified. Stuck-at faults in the address decode and the data input and output lines are also detected. Transition faults in the cells will probably be detected, but only one transition was explicitly exercised (e.g., if the first cell were incapable of transitioning from a 1 to a 0 but powered up in the 0 state, it would not be detected).

Counting only the DUT read or write operations gives a run time for the algorithm of 8N where N is the number of addresses in the DUT. If extra cycles are required by the ATE to prepare the internal registers, the run time will be longer. The 8N is derived by simply counting the one write per address in the background preparation, adding the three reads per address in the read and test section, and doubling the subtotal to account for the second pass with the complemented background. Notice that the routine is linear with respect to N. A 32 K memory array test takes twice as long as a 16 K test, and a 64 K test takes twice as long as a 32 K test.

March Test. A march test begins with an homogeneous array of data (either 0s or 1s) and proceeds from one end of the array by checking and complementing each bit. In an $N \times M$ array all the bits in a row are checked and complemented together. The flowchart of Figure 10.13 shows the arrangement for an $N \times 1$ memory device.

The march test detects cell transition from 0 to 1 and back, in addition to the static stuck-at faults of the array. Test time for the algorithm is 10N times the time for a DUT access if the internal ATE register operations are carried out transparently. March patterns are still linear with respect to the size of the array being tested. This time-saving advantage has encouraged numerous improved versions of the test which are intended to detect the more complex failure modes such as coupling faults between cells [3].

The improvements hinge on the nature of coupling between cells during tran-

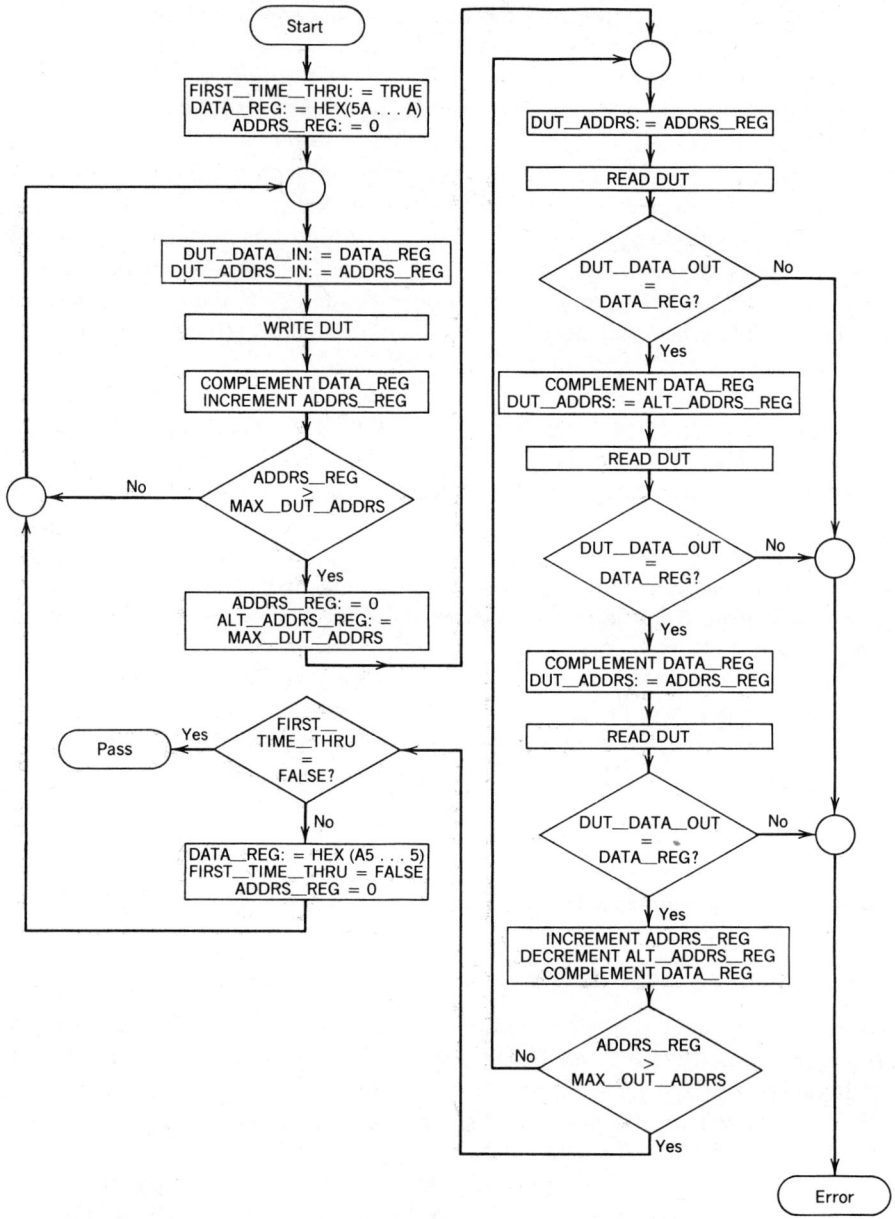

Figure 10.12. Flowchart of an address pattern memory test.

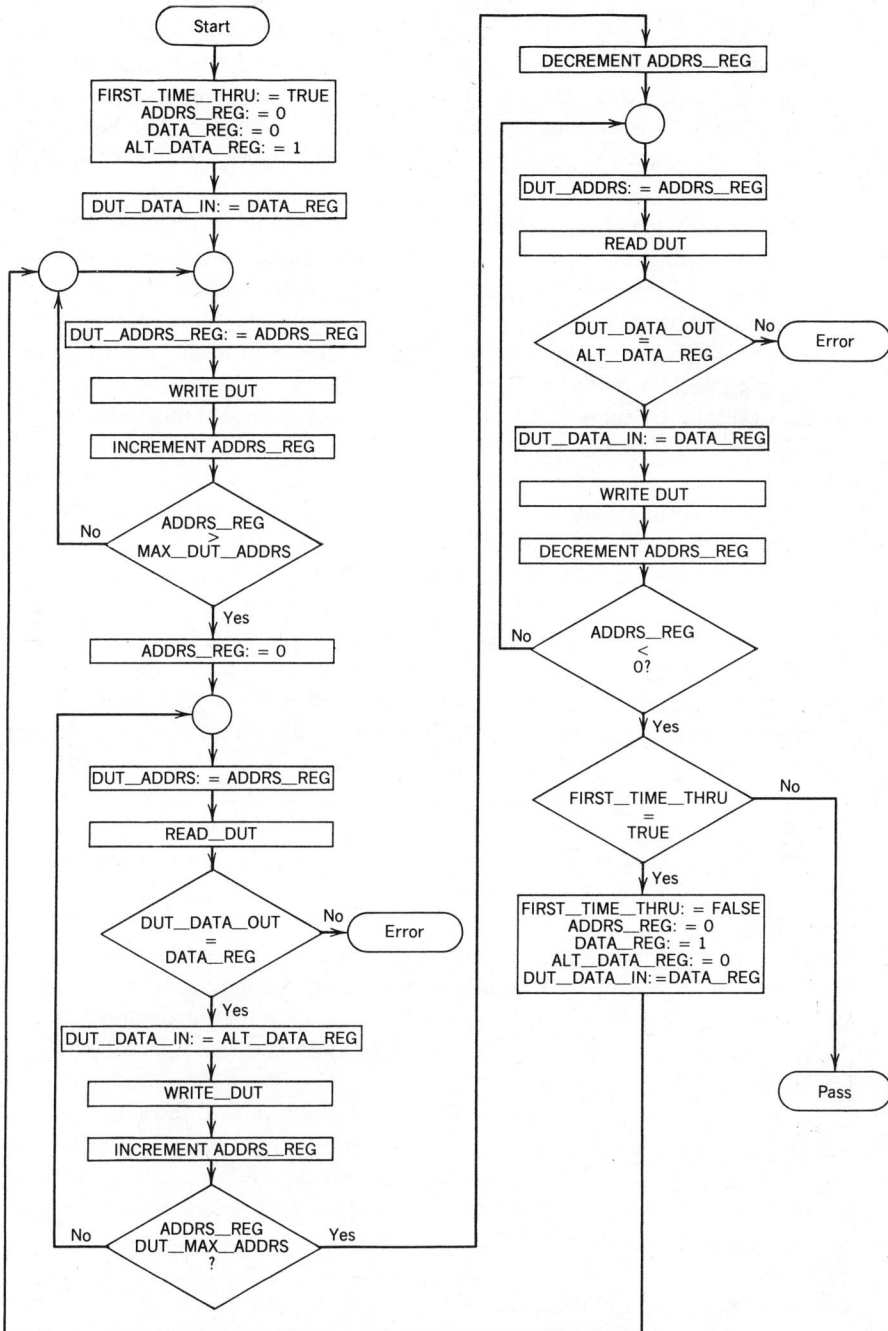

Figure 10.13. Flowchart of a march pattern memory test.

sitions of one of the coupled cells. Coupling between cells A and B may be bidirectional or unidirectional in effect. That is, a change in the state of A may disturb the stored state of B, but the reverse may or may not be true. In addition, the direction of change of the disturbing cell may be critical. A change in cell A from 0 to 1 may affect B, but a change from 1 to 0 may not. Finally, the state of the disturbed cell is critical in some faults such that a change in A will only affect B if $B = 0$ (or 1) at the time.

Although the $10N$ march pattern described above appears thorough and in fact detects single coupling faults, it does not detect multiple couplings which may occur in a different fashion between several cells. Adding more write operations to the basic march pattern can improve detection by providing additional transitions to force coupling. By changing the parity of the transitions, multiple couplings which cancel out under the basic march test can be detected. It has been shown that at least $14N$ steps will be needed to detect multiple couplings.

Galpat. Cell coupling and address transition faults have traditionally been detected by the "galloping pattern" or Galpat algorithm. Galpat exhaustively tests the relationships between each cell and every other cell. The popularity of Galpat waned rapidly as memory sizes grew, however, due to its excessive test time per cell.

Nested loops are prevalent in Galpat's flowchart (Figure 10.14). The N^2 characteristic is easily understood since the basic concept is to make a change in a cell and then read all the other cells to check for disturbed content. After repeating the pattern with the complement background the overall algorithm time is $2(2N^2 + N)$. Galpat tests for multiple address faults in the write operation and two-cell coupling in which writing a cell disturbs the value of another cell. Its advantages are its thoroughness and the universality with which it can be applied. Since no assumption is made about internal structure of the memory array, the test is equally effective on any geometry.

Column Tests. Algorithms such as those described above detect the stuck-at functional, transition, coupling, and multiple access faults without knowledge of the internal layout of the memory array. Tests for sense amplifier characteristics cannot be approached in the same manner since the time between operations on addresses within a given column is of importance.

Once the internal arrangement of an IC has been determined, the addressing for an algorithm can be sequenced such that a particular pattern follows either a row, a column, or a diagonal. A 64 K memory chip may be internally arranged as two banks of 32 K each with each bank in turn arranged as a 1 K-row $\times$ 32-column array, as suggested by Figure 10.15.

Each of the banks is treated independently since they each have separated decoders, drivers, and sense amplifiers. The procedure described is simply sized to a particular bank and repeated to test the other bank. Assuming that the most significant address bit controls bank selection, the lower address bits are considered alone in sequencing the pattern. Since there are 32 columns in each 32 K

Start

FIRST__TIME__THRU: = TRUE
ADDRS__REG: = 0
DATA__REG: = 0
ALT__DATA__REG: = 1

DUT__DATA__IN: = DATA__REG

A

DUT__ADDRS__REG: = ADDRS__REG

WRITE DUT

INCREMENT ADDRS__REG

ADDRS__REG > DUT__MAX__ADDRS — No → B

Yes

ADDRS__REG: = 0
ALT__ADDRS__REG: = 1

C

DUT__ADDRS__REG: = ADDRS__REG
DUT__DATA__IN: = ALT__DATA__REG

WRITE DUT

D

DUT__ADDRS__REG: = ALT__ADDRS__REG

READ DUT

DUT__DATA__OUT = DATA__REG ? — No → Error

DUT__ADDRS: = ADDRS__REG

READ DUT

DUT__DATA__OUT = DATA__REG — No → Error

Yes

INCREMENT ALT__ADDRS__REG

ALT__ADDRS__REG DUT__MAX__ADDRS — No

Yes

ALT__ADDRS__REG: =
ALT__ADDRS__REG = DUT__MAX__ADDRS

E

ALT__ADDRS__REG = ADDRS__REG — No

Yes

DUT__DATA__IN: = DATA__REG

WRITE DUT

INCREMENT ADDRS__REG
INCREMENT ALT__ADDRS__REG (twice)

ADDRS__REG > MAX__DUT__ADDRS — No

Yes

FIRST__TIME__THRU = FALSE — Yes → Pass

No

FIRST__TIME__THRU: = FALSE
ADDRS__REG: = 0
DATA__REG: = 1
ALT__DATA__REG: = 0

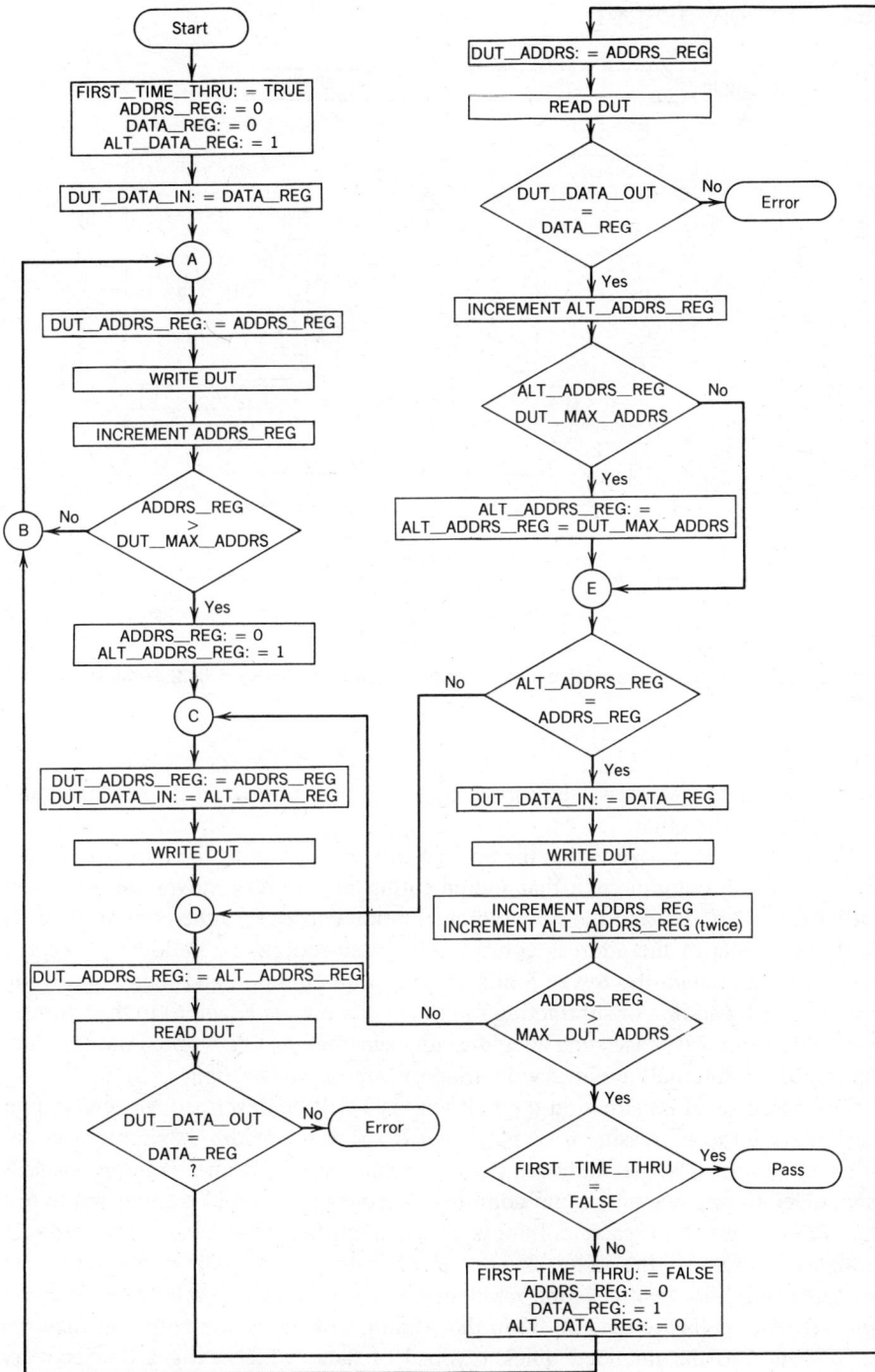

Figure 10.14. Galpat memory test flowchart.

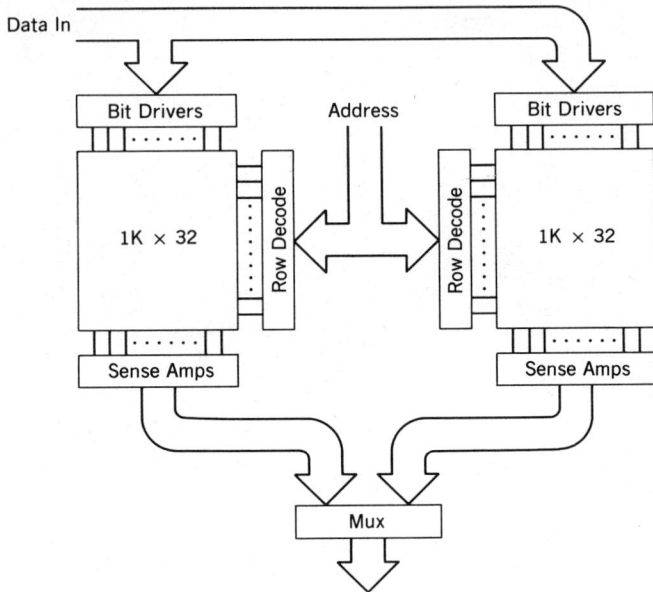

Figure 10.15. Block diagram of a 64K memory internally arranged in 2 sections with 32 columns each.

array, the address bits must be segregated to select one of 32 on the one hand and one of 1 K on the other.

For the example, assume that the upper 5 bits, not including the most significant bit, control the column such that incrementing the address moves up a column until 1023 is reached and the next incrementation carries into the column field. If the lower 5 bits of the address control column selection, the galloping diagonal test must manipulate the lower 5 bits to control the column and then control the row address by adding or subtracting multiples of a binary 32 (bit 6) to the address. In each particular IC case such an addressing scheme must be worked out to orient the test to the internal layout. Two variations are shown in Figure 10.16.

For purposes of explanation the column tests will be described for the simpler case, leaving the addressing for a particular RAM to the reader. A simple column-oriented pattern is a checkerboard pattern written and read on a column basis. It is intended to produce maximum column noise during the rapid reading and to test the read recovery of each column's sense amplifier. The concept is entirely straightforward once the addressing sequence is determined. The important aspect for sense amplifier testing is the rapid succession of reading each cell in the column. If other cells are read between the column cell, the sense amplifier may not be subjected to the intended quick reversal of data, which evokes the recovery problem.

A write recovery test requires a cell in the column under scrutiny to be written

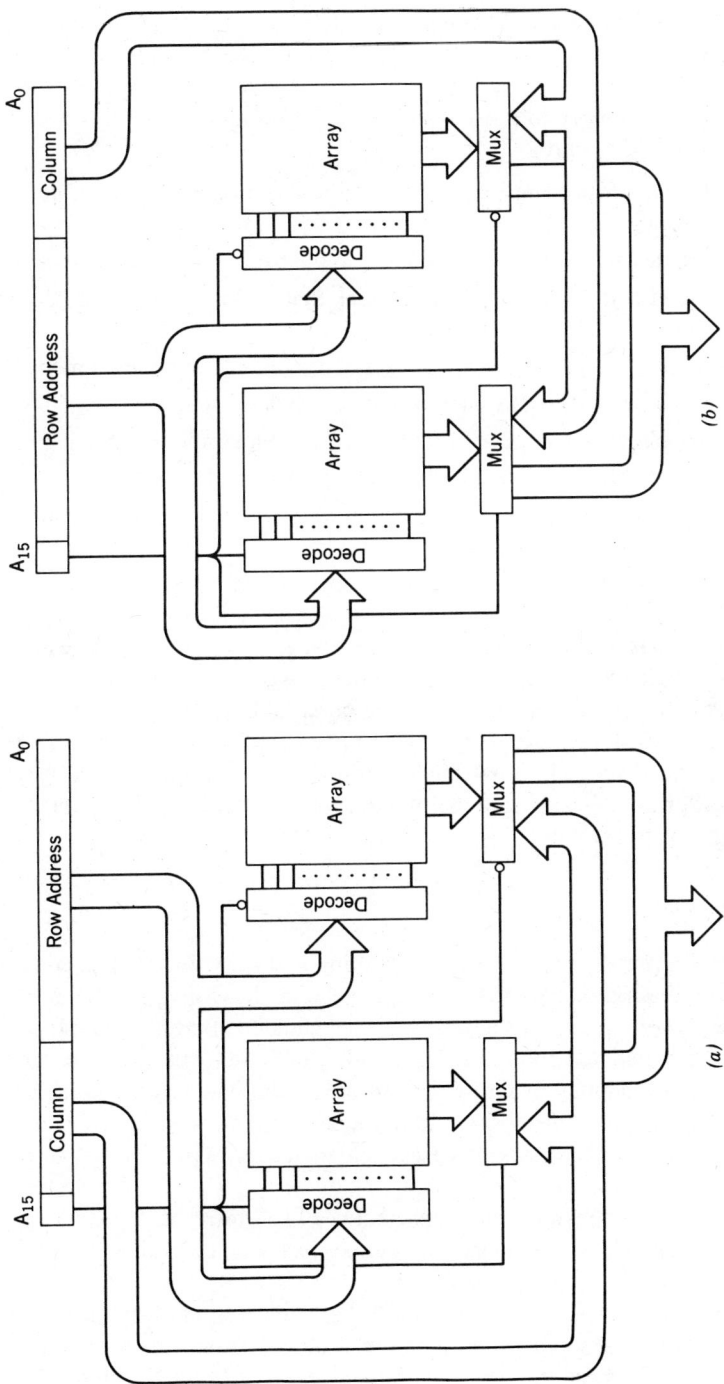

Figure 10.16. Block diagram of two similarly structured RAMs with differing address arrangements.

255

(some tests write the cell several times alternately to 1 and 0), and then some other cell in the column is read. An example of this type of pattern follows this sequence:

1. Write the memory to a background pattern of all 0s.
2. Set ADDRS = '0000'.
3. Perform the following loop N times where N = the number of cells in the memory array.
   a. Write cell ADDRS complement to the background (1 on the first pass).
   b. Read cell ADDRS + 1 and check that it still contains the background state.
   c. Read cell ADDRS and check that it contains the complement state.
   d. Write cell ADDRS back to the background state.
   e. Read cell ADDRS + 1 and check for background state.
   f. Read cell ADDRS and check for background state.
   g. Increment the target address and loop if it is less than the array maximum.
4. Write the background to all 1s and repeat step 3.

This pattern has the advantage of being linear in complexity with respect to N, but it does not check for column noise which might result from the first address being written and the last address (or some other address in particular) being read next. Pattern sensitivities are very device dependent, and a great many patterns have been derived for individual fault conditions in particular device types. The important criteria to consider are the maximization of fault types coverage and the minimization of test time.

Refresh Test

Dynamic memories depend on periodic reading and rewriting of data contained in each cell to refresh the charge. Rapid tests such as those given above accomplish the essence of refresh, even though it is not intended. Internal refresh for RAM chips is usually handled by a control circuit which steps through row addresses and refreshes all columns of a row at each operation. The sense amplifier of each column is connected to the write buffer for the column without the usual multiplexers intervening. A single read/write cycle then renews the charge in a complete row with one step.

The maximum time which can elapse between cell refreshes is a function of the cell size and the technology. Once the number has been determined, it is a simple matter to divide the time by the number of rows in the array to determine the refresh clock frequency which will just satisfy the minimum requirements. Remember, each refresh clock cycle refreshes only one row. Testing the maximum time between refresh requires slowing down the clock to the minimum frequency or disabling it altogether. If the clock is slowed to its minimum frequency, no reads or writes can be performed within a complete refresh cycle or the effect will be the refreshing of written or read cells.

An effective test can be created by disabling the refresh and performing the following sequence:

1. Write all cells to any given pattern of data.
2. Wait for the maximum rated time between refresh without allowing refresh or any memory operation.
3. Read the memory in the same order it was written and at the same operation rate to assure that no cells must wait longer than the specified maximum.
4. Repeat steps 1 through 3 using the complement of the data pattern.

CAM Algorithms

Due to its unique addressing mode the CAM requires an entirely different approach for testing. Although a checkerboard pattern could be written into the memory array, any attempt to read it would give an ambiguous result since half the words would match either pattern. A faulty bit in one of the words would be masked since any one matching word is sufficient to give a hit signal.

A simple test approach uses the hit/miss hardware as part of the fault detection system [6]. To avoid ambiguous comparisons, the first step writes each word with its position in the physical array. Then the data are presented again, and each word is compared for a hit. The data are rewritten in complement form, and the read sequence again verifies the results. Each of the storage cells has been tested for s-a-0 and s-a-1 faults. In addition, each cell's exclusive-OR has been tested for a successful compare of 1s and 0s.

The remaining task is to test for the miss condition. If all combinations of nonmatching inputs were tried, the test would be prohibitively long. From a structural standpoint the fault to be tested is in each case associated with a single bit. If it can be substantiated that each bit's comparator in each word does not erroneously give the hit signal, the test will be complete from a stuck-at functional viewpoint.

The "walking 1s" test accomplishes the needed result. After writing a background pattern of all 0s into the memory, a single 1 is floated across the input word. That is, the first bit is toggled to a 1 and the word is presented, then the bit is returned to a 0 and the second bit is set, and so on. Each bit position for the 1 tests for all the comparators in a column. All other positions of the words match, and only the one column inhibits the hit signal from being asserted. A final pass writes a background of 1s and floats a 0 across data inputs.

The test sequence described tests for stuck-at or functional tests only. Since CAMs are constructed using static RAMs, some of the algorithms described in the dynamic RAM area are inappropriate or unnecessary. The complexity of the test is linear with respect to N. The initial write of unique information takes N operations, and the hit test requires another N operations. Repeating these for complement data makes a subtotal of $4N$ operations. The background pattern write for the miss test requires N operations, and the floating 1s procedure requires M operations where M is the number of bits in a word and N is the number of word locations in the memory. Doubling this miss procedure to cover both complements results in $2N + 2M$, which gives $(6N + 2M)$ for the entire test sequence.

ROM Tests

Most of the complex RAM algorithms are inappropriate for ROM due to the obvious lack of write capability. Testing ROMs is therefore often reduced to reading and checking the contents for correctness against a known good example or a document. Access timing tests are applicable during IC testing since the same address decode and output buffer problems which occur in RAMs can occur in ROMs.

If the testing is for design or component qualification purposes, the retention characteristics of EPROMs should be tested by repeated reading. EPROMs are usually specified to withstand tens or hundreds of millions of reads without loss of data. As a manufacturing check this test is entirely too long.

TEST ENVIRONMENT

Several characteristics of memories result in a heightened interest on the part of the designer and the manufacturer toward the environment under which the memories must operate and their ability to cope with that environment. The massive number of gates involved in a typical memory array aggravates failure statistics since only one failure in a million gates will disable a memory device. Environmental factors such as heat and voltage margins bring out failures quickly which might occur only occasionally or after some time in operation. The two most popular forms of environmental manipulation will be discussed. They are typically applied toward different goals. Burn-in is a manufacturing screening procedure, while Schmoo plotting is a process check or design verification test.

Burn-In

In all cases mentioned thus far it is assumed that the faults exist at the beginning of the test sequence and are to be detected and located within the circuit. Memory circuits are at the forefront of complexity in terms of total gate count, although their regularity of structure simplifies testing. The massive number of transistors involved increases the concern about latent failures. Latent failures are the result of process irregularities which set up the conditions which make a fault likely without actually creating the fault at the time of manufacture. A poor lead bond might make contact at the time, but became disconnected after a few hours of use due to thermal changes, vibration, or corrosion.

Latent failures which become active only after a reasonable product lifetime has passed are usually taken in stride as part of the "wearing out" process. Failures in the first few hours or months of operation are regarded as infant mortality and usually result in claims against warranty. The cost of these claims and the desire to improve the customer's image of electronic products has resulted in a process called "burn-in."

Burn-in takes advantage of the distribution of latent failures over time, which universally seems to resemble the "bath tub" curve of Figure 10.17. As a product is placed in service it is exposed to a new environment which excites a relatively large number of latent failures. Once these failures are detected and corrected the number of failures per million units settles to a near constant rate for the product life. Subjecting the circuits to a harsher environment accelerates the life and shifts the curve toward the left-hand side. A burn-in procedure operates the circuits under harsher conditions long enough to force the infant mortality faults to occur before the product is installed for actual use.

The conditions for burn-in usually consist of higher temperatures and higher voltages than normal. A typical procedure would place parts in an oven at or near the maximum allowable operation temperature for the parts and apply the maximum allowable power and input voltages. During burn-in the inputs can be held in a given state for the entire 200 h or so, producing a condition called "static, power-on" burn-in. Alternatively, the inputs can be cycled through a test pattern to cause internal switching in a condition called dynamic burn-in. Neither case requires testing of the outputs per se since the object is to incite failure and not necessarily to detect it.

Testing can be conducted before and after the burn-in as a separate operation. Maximum use of the burn-in facilities can be made if early dropouts are detected and replaced during the operation. In a constantly flowing operation where parts are added and removed from the burn-in facility per individual time schedules,

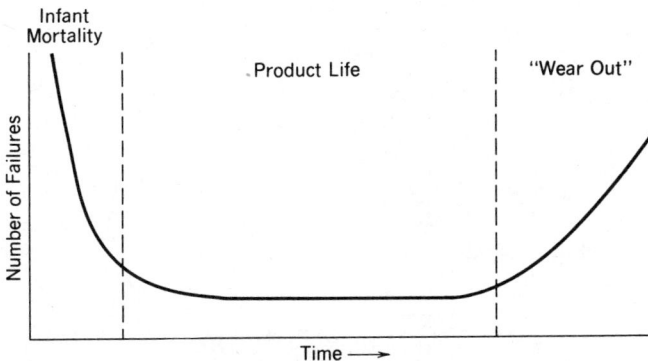

Figure 10.17. A graph of failures from a fixed population of circuits against time reveals the shape which gave it the name "bathtub curve."

early detection will improve throughput. A scanning monitor can periodically compare the output sequence of each part during its exercise sequence and log failures for removal. Devoting a tester to each part is seldom cost effective unless the parts are actually entire assemblies such as computers and the volume is a few units at a time.

Schmoo Plotting

When several environmental factors vary independently, the worst case is of greatest interest in testing. If the circuit can perform under worst case conditions, it is a foregone conclusion that it will work satisfactorily under any other conditions. Unfortunately, determining the worst case is not always easy or even possible. Different fault modes may become dominant under different combinations of the independent variables, leading to several "worst cases." In fact, failure rates may be higher for a combination of variables which does not include any one variable at its limit.

Multiple power supply voltages used in some memories are a good example of the multivariable worst case problem. Other variables investigated include clock or access timing, pulsewidths, or temperature. Statistical analysis can be brought to bear after a significant number of samples has been taken, but organizing the data to produce a recognizable result was an early problem. The Schmoo plotting procedure deals with the problem as follows. Select a test sequence which results in a pass or fail condition. For each of a statistically significant number of devices, repeat the test for each combination of two or more environmental variables and plot the test results. The test is easily visualized for two variables, but certainly possible for more.

At each point on the plane described by incrementally varying V_{CC} and V_{DD}, the number of passing parts is recorded. Plotting the numbers on a graph gives the shape shown in Figure 10.18, which resembles the cartoon character called a

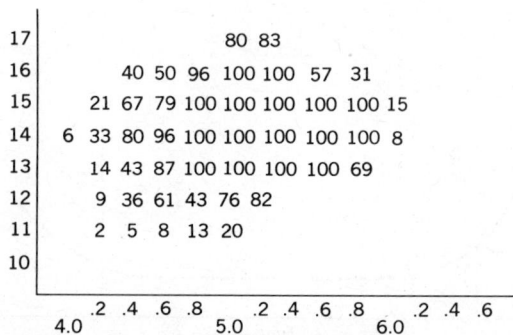

```
17│                    80 83
16│            40 50 96 100 100 57  31
15│         21 67 79 100 100 100 100 100 15
14│    6  33 80 96 100 100 100 100 100 8
13│       14 43 87 100 100 100 100 69
12│        9 36 61 43 76 82
11│        2  5  8 13 20
10│
   └─────────────────────────────────────────
      .2 .4 .6 .8   .2 .4 .6 .8    .2 .4 .6
     4.0            5.0            6.0
```

Figure 10.18. Plotting test pass/fail results against two independent environmental variables gives a blob-like operating region resembling the comic strip animals Al Capp called "Schmoo."

"Schmoo" in the long running series by Al Capp. Achieving sufficient samples for valid analysis is often a very long process requiring weeks of work. Characterization is therefore used only for design or process verification and not for production testing.

BOARD LEVEL TESTING

Several considerations become important as testing moves from the chip level to the PC board or subassembly level. The physical arrangement, the embedding of memory in other logic, and the economics of testing alter the applicability of many of the procedures described thus far.

Block Level Memory

As installed on a PC board, the memory chips appear as functional entities whose internal workings are not usually accessible. The availability of numerous interchangeable parts with the same functional specifications but different internal structure gives the manufacturer of memory boards flexibility in selecting supply sources and prohibits the test engineer from making any assumptions about the ICs below the functional specification level. Without such assumptions the column oriented, diagonal-oriented, and other structure-related tests cannot be precisely constructed, and their effectiveness may be greatly reduced.

A second influence on the application of complex patterns is purely numerical. The board level represents added complexity as potentially hundreds of memory chips are connected together to form larger arrays. Test procedures which run in greater than linear proportion to N time become infeasible, and the number of distinct algorithms which may be run economically is drastically reduced.

Taken as a whole the memory array of a board has different dimensions and different potential faults than the typical memory device. Although IC memories are often constructed as $N \times 1$ arrays, the boards have devices in parallel to form a data word of M width (8, 16, or 32 bits typically). The data lines routed across the board to interconnect devices and the associated input and output latches create new possibilities for stuck-at faults. Additional address decoders on the board select IC memory banks by enabling chips one row at a time. As seen in Figure 10.19 the structure resembles the internal structure of a memory chip, and similar patterns can be used, but with a different emphasis.

Writing efficient board-level tests for memory means taking advantage of any tests which may have been already performed at chip level. If the test engineer cannot assume that the ICs have been tested, the strategy shifts to an ordering of tests to find the board-level errors, such as solder shorts, first, followed by as many

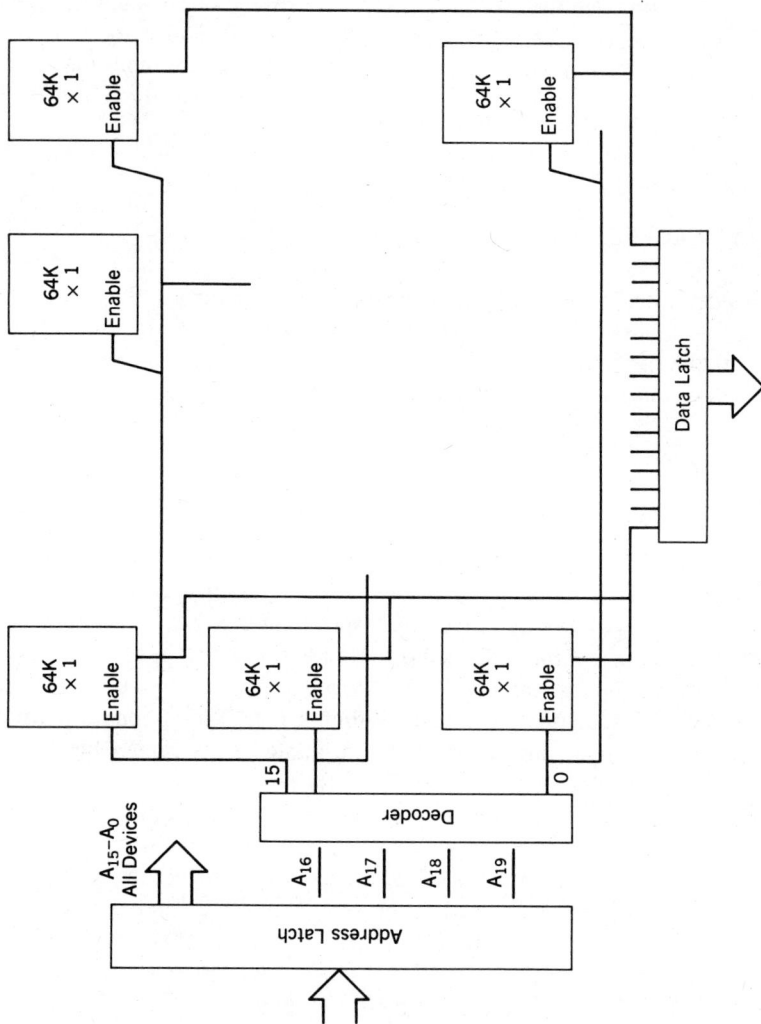

Figure 10.19. Block diagram of a board-level memory array resembles the internal arrangement of memory chips.

chip level tests as are economically feasible under the circumstances. An efficient board level routine might appear as follows:

1. Write an address pattern into the arrray at single binary bit address points.
    a. Write HEX(0000) into the address HEX(00000).
    b. Write HEX(0001) into address HEX(00001).
    c. Write HEX(0002) into address HEX(00002).
    d. Write HEX(0004) into address HEX(00004).

    Continue in this way at intervals of powers of 2 until (OFFFF) is written with (FFFF). Then write the complement of the address for the remaining address points.
2. Each of the addresses written in step 1 is read and checked.
3. The procedure of 1 and 2 is repeated, starting with complement data (that is, address HEX(00000) is written with HEX(FFFF), and so on.

The previous procedure is an abbreviated address pattern designed to detect stuck-at interline short faults in the address selection hardware of the board. It requires only $4 \log_2 N$ operations, and although it does not detect internal chip address faults very thoroughly, it locates solder shorts and open lines well. It should be followed with a walking 1s (0s) pattern which sweeps a 1 (0) across a data word of 0s (1s). Such a pattern should be written into and read from each bank of memory to detect possible shorts and opens in each chip's data interconnect with the board.

With the addition of the walking 1s pattern the entire test for board level stuck-at faults requires $4 \log_2 N + 2bM$ where b is the number of banks (chip rows) and M is the number of bits in a data word. For the example 1 Meg $\times$ 16 memory, the test requires only 576 steps. Further tests such as a march test may be desirable to test for individual cell function, but most typical manufacturing faults which occur at the board level will be detected by this short procedure.

The assumption of standard behavior of memory chips when viewed at the board level should be tempered. Although the specifications for intended operations such as read and write are usually thorough, unintended functions may be less predictable. A classic case involves the behavior of data outputs during a write cycle. In some implementations of a particular memory chip the outputs display the data being written into the cells in the so-called write through mode. In another manufacturer's rendition of the same part number the outputs may be disabled or show indeterminate data (unknown values). Care should be taken with respect to all assumptions.

SUMMARY

Although limitations of space and restrictions of focus make this an extremely brief coverage of the subject of memory testing, the methodology has been exposed.

Memory testing is distinct from the testing of other logic in that its orderliness allows algorithmic design of tests. The compaction thus achieved is required in turn by the sheer massiveness of memory structures which are in common use.

As for logic testing the thread of overall test design includes the following steps:

1. Identify the target fault set or sets.
2. Select a test application method including appropriate ATE and software support.
3. Chose stimuli that activate the faults and propagate their effects to observable points on the circuit. In the case of memories there is a shift in emphasis to selecting stimuli that can assure that only a well-defined set of faults has been activated in order to avoid masking. Writing and reading all 1s does not, for example, guarantee that all cells can be written to 1s if the address decoder has not been previously tested.
4. Establish the environmental conditions under which the tests are to be performed for maximum effectiveness and efficiency.
5. Compose the test and support software to assure a clear result to the test operator without requiring a complete understanding of device physics or testing engineering.

Use of these guidelines should be tailored to a particular circuit after a study of the architecture and technology. The goal of memory testing is to detect as many faults of as many differing types with as few test steps as possible.

REFERENCES

1. E. R. Hnatek, *A User's Handbook of Semiconductor Memories*, Wiley, New York, 1977, pp. 373–374.
2. C. A. Papachristou and N. B. Sahgal, "An Improved Random Method for Detecting Functional Faults in Random Access Memories," *IEEE Trans. Comput.*, **C-34**(2), 110–116 (Feb. 1985).
3. D. S. Suk and S. M. Reddy, "A March Test for Functional Faults in Semiconductor Random Access Memories," *IEEE Trans. Comput.*, **C-30**(12), 982–985 (Dec. 1981).
4. "Selected Test Patterns for 4K RAMs," Macrodata Corp., Woodlands Hills, CA, Appl. Note 139, 1977.
5. D. S. Suk and S. M. Reddy, "Test Procedures for a Class of Pattern-Sensitive Faults in Semiconductor Random Access Memories," *IEEE Trans. Comput.*, **C-29**(6), 419–428 (June 1980).
6. G. Giles and C. Hunter, "A Methodology for Testing Content Addressable Memories," in *Proc. 1985 Int. Test Conf.*, IEEE Comput. Soc., pp. 471–474.

CHAPTER ELEVEN

DESIGN FOR TESTABILITY

Nowhere in the test arena is there a more widely discussed and hotly contested subject than design for testability (DFT). The growth in the cost of developing and applying tests has grown to overshadow production of the circuits themselves in a number of cases. Multimillion dollar test capability acquisition projects focus the attention of high-level management on the problems of test and result in efforts to stem the increasing costs.

Test method research has created improvements in test generation methods, but continued pressure from complexity results in a reduction only in the rate of growth of costs, not in the dollars spent. The designers of complex ICs and pc boards are tasked to incorporate features into their circuits which are more "testable." Adding to the numerous criteria already to be considered during a design (speed, power, functions, options, manufacturability, reliability, etc.) has met with resistance in most design departments.

Testability is a difficult characteristic to define, and the softness of most definitions complicates acceptance of and compliance with DFT requirements. Before discussing alternatives for DFT implementation, a review of attempts to measure and define testability will be presented. The span of DFT is also difficult to define, as it includes such meager efforts as the addition of an auxiliary observable node or "test point" to a circuit and complete built in test (BIT) or self-test schemes. An attempt will be made to touch upon the most popular methods.

DEFINING TESTABILITY

From an engineering viewpoint the judgment of testability on the basis of the difficulty of developing and executing a test for an electronic device is unsatisfying. Yet such nebulous definitions persist in most test environments. Analyzing the factors that contribute to the subjectiveness reveals several supporting definitions which must be provided for each project which includes a test effort before testability itself can be defined. There are no universally correct answers, but the method for evaluation can be generalized.

Test Goals

In many cases the goals of testing and the goals of testability are confused. Requirements for comprehensiveness and diagnostic resolution are test goals and say little about the design of the circuit. Only when the test goals cannot be met with available test technology is the correlation an absolute one. Otherwise, the relationship is economic. Economics is the driver of testability, and it is therefore necessary to define the goals in order to place the economics of testing on a sound footing.

It is assumed that a careful job of planning the test approach has been completed (see Chapter 12). Peculiarities in the intended approach, such as the availability of particular test equipment and test generation systems, are the basis for setting numeric test goals. Compromise in the process of negotiating DFT measures to be taken may result in a later change in the test approach, but it is essential to begin with a clear test concept.

Comprehensiveness is often quoted as a percentage of faults occurring in the circuit which will be detected by the test. "Detected" means only that the test will not result in a pass or go indication if a detectable fault is present. A rigorous definition must also include a description of the fault class or classes to which the test applies. The ability of a test to differentiate between faults and give a clear indication of the nature of the defect is also important if the defective circuits are to be repaired.

Diagnostic resolution is a measure of the ability to differentiate between faults, but its numbers are stated from a different perspective. Referring to Chapters 3 and 4, it can be seen that when a path is sensitized from a particular active fault to an observable output, several faults along the path may also be active. The test detects all these faults simultaneously. There are several techniques for pinpointing the faults, as has been covered in Chapter 9. Recording the degree to which each attempt to differentiate the detected faults is successful is the diagnostic resolution. A histogram of the results appears in Figure 11.1.

The precise meaning of the histogram can be adjusted by the accounting routines used and is often changed to record the number of replaceable units (components) implicated by each failure message. In either case the histogram form is useful in showing the distribution nature of diagnostic resolution goals. An absolute goal of "all tests must locate the fault to within two ICs" is often proposed. A more realistic goal would be a set of numbers which characterize the histogram, such as "50% of the faults shall be located to within one replaceable unit, 90% shall be located to within three replaceable units, and no more than 5% shall be located to greater than 8 replaceable units."

Diagnostic capability may be stated indirectly in the "mean time to repair" or

Diagnostic Messages

Implicated ICs		
1	XXXXXXXXXXXXXXXXXXXXXXXXXXXXXXXXXX	34%
2	XXXXXXXXXXXXXXXXXXXXXX	21%
3	XXXXXXXXXXXXXXXX	16%
4	XXXXXXXXXXX	11%
5	XXXXX	5%
6	XXXXXXX	7%
7	XX	2%
8		0%
9	X	1%
10	XXX	3%

Figure 11.1. Distribution of suspect ICs per diagnostic message. Each unique faulty output condition implicates one or more possible faults.

MTTR requirements for the assembly being tested. MTTR incorporates both the diagnosis and repair operations and is dependent upon the level of repair expected. Replacement of a socketed part is quick, while unsoldering and replacing a large IC package is laborious. The time to diagnose must be extracted before becoming a meaningful part of the test goals.

Time is often a consideration in test application at the go–nogo level also. The maximum speed of the ATE used is fixed, so the test time is directly related to the number of test vectors needed to achieve the required test comprehensiveness and diagnostic resolution. Test length is not absolutely tied to testability, however, since the intelligence with which test vectors are created influences their efficiency. Many more random patterns than deterministically created patterns are needed for a given level of comprehensiveness for a complex circuit.

Test Techniques

The dependence of testability upon test techniques is easily stated, but not so easily evaluated. Use of a particular test generation method or a particular variety of ATE has more influence on the approach used in DFT than on the degree to which DFT is applied. Some DFT methods, such as LSSD, require capabilities which may not be present in all ATE to be used efficiently.

There is a general ranking of test tools and the degree of DFT needed. ATG systems generally work best if the amount of sequentiality is held down in the logic design, for example. During testability discussions the limitations and strengths of the test tools available should be kept in mind.

Economic Tradeoffs

Evaluation of DFT efforts is best made on the basis of cost. The estimate of test generation and test application costs for a particular approach is weighed against the cost of implementing DFT. Adding a connector has costs associated, beginning with the circuit layout and continuing through the life of the product in the form of component and assembly costs.

Test costs also include development and operating costs. The extra few minutes to execute a test on a relatively untestable circuit must be multiplied by the total expected production volume for the life of the product. The extra cost of troubleshooting is likewise multiplied by the production volume and then multiplied by the predicted failure rate. In general, the equation for cost evaluation of testability changes is

$$C_{IMP} + Q(\Delta C_{MAN}) < \Delta \{ C_{GEN} + [C_{APP} + (1 - Y)C_{TS}]Q \}$$

In words, the equation states that as long as the cost of a particular DFT measure is less that the savings in test cost, it should be implemented. The cost of DFT implementation (design cost) is added to the recurring manufacturing cost times the quantity of units to be manufactured to form the DFT cost on the left-hand side. The right-hand side is the difference in costs associated with test generation, test applications, and troubleshooting. On the right-hand side Y represents the manufacturing yield appearing at the test station. If 80% of the units pass the test, then 20% must be troubleshot.

Seldom are the costs calculated as accurately as is suggested in the equation. The labor for calculating the equation for each test point or gate added would be too great. Rather, the impact of a group of changes to the design is estimated and weighed against the estimate of test impact.

TESTABILITY MEASUREMENT

Adding objectivity to testability determination would help the analysis of DFT versus test cost tradeoff greatly. Rather than calculate the effect of each change on the circuit requested, a formula for determining the effect of each change on "testability" and another for the effect of "testability" on test cost could guide decisions quickly. The "testability figure of merit" of a design could be specified along with other design criteria and monitored by management.

Several methods for testability measurement have been proposed, and a few have been implemented in limited applications. Correlation of the resulting testability figure of merit and the actual test costs is not well established for any of the methods. The assumption is made in most methods that the test to be developed is a digital functional one and that an ATG program is to be used to derive test vectors. This is a worst case assumption since digital functional tests are the most difficult to create and ATG programs generally show less ingenuity in finding a test than manual methods.

Controllability/Observability

As introduced in Chapter 4, the logic-structure-dependent constants of controllability and observability are interesting and useful characterizations of a circuit. The relationship of these constants to testability is strong enough to result in assumed equivalence, although several other factors are involved. Intrinsic [1] measures of testability such as controllability/observability can be converted to extrinsic measures such as test coverage and diagnostic resolution if the factors for test generation and test application are well characterized and constant.

Controllability directly affects the ease of activating a fault and indirectly affects the ease of finding a propagation path for the fault effects. Observability has direct influence on the latter and refers to the controllability constants for a basis of calculation. A generally accepted set of measures for controllability and observability of each node in a circuit consists of six values divided into two classes [2]. The two classes are combinational and sequential values, and each class consists of the controllability of achieving a 1 at the node, the controllability for a 0, and the observability of the value at the node. Herein the notation CC1, CC0, C0, SC1, SC0, and S0, will be used, and the method of calculating the values as nonnegative integers will be adopted.

Combinational contollability functions must be derived for each gate or functional block in the circuit. The essential question to be answered for each output of each gate is "What is the easiest way to achieve a 0 (1) at this node?" For the NAND and NOR gates the functions are shown in Figure 11.2.

Each of the output node controllabilities is calculated in terms of the input node values. The controllability or a primary input to the circuit is defined as 1. The function "min" is defined as selection of the lesser of the controllabilities within brackets. The "+" sign denotes arithmetic addition of the controllabilities. MSI and LSI functional blocks can be treated without breaking them into constituent gates once the controllability functions have been derived and catalogued (Figure 11.3).

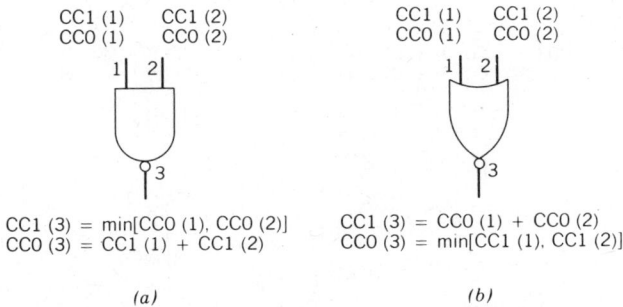

CC1 (1)   CC1 (2)
CC0 (1)   CC0 (2)

1│ 2│

3

CC1 (3) = min[CC0 (1), CC0 (2)]
CC0 (3) = CC1 (1) + CC1 (2)

(a)

CC1 (1)   CC1 (2)
CC0 (1)   CC0 (2)

1│ 2│

3

CC1 (3) = CC0 (1) + CC0 (2)
CC0 (3) = min[CC1 (1), CC1 (2)]

(b)

Figure 11.2. Combinational controllability calculation for basic gates. (a) NAND. (b) NOR.

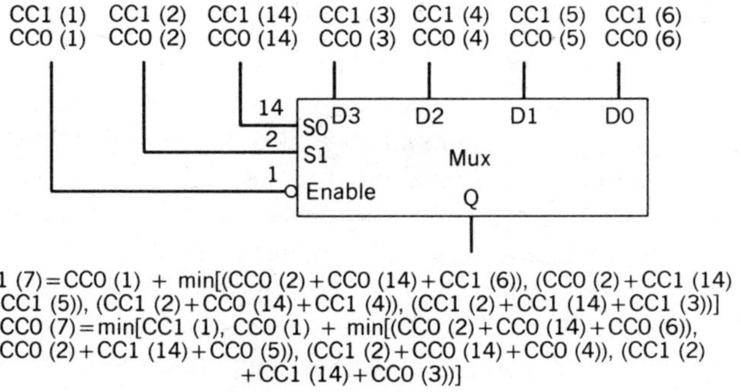
CC1 (1) CC0 (1)
CC1 (2) CC0 (2)
CC1 (14) CC0 (14)
CC1 (3) CC0 (3)
CC1 (4) CC0 (4)
CC1 (5) CC0 (5)
CC1 (6) CC0 (6)
14
2
1
S0
S1
Enable
D3
D2
D1
D0
Mux
Q
CC1 (7)=CC0 (1) + min[(CC0 (2)+CC0 (14)+CC1 (6)), (CC0 (2)+CC1 (14) +CC1 (5)), (CC1 (2)+CC0 (14)+CC1 (4)), (CC1 (2)+CC1 (14)+CC1 (3))]
CC0 (7)=min[CC1 (1), CC0 (1) + min[(CC0 (2)+CC0 (14)+CC0 (6)), (CC0 (2)+CC1 (14)+CC0 (5)), (CC1 (2)+CC0 (14)+CC0 (4)), (CC1 (2) +CC1 (14)+CC0 (3))]

Figure 11.3. Controllability for an MSI combinational chip derived directly from function descriptions.

The complex functions resulting from a relatively simple MSI IC serve to illustrate the computing challenge of analyzing a complete circuit. Reading the equations is not as difficult as it first appears. The CC1 of the output pin 7 is the controllability of enabling its output plus the easiest controllability of selecting an input and driving that input to a 1. In the case of CC0 the output can be driven to a 0 by disabling the output, which then is examined by the minimum function as an alternative to selecting and forcing a 0 through one of the channels.

Observability is calculated for each circuit node in a reverse fashion. That is, the combinational observability of the input to a gate (Figure 11.4) is dependent upon the observability of its output. The controllability of other inputs necessary to sensitize the output to the input node being examined is also vital to the calculation.

Note that the same observability applies to observation of either a 1 or a 0, which also means that the cases of a NAND gate and an AND gate are the same. The use of controllability in the calculations implies that all controllability calculations must be performed first.

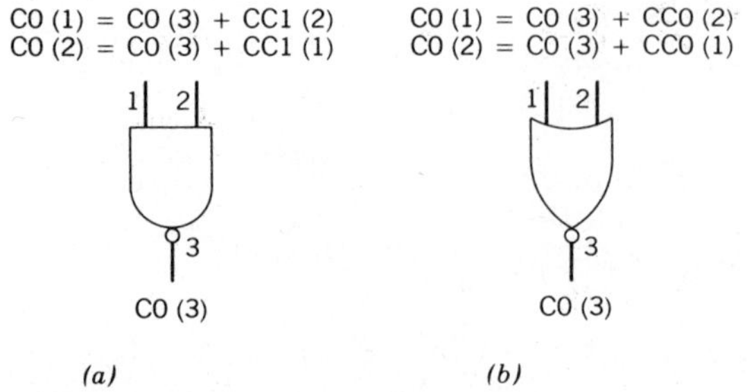
CO (1) = CO (3) + CC1 (2)
CO (2) = CO (3) + CC1 (1)
1
2
3
CO (3)
(a)
CO (1) = CO (3) + CC0 (2)
CO (2) = CO (3) + CC0 (1)
1
2
3
CO (3)
(b)

Figure 11.4. Combinational observability calculations for basic gates utilize the results of the controllability calculations. (*a*) NAND. (*b*) NOR.

Combinational controllabilities and observabilities obtained by the method described above do not reflect the depth of a node within a circuit. Depth is the number of logic ranks or gates that must be traversed to reach a primary input (or output, depending on which way the counting begins). Adding an extra controllability of 1 to each of the functions reflects the node depth.

The depth of a node becomes useful when feedback occurs in the circuit. Properly initializing the circuit values permits calculation of all node controllabilities by iteration. All primary inputs are given a combinational controllability of 1 at the beginning, and all other nodes are given a value of infinity (or a very large number). Controllabilities are then calculated iteratively until the values do not change or until an arbitrary number of iterations is reached.

The first pass through the left-hand gate in Figure 11.5 gives an output $CC0(3)$ = ∞ since the $CC1(6)$ = ∞ from the initial settings. The value of $CC1(3) = 2$ using the "1-extra-for-gate-depth rule." The value of $CC0(6) = 4$ since it is the sum of $CC1(3)$, which is now 2, $CC1(5)$, which is a primary input and thus 1, and the extra 1 for gate depth. $CC1(6) = 2$ in the same way $CC1(3)$ was calculated. The second pass at $CC0(3)$ now gives a value of 4, and further evaluations do not result in changes for any of the values.

Sequential controllability uses the idea of depth directly. In an attempt to characterize the difficulty of manipulating sequential elements a sequential depth constant is given to each gate or functional block. If the gate is combinational, its sequential depth is 0. A simple flip-flop might be given a sequential depth of 1. The assignment of constants to more complex circuits is difficult and usually involves evaluating a gate and flip-flop model.

As with combinational constants the sequential constants represent the easiest method of achieving the objective. If a flip-flop can be asynchronously set or cleared, it is more controllable than a similar flip-flop which can only be synchronously set or cleared. The sequential depth assigned to the first case might be 0, while the latter would be assigned a 1. The entire circuit is evaluated repeatedly until the constants converge for both combinational and sequential controllability.

Testability estimation programs perform the calculations based on a description of the circuit and tabulate the constants for each circuit node. Although the individual constants have no direct correlation to the cost of test generation, their relative values indicate areas of the circuit which need change to improve testability [3]. A value of infinity (or whatever maximum number is expressible) has

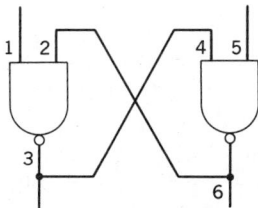

Figure 11.5. Cross-coupled NAND gate flip-flop requires several iterations of controllability calculation for solution.

the special meaning that the indicated logic state cannot be achieved (at least within the maximum number of loop iterations allowed during calculations). Such nodes would be prime targets for investigation.

IMPLEMENTATION OF DFT

Whether the testability of a circuit is estimated automatically by a computer algorithm or manually by a test engineer, there is no value in the effort unless the feedback can be used to decrease the cost of designing, manufacturing, or testing the circuit. The economic pressure of rising test costs have resulted in the invention of several elaborate and rigorous design methodologies, but design for test is not restricted to the application of one of these schemes. Many straightforward guidelines have been developed. These ad hoc testability improvement techniques are in more widespread use due to their lower impact on design cost and general applicability to MSI and LSI designs which include off-the-shelf parts.

The ad hoc testability improvements consist of minor circuit changes which improve controllability or observability. A test engineer may suggest these changes to a design after the basic work has been completed without throwing the entire circuit back to the beginning. Effective administration of ad hoc techniques usually requires that the test engineer look over the designer's shoulder occasionally and suggest changes before the process of design is complete. The most efficient administration method of the design process involves training the designer to be test engineer and thus eliminating the middleman. Training in test engineering is a diverse, time-consuming proposition and due to changes in the ATE and test generation systems would become ineffective if not maintained. Separation of the two tasks allows better efficiency in each and maintains a clear set of priorities which might otherwise get lost in the effort to complete the design.

The emphasis of ad hoc DFT varies with the style of design being implemented and the test technology to be used. General rules will be given first based on a design methodology using off-the-shelf parts and a test approach using static functional testing for a pc board test. These least expensive methods for design and test give common denominator results which can be embellished for more complex situations.

Basic DFT

Beginning with the basics means thinking of DFT in terms of three objectives: initialization, control, and observation. These themes sound familiar after the previous chapters. Initialization is the process of forcing all circuit nodes to a known value. Control is the process of achieving a goal logic state of a given node and is associated with the activation of faults. In reality, control plays a part in both initialization and observation. Observation is the process of sensitizing a path from

a given node to a primary output to enable comparison of the actual value against
an expected result.

Initialization. The difference in environment between testing and operation of a
circuit is greatest in the area of initialization. The scale of time and the importance
of synchronization with external clocks are exaggerated during static functional
testing. At test application speeds of a few thousand per second, cycling a counter
string of 16 bits is inconvenient, and cycling a string of 32 bits is impractical. The
designer may depend on the first few seconds of circuit operation at 10 MHz to
home the circuit values to known states, but the test engineer cannot wait.

   A simple implementation of a clock generator which produces a base frequency
(assume 10 MHz) and a second frequency 24 times slower (416.7 KHz) is shown
in Figure 11.6. The circuit uses MSI logic IC and a crystal-controlled oscillator.
Although the design apparently obeys good design guidelines, it is untestable by
static functional means. The oscillator output cannot be interrupted or even observed directly. Even if a dynamic tester were used, the task of synchronizing the
tester to the internal oscillator would be formidable since the oscillator output state
would have to be inferred from secondary effects elsewhere in the circuit. An in-
circuit tester could access the oscillator output via a probe contact and overpower
it to substitute the tester's own clock.

   Beyond the oscillator problem lies a second impasse in the form of the divide-
by-two flip-flop. The designer was unconcerned about the phase relationship of
the slower signal to the faster one, so the initial state of the flip-flop is unimportant.
The set and clear inputs have been pulled up to the same resistor in the interest of

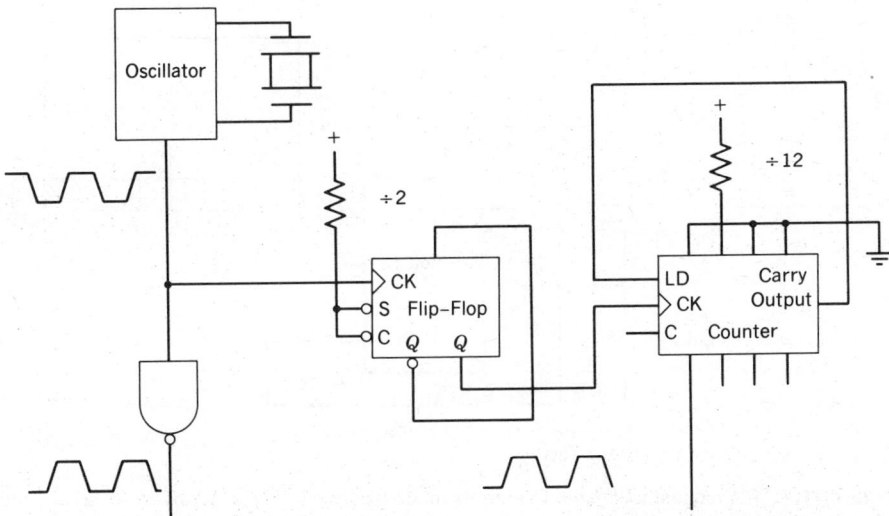

Figure 11.6. Schematic of a multiple-frequency generator exhibiting poor testability.

saving resistors. Again, the tester cannot directly test the output of the flip-flop in order to determine its state, and a complex distinguishing sequence will have to be programmed to resolve the circuit condition. In this case the in-circuit tester also has difficulty in dealing with the situation. If the set or clear is forced with a contact, the other is also activated. The output of the flip-flop will become (1,1), while the set and clear are held but will revert to an unknown when they are both released simultaneously. The output state could be sensed directly through a contact in the in-circuit case, which would result in a single-step distinguished sequence.

The counter is simply an enlargement of the flip-flop problem with the carry output connected to load the counter with (0,1,0,0) after its output reaches (1,1,1,1). The only remaining component shown is a buffer gate used to supply the current necessary to drive a large fan-out of the high speed clock. In the world of the static functional tester all outputs of the circuit will be unknown at power up. Actually, no matter what values the flip-flop and counter assume upon power up, it can be proven that their values will begin a sequence resulting in the counter passing through counts of 4, 5, 6, 7, 8, 9, 10, 11, 12, 13, 14, 15 repetitively.

The problem at hand is to modify minimally the circuit to permit single stepping of the clock and combinational initialization by the tester. A possible solution is shown in Figure 11.7 which uses only two resistors and three connector pins (or test points) to achieve the result. Many other solutions are possible.

In place of the totem pole output TTL NAND gate used as a buffer for the high-

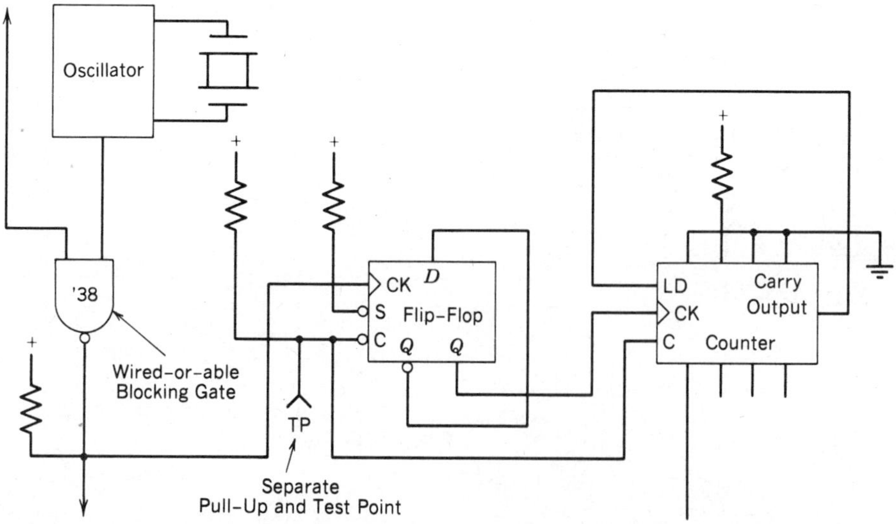

Figure 11.7. Ad hoc testability improvements made by simple modifications of the schematic of Figure 11.6.

speed clock, an open collector '38 NAND gate has been used. The input of the flip-flop has been moved to the output of the buffer rather than feeding directly from the oscillator. The second input of the '38 gate has been connected to an external input, allowing the tester to force the input low and disable the output. The high-speed clock net is also connected to an external pin. When the '38 gate output is disabled, the tester clock can drive the net directly, allowing stepping or synchronization.

A pull-up resistor has been added to the net since the '38 does not contain an active pull-up. The connection of the tester driver and the '38 is characterized as a wired-OR (although it exhibits the truth table of an AND), but the '38 output will probably be held off except when measuring its frequency. The second resistor has been used to split the pull-up functions for unused gate inputs so that the set and clear of the flip-flop can be separated. A third connector pin (in this case a test point) enables the tester to force the clear input to a low. This action clears the flip-flop asynchronously and assures a known state in a single step.

The clear of the counter which was unused before has been added to the test-pointed clear net. The circuit can be brought to a completely known state in the first input vector without specialized homing or distinguishing routines. Alternative solutions depend upon the priorities of the design. If the oscillator is mounted via a socket, it can be removed during the test, and a plug which matches the socket could be used to control the net. It is not the method used but the goals achieved that make the circuit initializable.

In the example described, the changes allowed initialization in a single step. Such an ideal solution may not always be available due to design restrictions. Use of a RAM embedded in a circuit precludes single-step initialization, for example. Replacing a RAM chip with separate flip-flops would allow direct clear of all cells, but would be ridiculous as a solution due to the much greater number of ICs involved. A more realistic testability feature might involve providing an alternate method of activating the WE and manipulating the address lines to reduce the initialization time to one or two steps per address.

Initialization is the process of determining the states of all nodes of a circuit or forcing a known value to occur at each node of a circuit. When a tester can be synchronized to the circuit's internal clock, the internal states must be determined in reference to that clock. Otherwise, the circuit must be forced to known states in reference to tester inputs. Disabling oscillators, opening feedback loops, and providing combinational paths for setting or clearing registers are the usual techniques of initialization testability improvement.

When LSI or VLSI technologies are encountered, the direct approach to initialization cannot always be taken. The sets and clears of a microprocessor's register set are not usually accessible directly, but a reset signal is usually provided. Initializing circuits around the microprocessor requires either a built-in micropro-grammed initialization routine or external access to the address and data buses along with associated interrupts and controls. Tester access to the microprocessor's control signals should therefore include the ability to interrupt the microprocessor and force it to relinquish bus control. Further initialization capability can then be pursued as bus-level controllability issues.

Controllability. Initialization is merely a specialization of control. The ability to control a clock circuit is useful beyond achieving a known state for all nodes. The specialization of control that is termed initialization does not concern itself with forcing or achieving specific values, but only known values. Activating the stuck-at faults of a circuit requires making each node either a 0 or a 1 and observing the state and then reversing the node state and observing its value.

In the example of Figure 11.7, if the buffer gate were not replaced but its other input were brought out through an input pin, the circuit would be initializable but not controllable. The clock signal could be blocked by driving the extra input to a 0, and a known value 1 would be forced on the clock net. Since the buffer in place cannot be safely overdriven by a functional tester and no external access was provided on the net in the first case, the value 0 cannot be achieved on the bus and the circuit is untestable.

Control has a functional meaning to the test engineer, not the circuit architecture meaning which the logic designer commonly uses. Achieving control of a data path is of importance to deriving a test. An internal bus which is used extensively between a microprocessor and its memory is not very controllable if the only access from outside the circuit is an RS-232 port. A test engineer uses the term controllable to mean that a circuit's values are easily or at least reasonably easily changed with the context of a test generation and application scenario.

Circuit structures which are normally considered control circuits by the designer are key to influencing data paths by nature and are prominent in the controllability review. At each MSI or LSI chip input, signals labeled enable or select are excellent choices for scrutiny in a testability review. The typical hierarchy of circuit feature importance to control is as follows:

Clock circuitry. This refers to circuits which generate or control synchronizing signals, particularly those with large fan-out.

Enables and Selects. These are signals and associated circuits which direct data flow and activate functions. Instruction buses and interrupt structures may be thought of as enables.

Data. Manipulation of data buses and storage mechanisms is required to achieve control of the states of many nodes within a circuit.

Combinational circuits provide no challenge to controllability. Other than the possibility that a large number of primary inputs may have to be chosen correctly to activate a given fault in a single test vector, the task is uncomplicated. Sequential circuits add the dimension of time and the complexity of potential control procedures with a large number of test vectors.

The microengine in Figure 11.8 is a common example of an initializable circuit which is nevertheless not controllable. The counters which constitute the next address register can be cleared by an external line. Once the sequence is started, however, the address selected is either the next count or a jump directed by the contents of the microprogram ROM. Since there are three 4-bit counters, there may be 4096 addresses. If the program is a lengthy one which repeats most of its sequence but differs slightly depending on data, most of the steps will not add to the test after the first data example is presented. Yet there is no way of jumping quickly to the portion of the program which depends on data.

Again, there are several possible improvements to the circuit from the standpoint of controllability. A multiplexer could be inserted into the address bus with alternate inputs connected to an external port. This would allow immediate access to any address in the memory. Unfortunately, the addition of a mux would also slow the maximum cycle time of the microengine. Placing the mux in the next address loop and allowing external alternate control of the jump enable line would

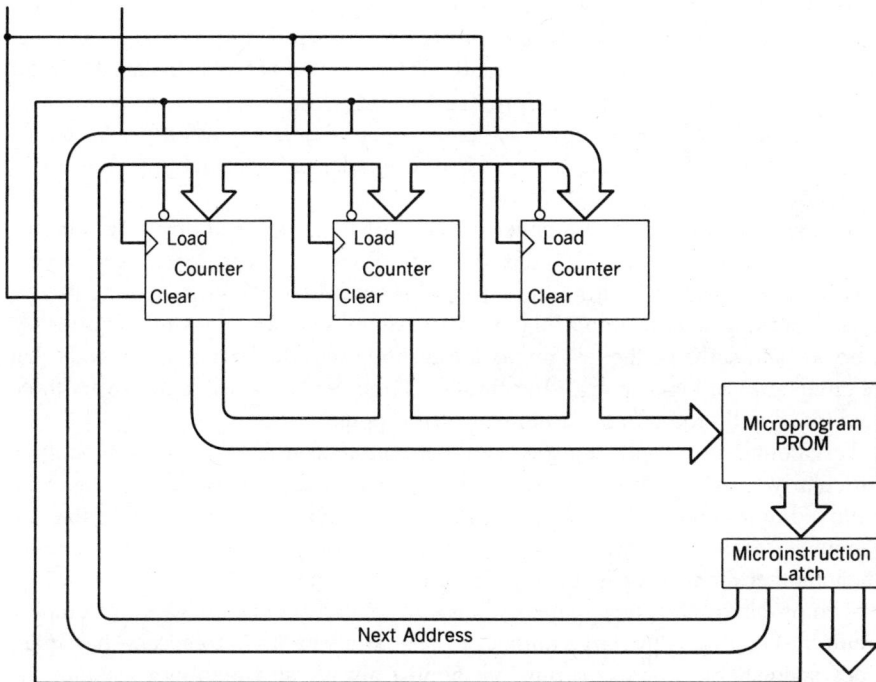

Figure 11.8. Schematic of a microengine. Deep sequentiality renders the circuit uncontrollable for test purposes.

create the same effect while slowing only the jump instructions. Other drawbacks include the relatively large number of primary inputs needed to feed the mux in any case.

An alternative which uses no hardware but may restrict available space for software is the direct use of instructions written into the microprogram and stored in ROM. A simple set of instructions which perform conditional jumps to addresses which are derived from data would suffice to check out the circuits. The conditional jump hardware is not shown in the figure for simplicity, but it is not hard to imagine. If some form of alternative addressing is not allowed, thousands of clock cycles will be needed to test the circuit. In the case of a static functional test this translates into thousands of test vectors. These are expensive to derive and expensive to perform.

Another common example is a communications circuit which performs serial interface. Multiple operating modes are often used, and the circuit may be required to change protocol or role between messages. If the controls for forcing a role (receiver, transmitter, bus controller, bus monitor) and a mode (1 stop bit, 2 stop bits, even parity, odd parity, etc.) are not brought to a primary interface in a straightforward combinational manner, lengthy sequences will be required.

Observability. A review of the path sensitization process discussed in Chapter 3 will show the importance of observability. Once a fault has been activated, its effects must be propagated to an externally observable point. At each gate or circuit block which must be traversed the problem of sensitizing the gate to the path which as a fault effect must be faced. Sensitizing a gate requires controlling the inputs that do not have fault effects present. Therefore, a loss of observability adds to the burden of controllability.

As a simple guideline the number of ranks of logic between observable points should be controlled to increase observability. Each rank represents a gate which must be sensitized, so limiting the number of gates between primary outputs is a commonsense generalization of the problem. Suppose that a block of test points is reserved, and some of the test points are inserted into the wiring at intervals that are three ranks of logic apart. If the circuit is large, inserting test points every three logic ranks will require a large number of test points.

Test points are usually clustered or concentrated around complex functions, particularly those residing in VLSI. Since the insides of these chips cannot be improved in testability, the strategy is to get as close as possible to assure that the diagnostic resolution is adequate to select the chip only when it is at fault and not when another member of its uncertainty group is named.

Two problems become apparent when test points are to be assigned in complex circuitry. The first is the large number of test points required to access the circuit nodes adequately. Connector pins are almost always at a premium whether the design is an IC or a pc board. Other forms of test points require components such as sockets or at least takeup area on the substrate which might be used for logic.

Additional concerns are often raised over the effect of connecting unterminated wires to internal circuit nodes. Although the line lengths are usually small, the effect could be a problem in very-high-speed circuits such as ECL or gallium arsenide technologies. The test point connectors themselves present an environmental problem for military designs since they are not sealed in insulative material. Corrosion can cause high-resistance leakage paths between test points, and there is the possibility that a field technician could inadvertently short out an open test point, damaging the circuit.

The second problem encountered is the tendency for logic to be concentrated within VLSI components. If the IC designer has not included accessible test points or usable testability circuitry, it cannot be added later. The result for the pc card test engineer is a lack of test points where they are needed most.

Several mechanisms exist for creating virtual test points within a circuit at either the pc or IC level. It is not necessary to examine all the test points at each test step. The important points are closest to nodes which carry fault effects from a presently active fault. Adding a multiplexer and allowing the test generation process to control the selection of sources multiplies the effect of a few primary outputs. The extra component (multiplexer) added as shown in Figure 11.9 is a tradeoff for a savings of 10 primary output connections. The 16-to-1 multiplexer has

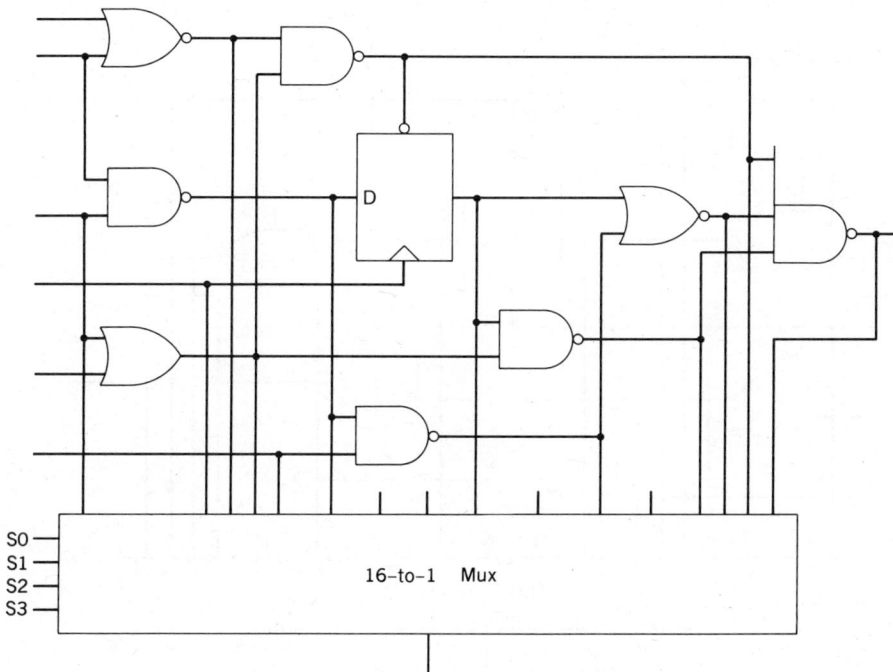

Figure 11.9. The multiplexer at the bottom of the schematic allows direct observation of many internal points through a single physical pin.

no function in the circuit other than the enhancement of testability. Its select lines are connected to primary inputs to permit the test generator control of which internal point is being monitored by the single multiplexer output at any given time. Five test point connections (four inputs and an output) can thereby access sixteen internal nodes. The example shown is not an advisable distribution of those 16 nodes since maximum use would be made if each monitored node in a line were several layers of logic from the next. The compactness of the drawing requires the simplification.

Multiplexed test points achieve a $2^n - 1$ expansion in the number of points accessible with n primary connections, but the gate cost is greater than n gates, and the gates are not used except for test. Extra gates require space, consume power, and lower reliability since they may develop faults themselves. Such failures are sometimes not counted as true faults since they do not impair the function of the circuit. A shortened input to the mux would cause an s-a-0 at the node which it was intended to examine, however, so some faults count in any case.

A different approach to the same end of using extra gates to multiply the effect of a few primary outputs involves the use of shift registers. Replace the multiplexer of Figure 11.9 with a parallel loading serial shift register as in Figure 11.10, and the number of actual primary connections needed to access 16 internal points drops to 3 (load, shift clock, and serial output). The test generator must control the load

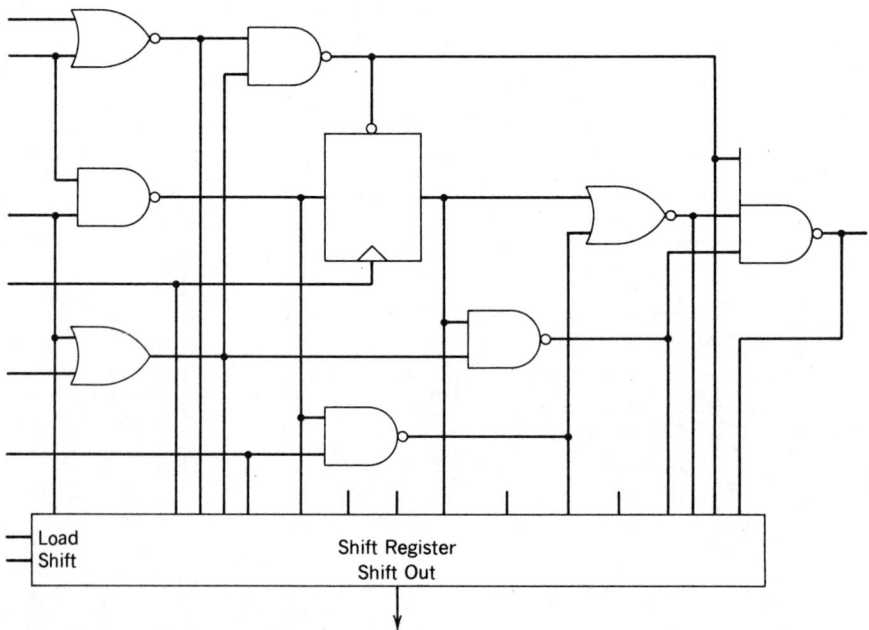

Figure 11.10. Substitution of a shift register for the multiplexer in Figure 11.9 saves control pins, but slows the test routine.

and shift clocks in order to trap the node states being monitored at some appropriate time and shift out the results later. Once the information is loaded, all access points can be examined during the shift operation. The correspondence of the shifter output with the clock count during the shift out allows deduction of which circuit nodes are represented by the current output.

If the shift register is loaded and shifted out between each assertion of a new test vector, the state of all monitored nodes is known for each test step, and the internal points can be considered virtual test points. Note that the relationship between the number of virtual test points created and the number of external connections needed is irrelevant. Only the three test points will be required, no matter how long a shift register is used. The gate count added to augment testability rises linearly with the number of virtual test points. Lengthy tests may also result as the long shift register must be emptied after each pattern. The repetition of a simple fixed routine between each test vector can be directly coded into the ATE language for transparent operation as far as the test generator is concerned.

Rigorous Design for Testability

Rigorous methods are techniques for improving controllability or observability which must be engrained into the design. As opposed to the basic techniques which were described as add-on circuits or circuit changes which could be made after the fact, rigorous methods require consideration during design and influence the circuit architecture heavily.

The most widespread technique in use incorporates the shift register idea, but uses it for control as well as observation [4, 5]. In addition, the shift register chain is incorporated into the design function, reducing the added gate cost of testability. The premise of LSSD is that any circuit can be realized using combinational gates and a simple latch circuit which doubles as a shift register in the test mode. In Figure 11.11 the latches labeled LA and LB are simple cross-coupled flip-flop-type latches, not edge-triggered devices. LA latches are fed by either the parallel inputs from other combinational logic or from the previous LB latch in the string. All LB latches are fed from their respective LA latches. The latch enable called PAR1 allows data from the parallel inputs to determine the state of the LA latches. When the PAR1 enable is closed, data are held in LA. The second rank enable is called SHIFT2 and permits data to transfer to the LB string. Cycling these two enables in that order places the data from the previous combinational logic stage onto the latch parallel outputs and makes it available to the next combinational logic. During normal circuit function this sequence will be used.

During test the sequence changes to take advantage of the shift characteristic created by the linking of LB's output back to LA of the next bit in line. The enable called SHIFT1 allows this to take place. Data can be parallel loaded from the previous stage and shifted out serially or shifted in serially and applied to the next combinational logic stage. No other flip-flops or storage cells are allowed. Testing

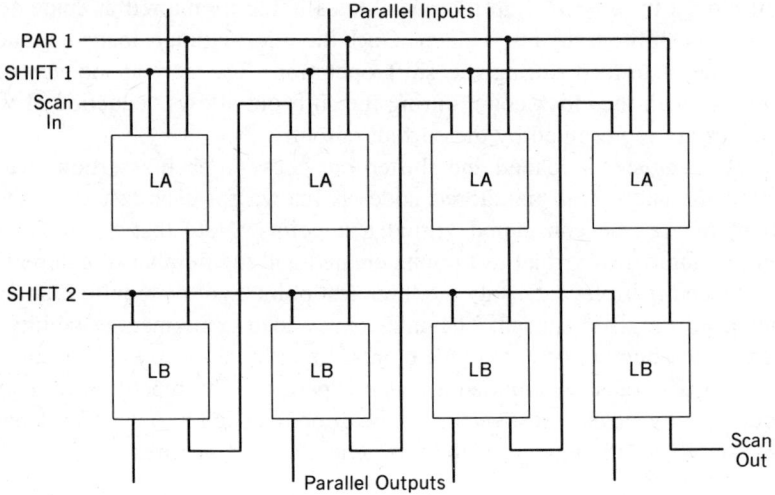

Figure 11.11. A dual mode parallel register–serial register made up of two ranks of latches.

an amorphous block of logic between two scan registers is a matter of shifting in the required drive signals (see Figure 11.12) and shifting out the results. The sequence is as follows:

1. Cycle SHIFT1 and SHIFT2 to move the next test vector for this block of logic into the upper register serially.

2. After the logic values have had time to settle, use the PAR1 clock of the lower register to trap the results of the test step.

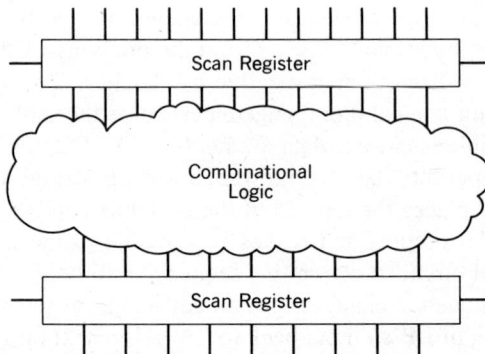

Figure 11.12. Using scan registers as the only sequential element of a design reduces testing to a series of tests for the logic between ranks of registers.

3. Use SHIFT1 and SHIFT2 of the lower register to shift the results out serially and compare them bit by bit with the expected results.

4. Repeat 1 through 3 for each test vector in the sequence.

Notice that while data are being shifted into the upper register the same clock sequence can shift the results of the previous test step out of the lower register. In fact, all registers are loaded and unloaded simultaneously. The scan clocking can be carried out autonomously since it is not test data dependent. The shift cycle is implemented between each test step and is usually performed by a subroutine in the ATE language or a hardwired controller in the interface adapter.

Test generation is performed on each of the combinational circuits between scan ranks separately. The shift path is tested by a simple algorithm such as sending a "goal posts" pattern of 0s bounded by two 1s through the shifter from the serial input to the serial output without cycling the functional clock. As modeled for fault simulation and test generation each of the combinational circuit sections is bounded by "virtual test points." The outputs of a circuit section are the parallel inputs to the scan path, and the inputs of the next circuit are the parallel outputs of the scan path. Since the results are shifted out as the next rank's inputs are shifted in, there is no interference, and each circuit section's test is fully independent.

Although the test engineer's task is simplified, there are several caveats for the designer. The rule concerning only combinational logic between scan paths is a difficult one to follow, but the consequences are serious. During the shift operation the outputs of the scan path latches change as data shift past. The effect is tantamount to cycling the inputs to the circuit section fed by the virtual inputs which the scan path supplies. Any asynchronous set or clear is likely to be triggered, and any stored data will be altered.

It is possible to use storage elements which have synchronous clocks, but only with care to restrict their clocking to non-scan-path sources. Multiple clocks must be controlled carefully to assure that the scan shift action does not occur when the clocks are active. Several hazardous situations are depicted in Figure 11.13 including a gated synchronous clock.

The scan path gated clock requires the outside clock to be held in a state that desensitizes the gate while the shift sequence is performed. Exceptions to the combinational circuit rule add complexity to the test clocking procedure and require special handling. The concept of scan path design is the creation of a transparent mechanism for simplifying tests. Many of the benefits are negated when each circuit must be separately treated.

When properly implemented, the LSSD or scan path design method results in nearly ideal circuit sections for test generation. Each section is combinational, which allows methods such as PODEM or the 9-V algorithm to work efficiently. In addition, the scan registers form natural boundaries which partition the circuitry into reasonable-sized segments for fault simulation. Available tools create func-

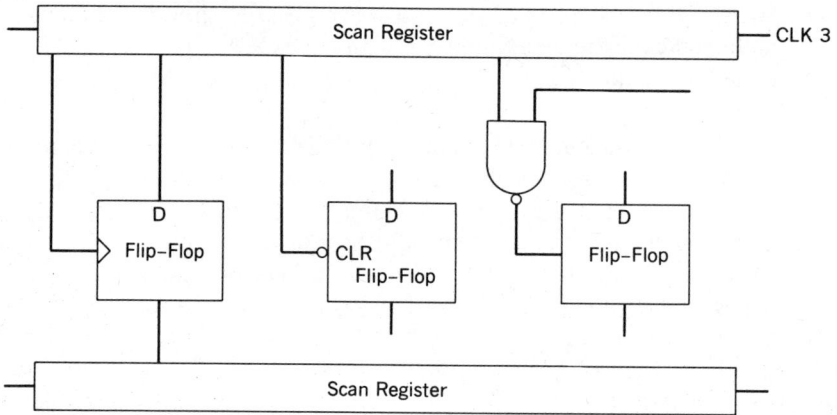

Figure 11.13. A simplified schematic shows forbidden sequential elements between the ranks of scan registers.

tional tests for each of the subcircuits, and the tests are then combined using the scan subroutines at the tester to make a complete test.

Scan path design lends itself particularly well to IC design since the gate-level design usually encountered in ICs can be manipulated to include the scan registers. Printed circuit design utilizing scan-designed chips can easily be adapted to make use of the same mechanism if the rules for design carry uniformly throughout all chips on the board. In theory, all the scan paths from each IC could be connected together in a long string and access in one cycle. The shift process would be too long to be practical in that case, so several shorter strings are accessed as separate entities.

The shift inputs and outputs could also be multiplexed if necessary, allowing the tester to switch between tracks of virtual test points. A tester which is to be used for scan path design board testing can be fitted with an interface which includes a multitrack serial–parallel converter. Each converter on the interface adapter must be at least as long as the scan path to which it is connected. Since most interface adapters are designed with a family of DUTs in mind rather than just one, it is advantageous to maintain a somewhat consistent length to the scan path strings.

An elegant but uncommon approach to scan path board design makes use of the hierarchy of testing to reduce test generation at the board level to a continuity test. The method requires all input or output pins of an IC to be connected immediately to a scan path on the inside of the chip. The key assumption asserts that all ICs have been tested before assembly into a pc board. The faults which remain are associated with the interconnection of those chips. The interconnection can be tested by loading a scan path at the output of a chip and reading the input scan paths of all chips which it feeds [6]. (See Figure 11.14).

A number of unique patterns can be used to detect interline shorts as well as

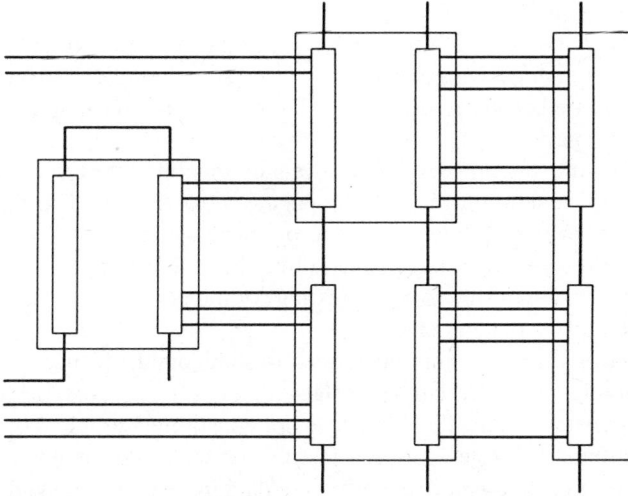

Figure 11.14. Scan design block diagram with scan registers at the input and output of all ICs. Test is reduced to individual IC tests and a continuity check between chips.

grounded lines and open lines. Even with sophisticated continuity algorithms there is no need for fault simulation or test generation since the actual gates within the chips are bypassed. Unfortunately, this method has the serious drawback of being unable to test for functional failures which may occur in the gates after some time in the field application.

The multiplexed chaining of scan paths can be continued to the system level to support fault tolerance and diagnostic programming. Complete design of a system using only gate array logic is certainly possible, but not the normal design for most cases. As costs for IC fabrication decline, the use of ASICs is increasing, but the exclusion of "glue chips" from between the ASICs is seldom practical. Inclusion of LSSD or its equivalent within ASICs does not seem to be a part of the trend.

Off-the-shelf parts are the nemesis of scan design since there are no major manufacturers of ICs that build custom VLSI or even LSI with scan path architecture. A discrete scan path register could be placed at the perimeter of each off-the-shelf microprocessor or other part, but the added circuits would be prohibitive and the benefit would be minimal due to the clustering of faults within the VLSI parts. Added parts count is a problem which plagues all scan path design. Typical scan path designs lose 20–30% of their gates to the testability improvements.

FAULT TOLERANCE

Circuit designers are barraged by requests for manufacturability, testability, reliability, maintainability, and a number of other "ilities." Satisfaction of one of the areas of concern is not necessarily helpful to the others. In addition, there is often confusion in the goals, which may lead the designer to think that adding a feature satisfies all requirements, resulting in neglect of some concerns. Many fault tolerant features are perceived as testability features, but in fact they result in problems for the test engineer.

The goal of fault tolerant design is to reduce or eliminate the propagation of fault effects. Test generation seeks to propagate those effects. The diametric opposition of these goals is indicative of the problem which fault tolerant circuitry poses for testability. Methods of circumventing the action of fault tolerant circuits must be included in the design to allow testing, or the confidence in the operability of the fault tolerant features is eroded.

Error correction circuits commonly added to semiconductor memory data paths are prime examples of potentially untestable fault tolerance (see Figure 11.15). Simple parity generation and checking does not constitute a problem other than the difficulty in automatically generating a test vector set as discussed in Chapter 3. The inclusion of a mechanism for altering the data in response to syndrome bits is the point at which fault effects are masked. Error correction circuitry (ECC) depends on interdependent parity bits which represent partial data word parity and are stored in memory along with the data word [7].

When the data are recovered, the ECC code bits are checked by a network which detects errors and deduces the faulty bit by the relationship of the code bits and the data bits. The faulty bit is then complemented and thus corrected before the data are passed to the requester. Most ECC systems correct all single-bit errors and detect all 2-bit errors, although the concept can be extended to other levels of correction by increasing the number of code bits that accompany each data word.

From a system point of view, ECC masks single errors. In order to prevent a buildup of single bit errors, which eventually would result in a double bit error occurring in a single word at some address and therefore in a failure, maintenance features such as an error log or single bit failure indicator lights are often included in the design. Unfortunately, ATE seldom includes the ability to sense light, and the test must then include operator intervention. The memory array may be tested by the usual means if the capability for reading the log is included as a subroutine at the end of the test program.

Testing the ECC feature is not at all straightforward. The syndrome bits are usually stripped off the data word before it is passed to the user. Probably, a fault in the ECC would result in a data failure at the user level. Positive identification of a fault and location within the ECC network requires observability of the code bits. Testing the ECC circuitry requires access to the data bus on the memory array side of the network to allow injection of faulty data. Testability features needed for ECC memory testing include

1. The capability to disable the error correction or observe memory array data before correction

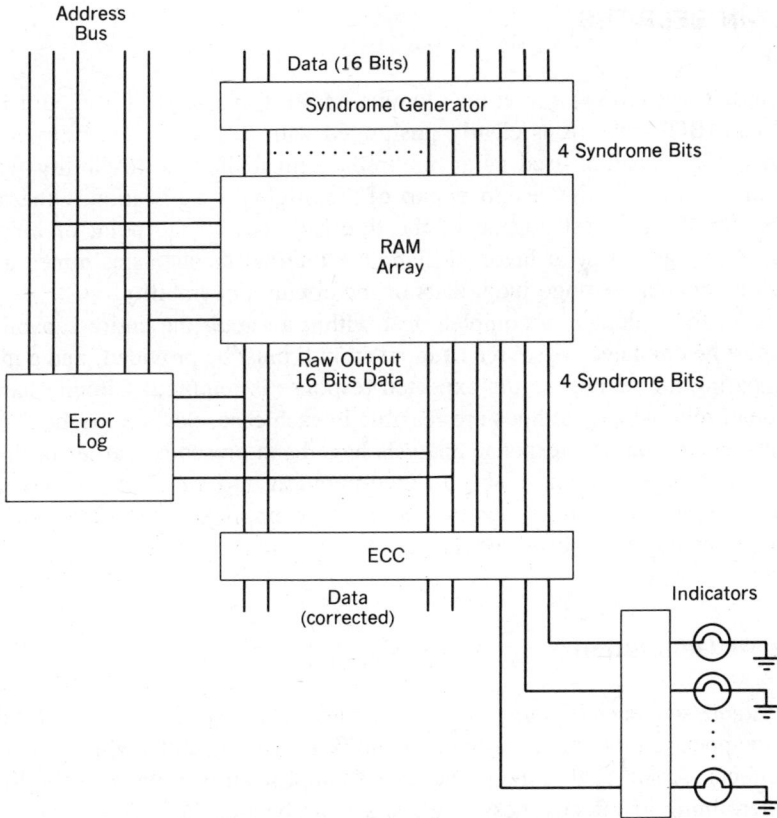

Figure 11.15. Block diagram of an error-correcting memory.

2. The ability to inject faulty data into the ECC circuitry in place of normal memory data
3. Observability of error correction syndrome bits
4. Direct access for both input and output of error log data and indicator light control

When fault tolerance is a high priority of a digital design, redundancy and voting schemes may be applied to the computational elements in addition to parity and error correction for memories and buses. Hardware redundancy must be made partitionable in a such a way that a test mode can deactivate the redundancy and exercise each unit separately. Then access to the voting or checking mechanism must be allowed to permit data which simulate a unit failure to be injected into the mechanism. From only a few examples it becomes obvious that fault tolerance and testability are not the same and are often opposite in effect.

BUILT-IN SELF-TEST

Built-in self-test (BIST) sometimes known as built-in test (BIT) or built-in test equipment (BITE) is not as clearly positioned with respect to testability as fault tolerance. BIT is often used as an umbrella term to include testability features which are included in the design as part of the original specification. Other applications take the self-test portion of the title seriously to the point of hindering outside ATE-based test in favor of elaborate internal mechanisms aimed at providing on demand go–nogo judgments of the circuit's operability.

In order to implement a complete test within a circuit the entire capability of ATE must be emulated. A source of input stimuli must be provided, and a method of comparing the outputs to the expected response is required. Circuit characteristics determine which methods are feasible in each case. Several mechanisms for stimulus generation and response analysis have been presented earlier in this text which lend themselves reasonably to incorporation into a circuit. In ideal applications, normal circuit functions can be used to supply one or both of the ATE functions, leading to reduced overhead.

Stimulus Generation

A repeatable sequence of data or control sequences suffices as stimulus if it can be made to appear on the input to the CUT and if it activates faults while propagating their effects. Assuring that the sequence will appear on the inputs is not difficult, but determining its effectiveness as a test set can be a challenge.

Integrated sources of stimuli include the contents of a microinstruction ROM or the output of a control sequencer. These sequences are repeatable and represent the entire set of possible input vectors for the decoders and other circuitry they drive by definition. Unfortunately, the microprogram usually includes jumps and seldom executes a signifiicant fraction of its instruction list in a reasonable amount of time. A method of disabling jumps and simply counting through the program counter must be installed to efficiently cycle through the sequence.

For circuits which are not directly driven by a source such as the microprogram ROM a pseudorandom pattern generator may be needed. The LFSR (Figure 11.16) covered in Chapter 3 is an adequate choice if the circuit to be tested is combinational.

The question which remains unanswered for the use of an LFSR is "How many times should the register be clocked to assure adequate test coverage?" Since the patterns generated by the LFSR are repeatable if the starting point is the same each time the sequence is generated, the patterns can be trapped from an identical LFSR and used as input to a fault simulator or other fault grader. Effectiveness of the test sequence will be reduced greatly if sequential elements exist within the CUT.

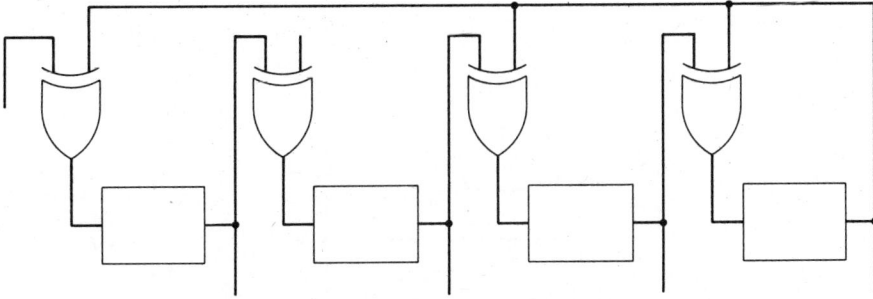

Figure 11.16. LFSR in the pseudorandom pattern-generating configuration.

Response Analysis

Simple storage of a sequence of expected outputs of a circuit for an entire sequence of inputs is far too cumbersome. A compression method must be used to make the data tractable. The LFSR again becomes useful, but in this case it is connected as a parallel input serial output register, as shown in Figure 11.17.

An LFSR used to monitor and compress the content of a set of parallel lines is called a built-in logic block observer (BILBO) [8]. After any specific number of patterns has been monitored, the contents of the LFSR contain a signature which has a high probability of being unique. The pattern is easily stored in a single address of ROM, a set of switches, or even as a hardwired pattern in the chip or board on which the circuit resides. The test consists simply of activating the sequence in the stimulus generator, enabling the BILBO to analyze the results, and comparing the signature to the prerecorded answer.

Limitations in the ability to cope effectively with sequential logic are the greatest drawback to the pseudorandom technique. BILBOs can be combined with other sources of stimuli to improve the sequential performance. The drawbacks which remain are classic in the area of self-test.

1. Added circuitry used for the self-test consumes space, consumes power, and reduces the overall reliability by providing more possible faults.
2. No explicit means of establishing confidence in the operability of the self-test circuitry is provided in most cases.
3. Test results are usually limited to a go–nogo determination. Diagnostic information is scant or nonexistent.

As a testability scheme usable by external ATE, BIST has some definite hazards. When BIST uses completely internal mechanisms, the impact on conventional test is negative. Even if the logic being tested internally is combinational, the LFSRs are sequential and the net result is inaccessible sequential logic. Exter-

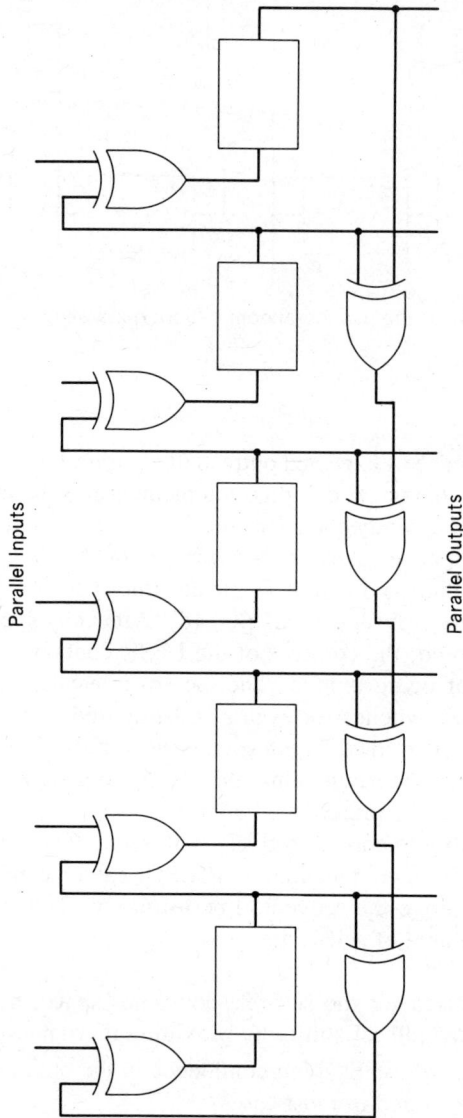

Figure 11.17. Built-in logic block observer (BILBO) consists of an LFSR in the parallel-in–parallel-out configuration.

Parallel Inputs

Parallel Outputs

nally accessible ports and special test modes which are sometimes added as BIT hardware enhancements have a positive impact. When the ICs containing such features are correctly incorporated into a design, the test features are still accessible at board or system level, and the overall improvements can be large.

SUMMARY

The Motorola 68020 microprocessor is an example of the application of several testability enhancement steps as a conscious strategy during the design process [9]. As such, it summarizes the potential harmonious use of different techniques of DFT.

The partitioning of the chip into circuit sections or blocks of differing function and differing characteristics allows appropriate test techniques to be applied to each subcircuit. Natural partitioning was aided by a few muliplexers and a small amount of added microcode to control them. Test mode features were added to reroute internal nanoinstructions to external ports, thereby improving observability for external testing. Inputs were redirected in the test mode also, improving controllability and shortening the test sequences needed to test the instruction store ROMs.

Internally generated algorithms self-test the cache RAM by providing and controlling a feedback path from its outputs to its inputs. Similar short-circuit feedback paths segregate the instruction decode PLAs, allowing them to be exercised without actually carrying out all the instructions fully. Other circuitry was tested externally after the internal test mode disabled internal cycling and made the buses more accessible.

The results of careful DFT for the MC68020 are a high fault coverage test with less than one tenth the normal test generation effort. Structured portions such as the PLAs and the cache memory were 100% tested. Yet the design tradeoffs were minimal. Only 3% of the logic and 2% of the microcode on the chip were devoted to test.

Similar principles are applicable at pc board level as well as in the ICs. DFT is an ideology and not a particular method. Many viable techniques exist for enhancing testability, and the task of the test engineer is to aid the designer in selecting the most appropriate, cost-effective methods for a given circuit.

REFERENCES

1. W. Keiner and R. West, "Testability Measures," in *1977 Autotestcon*, IEEE, pp. 49–55.
2. L. H. Goldstein and E. L. Thigpen, "Scoap: Sandia Controllability/Observability Analysis Program," in *Proc. 17th Des. Automat. Conf.*, ACM, 1980, pp. 190–196.
3. E. Archambeau, "Testability Analysis Techniques: A Critical Survey," *VLSI Syst. Des.*, **VI**(12), 46–52 (Dec. 1985).

4. M. J. Williams and J. B. Angell, "Enhancing Testability of Large Scale Integrated Circuits via Test Points and Additional Logic," *IEEE Trans. Comput.*, **C-22**(1) (Jan. 1973).

5. E. B. Eichelberger and T. W. Williams, "A Logic Design Structure for LSI Testability," in *Proc. 14th Des. Automat. Conf.*, IEEE, 1977, pp. 462–468.

6. E. J. McCluskey, "Built-In Self Test Structures," *IEEE Des. Test Comput.*, **2**(2), 29–36 (Apr. 1985).

7. J. J. Stiffler, "Coding for Random-Access Memories," *IEEE Trans. Comput.*, **C-27**(6) (June 1978).

8. B. Könemann, J. Mucha, and G. Zwiehoff, "Built-In Logic Block Observation Techniques," in *Proc. 1979 Int. Test Conf.*, IEEE Comput. Soc., pp. 37–41.

9. J. Kuban and J. Salick, "Testability Features of the MC68020," in *Proc. 1984 Int. Test Conf.*, IEEE Comput. Soc., pp. 821–826.

CHAPTER TWELVE

TEST PLANNING

It is not within the scope of this text to explore all the options and trade-offs pertaining to the planning and establishment of a test operation. Rather, an attempt will be made to expose the primary issues and provide some guidelines toward an organized attack. The reader will seldom find the ideal situation in which someone of great authority has declared that a new test facility must be established and the funds will be available. More often, the situation calls for a revision of existing test methods in a complex and constantly changing environment. The ideas discussed in this chapter must be weighed against a given situation's resources and needs.

TEST PHILOSOPHY

The first task in test planning is to secure an understanding of the prevailing test philosophy. Frequently, there is no formal test philosophy since the tradition of management is to treat test as an afterthought. An informal test philosophy may exist in the form of relating policies such as quality policies, manufacturing methods, logistics plans, and warranty policies. Researching and formalizing the policy or establishing a new policy is not an end, but a means for changing reaction management to planning management.

Policies should not contain details such as how products will be tested. Test engineering is a dynamic discipline, and any such policies will need revision as soon as they are formed. The center of a coherent policy on test must be the process by which test decisions are made. Subjects for a potential policy include the method for establishing requirements for a particular test project, the relationship of test projects to the design and manufacturing environments, and the criteria for controlling the test as an end product.

Requirements are important, underrated, and variable in the test arena. Their importance stems from the use of tests to judge the quality of a product or subassembly. Quality is becoming understood as the conformance to requirements[1], and seldom are the requirements and the product so objectively compared as in an automatic test. The test must itself carefully comply with requirements in order to maintain the value of the comparison.

Test requirements can, in some environments such as military electronics production, be externally imposed. In commercial environments the requirements must be derived from the requirements of the product and the overall quality policy of the company. In either case a translation must take place to convert the general goals and isolated measures usually included in either of these sources into a test process guideline and control policy which can account for the probable variations in environment from project to project.

As an example of the range of environments that may be encountered, consider the contrast between a military repair depot and a Japanese-style commercial manufacturing facility. The uncertainty of personnel training and the variety of electronic assemblies that must be repaired at the military electronics repair station place emphasis on uniform test methodology and comprehensiveness of the tests. A completely automatic test which detects and locates 95% of all functional faults is a frequently sought goal in this environment. On the other hand, the "do-it-right-the-first-time" factory seeks to minimize testing by providing feedback about customer requirements and by controlling the process. Tests are likely to be statistical and analytical rather than formal. Most situations will fall into an intermediate category with a number of differing tests at different levels of the manufacturing and field support process, each with its own goals and specifications.

The contents of a policy on test capability might contain statements on any or all of the following subjects.

Design for Testability.   Provisions should be made for review of all newly designed subassemblies and systems for compatibility with available test equipment and test generation methods. Inclusion of a test engineer on the project team is preferable, with formal review and approval of drawings or schematic files as a substitute.

Capital Equipment Acquisition for Test.   If the capital allocation and approval system of the organization is not clear and sufficient to promote proper decision making concerning ATE, it should be augmented or modified. The circumstances for treating ATE as a general-purpose machinery asset rather than a sole project-justified tool should be specified.

Test Generation Process.   Responsibility for test generation (test programming) should be clearly defined, and support responsibilities should be identified. If software maintenance is required for fault simulation or test generation programs, for example, the organizational relationship should be defined. Documentation and acceptance methods for test programs should be delineated.

Test Operations. Product scheduling responsibility, equipment maintenance responsibility, operating personnel, and technical support must be identified. Control techniques for revision or updating of test programs and the responsibility for achieving revision must be defined. Management information reporting methods may be included.

Even though a well-thought-out test policy exists, there is little guarantee that it will endure unchanged for more than a few years. New design and manufacturing methods change the inputs to the test environment both in information and in hardware. As tools and techniques in the test arena develop to combat rising complexity and cost, the policy must be adjusted to incorporate improvements. Most of the changes in policy or procedure that must be made are driven by economic pressures. In view of this reality much of this chapter's discussion will be in terms of cost effectiveness.

TEST ENVIRONMENT EVALUATION

Most test planning guides suggest a sequence of events which begins with establishment of the requirements, continues through equipment acquisition and technical staffing, to end at documentation and control. The ends of that list need to be extended. Before requirements can be meaningfully established, the environment in which the test is to operate should be assessed. Not only will specifying requirements for a particular test project be simplified, later steps such as equipment acquisition and personnel training will be aided.

Manufacturing Parameters

Manufacturing is often the environment in which test is enveloped. In any case, the results of manufacturing are the subjects of the test, and a Test Engineer has a vital interest in the manufacturing process and its foibles. For purposes of analysis it is convenient to divide the manufacturing environment into the four categories of product, process, quality program, and management.

Product Characteristics. Even within the limitations of digital electronic products there is a myriad of variations. Many products follow the hierarchy of Figure 12.1 although the nature and number of units, subassemblies, and components vary. Adopting the assembly process hierarchy as a starting point, the levels of the hierarchy are then partitioned according to potential test technologies that may be applied.

At the top level the finished product might be a computer, a communication system, or an original equipment manufacturer (OEM) module to be further integrated into a system by the end user. In the middle of the diagram are the subassemblies. Usually, they are pc boards, surface mount ceramic modules, power supplies, or electromechanical modules such as disk drives, displays, keyboards,

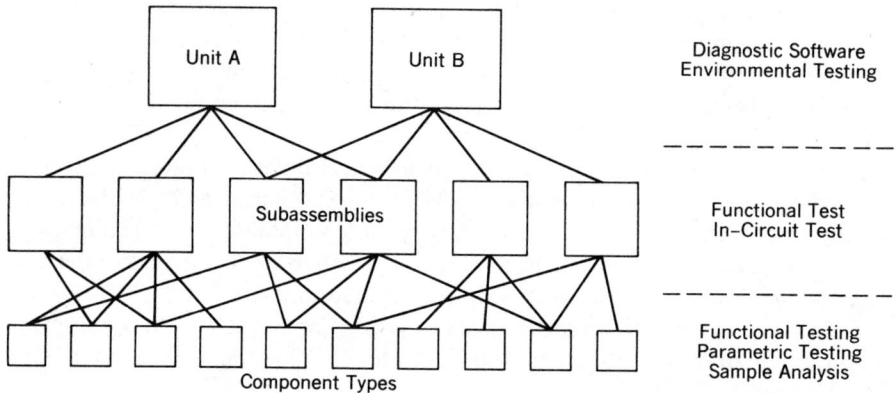

Figure 12.1. The physical hierarchy of an electronic device and the appropriate tests for each level.

and so on. The lowest level consists of components. In addition to ICs, transistors, resistors, and other electronic components there will be mechanical and electro-mechanical components as well.

Before dismissing all but the usually thought-of electronic components as un-testable, consider bare circuit boards and electromechanical components carefully: they may be testable by inexpensive continuity checks which may be cost effective. For discussion in this text the concentration will be on digital components. Test planning applies to all functional components and subassemblies, but the test meth-ods for analog or electromechanical components and assemblies have not been covered in this text and therefore will be largely ignored here.

The information to be gathered about components, subassemblies, or assembled units is nearly the same. The volume, expected manufacturing yield, and type of circuitry involved in the part must be assessed for each item before a test strategy can be formed. A table similar to Table 12.1 is helpful. In this case it has a few added columns to address digital pc module characteristics.

Information placed in the table may be rough estimates and will certainly be revised several times in the planning process. Items such as yield are dependent on the test strategy, which is partially a result of this table, leading to an iterative solution at best. A first cut will reveal areas which need further data, and the plan will converge rapidly afterward.

Not all of the columns will be needed in any one decision, but sooner or later the information will come in handy, so making as comprehensive a table as pos-sible is advisable. The second, third, and sixth columns are used in determining the test equipment strategy and capacity requirements. The connector configuration information will be useful in specifying test equipment and in determining the interface adapters needed. Gate count, memory type and quantity, and major ICs that are presently in the test generation model library will help estimate the test generation effort required.

TABLE 12.1 Product Characterization Table

Part Name (Designation)	Volume per Month	Estimated Yield	Number of Gates	Connector Configuration[a]	ROM/RAM	Models Needed
Microcontroller	2000	60%	75,000	2–120 m	8K ROM	80186 8259
Arithmetic unit	1000	65%	50,000	2–120 m	—	8087
Memory array	8000	80%	—	2–120 m	4M RAM	
Serial I/O	3750	75%	25,000	1–60 m 2–25 f	4K ROM 64K RAM	8250
Parallel I/O	1500	90%	15,000	1–60 m 2–34 f	4K ROM 32K RAM	8255
Graphics controller	2000	60%	60,000	2–120 m 1–96 m	1M RAM 4K ROM	6845
Clock–Keyboard	2000	85%	10,000	1–60 m	4K ROM	

[a] m = male, f = female.

297

Process Characterization. Characterizing the manufacturing process (or the repair process in the case of a field maintenance shop) is as important as studying the product. A flow diagram of the process should be obtained or constructed noting which subassemblies or components are received from vendors, major steps in the fabrication or assembly process, and shipping or product acceptance points.

The flow shown in Figure 12.2 is extremely optimistic! Although it contains most of the major manufacturing steps, there are no inspections, tests, or rework steps in the flow. Theoretically, a perfect process with perfect vendor components and materials and perfect assembly operations can work this way, but few manufacturers want to let the customer be the first to find out that their system is not perfect. The fabrication operations for the pc substrate itself often contain a number of visual inspections, but for the sake of simplicity the process will be lumped into a single block and largely ignored from here on.

Each of the assembly steps is investigated to determine what faults are likely to be caused at that point as shown in Table 12.2. Of course, without tests all the earlier errors would be inherited by a later operation, but the important consideration at each step is to discover what new faults will be introduced.

The cabinet assembly has been ignored under the assumption that no electronics is embedded in the cabinet, and therefore only inspection will be needed. Although the tables and diagrams in the last three figures and tables have suggested a single product, the same procedure can be used for multiple product operations either by repeating it for each product or by considering the variations of all products in a slightly more complex set of diagrams. In either case all products to be tested must be involved in the capability and capacity studies.

Quality Program. In the previous two sections the inputs to the planning process from the manufacturing environment were investigated. Before progressing further, the requirements for the output of the planning process should be checked. A policy with regard to quality control probably exists which could include requirements for inspection and testing ranging from simple guidelines to precise definitions of acceptance criteria for products and processes. If a policy does not exist, there are several subjects that should be thought out before proceeding anyway.

Key facts to be uncovered in the policy or discussed are as follows:

Vendor Specifications. Purchasing and receiving of components should take place according to an agreed-upon set of component specifications. The quality policy may precisely define the standards, or it may simply state that a specification must be defined and documented for each component. The acceptable quality level (AQL) or the parts per million (PPM) defective prescribed in the policy or in the individual specifications is vital to calculating an accurate yield for more complex assemblies.

Product Qualification. The method of criteria for judging acceptability of the

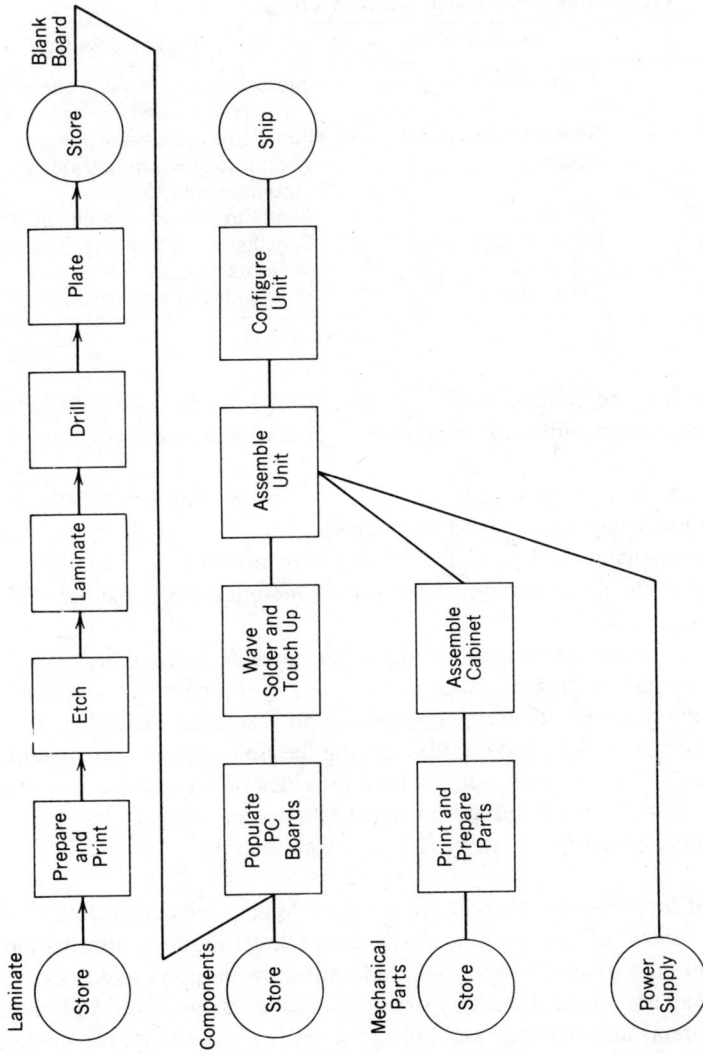

Figure 12.2. An idealistic manufacturing flowchart with no test or repair.

TABLE 12.2.   Typical Fault Sources in Manufacturing

Item	Process Block	Probable Faults
Board	Fabrication	Intercircuit shorts, open circuit paths, shorts to ground
Components	Receive from vendor	Functional, parametric, timing
pc assembly	Assemble	Wrong, missing, or misoriented components
pc assembly	Solder	Shorts to ground, intercircuit shorts
Power supply	Receive from vendor	Functional, parametric
Final unit	Assembly	Miswires
Final unit	Configure	Wrong jumper installation

final product may be spelled out in the policy. If it is not directly defined, the organizational responsibility for assessing product acceptance should be given.

Warranty Program. Replacement or repair of subassemblies and the requirements for requalifying the repaired items have direct bearing on the levels of test needed for components and pc boards. Spare part provision to the customer or to repair centers in the field is a typical area which includes specific subassembly test requirements.

Even if the specific items mentioned are not included in a quality policy, goals and guidelines may be present which can be used to compare against the expected outcome of the test plan. If the quality goal is to assure no more than 1% return from the customer for functional failure during the first year, the test requirements will be substantial. A warranty plan which provides on site maintenance and full spares may allow for a 3% fallout during the first year and that few percent difference can change the test strategy and cost significantly.

Management Information. Achieving a "zero defect" manufacturing system or maintaining control of the process quality at any level requires information feedback. Errors in the process cannot be fixed until they are pinpointed, and test and troubleshooting are an excellent way of obtaining the information. ATE networks, bar code readers, and manual tracking are a few of the alternative methods for gathering data for management reports. Mechanisms for gathering, storing, and formatting the pass/fail statistics and fault diagnoses are part of the test plan.

Simple statistical pass/fail data do little to aid the correction of problems. They only inform of the presence of process difficulties. A classification method for defect types and the collection of the results of diagnosis according to the system provide direct and useful information for the location and correction of problems. The categories should be chosen to reflect steps or variables in the manufacturing process. Table 12.3 is typical of a subassembly test and troubleshooting results classification.

Not only are the overall test yield of 70% and the individual yields of 80, 65, 80, and 71% easily obtainable from the information, but the culprit can be spotted

TABLE 12.3. Test and Troubleshooting Results

Part Number	Quantity Tested	Wrong/Missing Component	Solder Open or Short	pc Board Open or Short	Component Failure	Total Defects
205831	25	2	2	1	0	5
205846	136	8	26	3	11	48
205900	41	0	6	0	2	8
210030	228	32	17	8	9	66
Totals	430	42	51	12	22	127

by examining the column results for each category. Both overall and individual results are available, and the details should not be glossed over. Although the highest column total of faults was under the solder operation column, the 32 wrongly installed components on the 210003 part signals a serious misunderstanding (possibly caused by unclear instructions) on the part of the assembler.

A follow-up procedure usually called failure analysis will help avoid the propagation of management misconceptions due to errors in the data. It is easy to decide that the problem in a pc board must be due to a failed component, for example, only to find after the part is removed and discarded that a solder short was hidden under the package and was the real fault cause. In a production line atmosphere the first diagnosis will probably be recorded, and a different person may repair the board and add a note to the record. If the removed part is retested by a failure analysis technician, a "no defect found" condition will be uncovered and further investigation will find the real cause in the record and result in correction of the first entry.

Failure analysis also serves an investigative purpose in adding depth to the troubleshooting results. Knowing that the defective components were in most cases internally contaminated or that many have been overstressed will allow a precise corrective action to examine the incoming component test or discuss problems with the vendor. Failure analysis is rather complex and sometimes costly, however, and it may be reserved for investigation of specific problems rather than blanketly applied to all test failures.

Customer Requirements

External factors influence the test plan also, and prime among them is the formal or informal customer requirements. Informal requirements may be in the form of marketing information, while many businesses such as OEM supply to system integrators or sale to a government organization have formal contractual requirements.

If the requirements are levied at the final product level, the immediate task is to reflect those requirements back to the subassemblies and components as part of the test plan. The flexibility allowed in this process is the tradeoff of test and troubleshooting expense at each level versus that expended at the next. Lack of component tests will result in higher cost at a subassembly test, while lack of a subassembly test merely passes the cost to the final assembly. If all tests are removed, the cost of having the customers test to their own specifications may be the loss of customers!

Customer requirements may contain specific requirements for subassemblies or components. Production and distribution of spare subassemblies for field repair of units are a concern in many cases. Using untested spares is risky, as it may cost considerable troubleshooting time and requires the storage and distribution of a larger quantity of spares to assure availability of good ones. Spare parts are therefore usually specified exactly with respect to test and quality aspects.

Field support of the finished product may alternatively be required of the manufacturer as part of the warranty program. The costs of spares can then be weighed against the cost of maintaining a repair depot or portable test capability. Tests for subassemblies should be planned to allow rehosting of the test program on the chosen portable field tester, if possible, in that case.

Design Constraints

Test costs and test planning are driven as strongly by the methods and restrictions of the design as by manufacturing or the customer. Before attempting to consolidate the information gathered in the factory, the pertinent features of the design process should be considered. Test generation methods and costs are usually linked tightly to the design process, but test application methods are affected also.

DFT, as described in a previous chapter, reduces test generation costs, which in turn lowers the cost of the entire functional testing approach. Improvements in observability result in better diagnostic resolution in functional tests, which in turn reduces the troubleshooting time. Controllability improvements shorten the test sequence and reduce the test application time. Each of these affects the cost of test and could sway the overall test strategy.

A more subtle influence of the design process is a product of design stability. A design process which fails to thoroughly evaluate a new product for all operating conditions and for potential options or enhancements produces unstable designs which will require modification frequently in response to customer complaints or requests. Frequent design changes add cost to the test operation as test generation is repeated or test sequences are modified. Maintaining the correct version of test programs for the different versions of a product that may be returned for repair is costly and confusing. Test methods with quicker reprogramming such as in-circuit testing may be preferable to functional testing in such an environment.

Of course, the technology employed in new designs will influence the choice of test generation and test application tools. Dense custom VLSI or ASICs inherently increase the test and troubleshooting time as well as the test generation time. Utilization of surface mount devices makes in-circuit testing extremely difficult, if not impossible. Since most manufacturing and testing operations are planned as continuing ventures, and since ATE and other major test tools are major long term investments, looking toward the future in the design department can reduce long-term costs.

TEST STRATEGY

Having gathered information about and evaluated the impact of the environment in which the test strategy must exist, the central task of proposing and evaluating alternative strategies is at hand. Test planning takes place on a number of levels, with the later questions and alternatives depending heavily on the choices made early in the process. Although there are conventional methods of structuring the test capability, changing factors in design and manufacturing as well as test technology itself both create and demand flexibility in the planning process. The emphasis herein will not be on the right way to test (for there is no single right way), but on possible approaches and the process of deciding among them.

In a typical digital electronics manufacturing process there are five levels of testing [2]. The first is used by the designer to evaluate the suitability of a newly designed product. The last level is the field support or repair depot testing which is used to maintain the product in its operating environment. The remaining three middle levels are an integral part of the manufacturing process. According to the conditions and specifications as discussed earlier, these three may vary in complexity and coverage, and tradeoffs may eliminate one in favor of strengthening the others. The three are: incoming component testing, subassembly or production check testing, and finish product testing (sometimes known as final acceptance testing). Due to space limitations and in keeping with the scope of this text, the middle three test situations will be emphasized, and in particular, the subassembly test will be the focus.

Before proceeding, the time-honored test funnel should be introduced to place the subassembly test in perspective. The essence of the funnel is the selection of test goals to achieve detection of only those faults which are likely to have been caused in the operations which have followed the last level of testing. Retesting for faults which should have been detected and removed in an earlier test is wasteful. Delaying the test for a possible fault class has a cost associated which generally follows the "rule of 10" [3].

The cost of removing a defect from the product rises by a factor of 10 (roughly) at each of the assembly or distribution points in the process. Throwing away a component may cost $1 if the parts are purchased in quantity. Removing the same component from a pc board after it has been soldered takes more time, and replacing it with a good one is a wasteful repeat of the original installation labor. Altogether, the cost of removing and replacing the part will be about $10 when performed by skilled technicians in the factory atmosphere. Finding and replacing the faulty component in the final product involves considerably more troubleshooting and the necessary removal and reinstallation of the pc board from its place. $100 is a reasonable estimate of the total cost to remove and replace the component at this level. Finally, if the product is installed and operating in the field, a representative must be sent to the site or maintained there full-time to service the unit, remove the failed component, and replace the component after having troubleshot the system. Total costs of about $1000 per fault occurrence are probably quite reasonable.

Installing a series of tests along the manufacturing process combats the cost of repair. Narrowing the fault domain of each successive test focuses the tests and reduces the test cost needed to maintain confidence in the product without incurring the escalating repair cost unnecessarily.

Manufacturing Test Alternatives

As a prototype for test strategy in the manufacturing environment, consider the process flowchart of Figure 12.3. The chart contains a condensation of the blocks from Table 12.2, and the power supply and pc board fabrication processes have been left out for simplicity.

The upper left-hand tests labeled "continuity test" and "functional and parametric test" are intended to detect bad pc boards (without components) and defective components. The component test is designed to test as much of the vendor's specifications as is economically feasible. To reduce testing costs further, the test may be performed on a statistical sample rather than on 100% of the components to be used. The necessity for component test depends on the relationship with the vendor to some extent. If the vendor performs a 100% functional and parametric test just prior to shipment, the test results at the incoming component test will only detect shipping and handling damaged parts.

Component Test. Many vendors use only a sample test after packaging of the components, and the incoming test will therefore detect functional failures, mismarked components, and other manufacturing defects. The decision of how much to test is made on the basis of the cost of testing versus the projected cost of finding and repairing the failures at a later stage.

As a demonstration of the decision-making process, assume that the average failure rate for digital ICs at the incoming test is 0.8%. This is not an unrealistic number, although the recent emphasis on quality in the semiconductor industry may result in making it so. If no incoming test were performed, but a comprehensive test at subassembly level were expected to detect the failures, the parts would be mounted onto a pc board and tested. The probability of building a good pc board using probably good parts is given by P^n, where P is the probability that a given part is functional and n is the number of parts on the board. If the board in question has 200 of the 99.2% good parts installed on it, only $(0.992)^{200}$ or 20% of the pc boards will be failure free.

If the volume of pc boards used in the earlier example of manufacturing characteristics is applied under these circumstances and the average cost of component replacement is the rule-of-10 example's $10, the cost of not testing at the component level can be calculated.

$$\$ = (T + R) B(1 - P^n) + FBn(1 - P)$$

where F is the repair cost for a single component replacement, P and n are defined as before, B is the volume of pc boards, T is the cost of troubleshooting a defective board, and R is the cost of retesting the board after it has been repaired. Assume a technician whose employment cost (including overhead costs) is $25 per hour performs a test (including loading and unloading the ATE) in 0.1 hours and re-

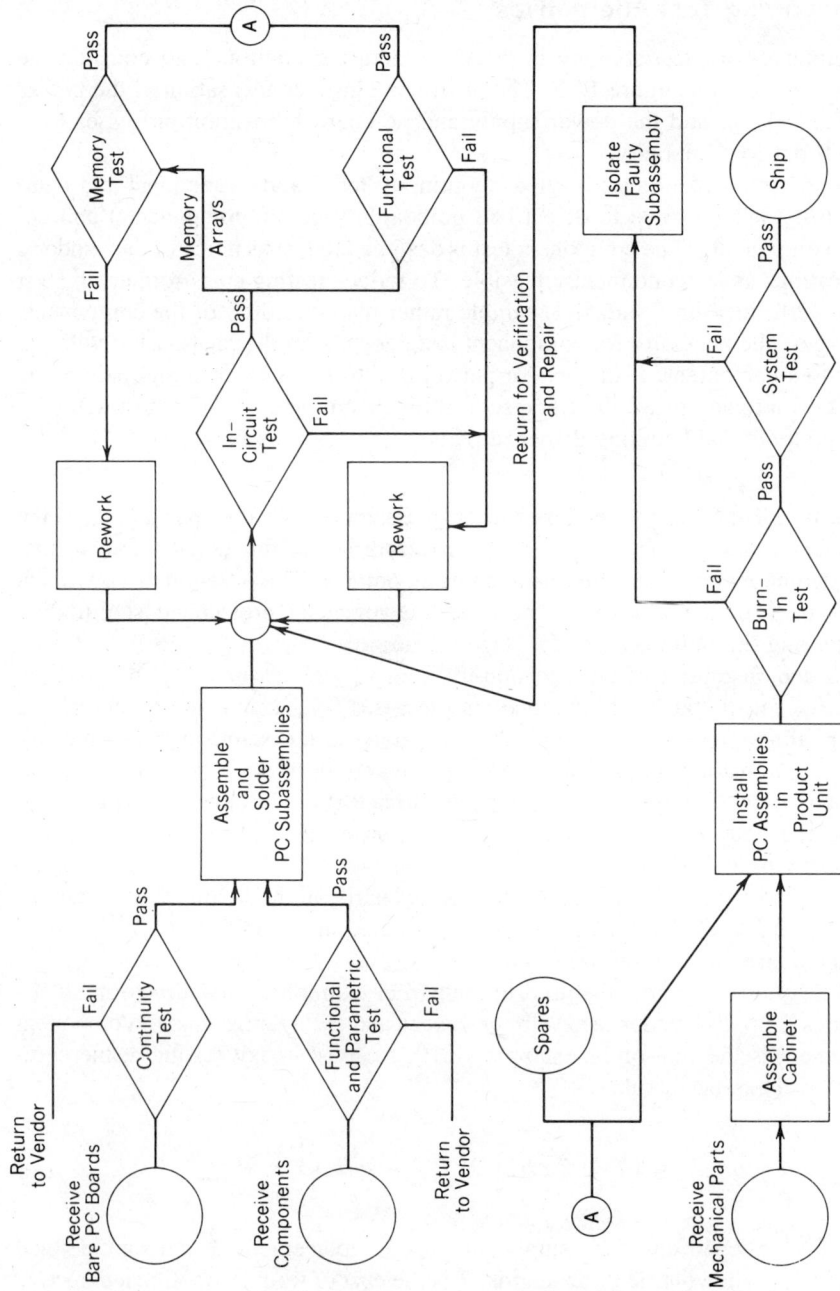

Figure 12.3. Manufacturing flowchart. Test operations appear as diamonds.

quires 0.3 hours to troubleshoot these complex circuits. The result using the 20,250-card per month total volume and the failure rates from above is a cost of

$$\$ = (7.5 + 2.5)\, 20{,}250(1 - 0.2) + 10(20{,}250)\, 200(1 - 0.992)$$
$$= 162{,}000 + 324{,}000 = \$486{,}000 \text{ per month!}$$

A potential savings of $5.8 million per year can be weighed against the cost of providing 100% test of all incoming components. If a sample plan for testing components is adopted, the savings will be less since there will remain a statistical probability of finding component test detectable faults at the pc board test. Note that the calculations include only those failures which are caused by the components. Implementing a component test does not mean that the pc board test can be eliminated. That analysis is based on a new set of probable faults induced by the assembly process.

Before a tradeoff decision concerning component testing can be made, the cost of testing must be calculated. There are two major types of cost associated with testing. The first category includes the costs of establishing a test capability. These one-time or "nonrecurring" costs include the cost of the equipment, any expenditures for preparing space in the building (air conditioning, electrical installation, etc.) and the costs of generating the test programs (labor, computing resources, etc.). The costs of applying the test to components is a recurring cost that must be paid as long as the ATE is in use. Recurring costs include maintenance, labor of the operator or technician to run the tester, and overhead (power, building maintenance, cooling, management).

Some of the costs are easily assigned to a particular DUT, and others must be distributed across a number of DUTs. The depreciation cost of the ATE and the recurring maintenance cost must be distributed if a per-DUT cost is to be established. It is generally easier to calculate a monthly or annual cost under an assumed load or throughput for the test operation as a whole. The catch in this approach is the prorating of nonrecurring costs. A product life estimate or an acceptable depreciation formula will be required. Since most depreciation formulas used in business are nonlinear, a table of yearly costs which are greater for the first year and declining thereafter is a convenient method of displaying the situation.

The table of costs presented in Table 12.4 includes equipment depreciation based on a $1 million purchase price (including installation charges), a 6-yr depreciation life, and the use of the "double-declining-balance" method of calculating depreciation [4]. The nonrecurring test generation costs are not depreciated since there are no precedents for depreciating labor. A recurring test generation charge is added to later years on the assumption that some enhancement of the tests may be required to correct deficiencies discovered through the analysis of management information from test results.

The savings of $5.8 million/year seems to be favorable compared even with the first-year cost of $1.2 million, but there is another catch. The capacity of the ATE available to perform component testing may not be enough to test all the

TABLE 12.4. Component Test Cost Example

	Non-recurring Costs		Recurring Costs			
Year	ATE Depreciation ($ in Thousands)	Test Generation ($ in Thousands)	Maintenance ($ in Thousands)	Operator ($ in Thousands)	Overhead ($ in Thousands)	Total
1	333	500	60	120	200	1213
2	222	80	60	120	200	682
3	148	60	75	120	200	603
4	99	40	75	120	200	534
5	66	20	90	120	200	496
6	44	0	100	120	200	464
					Total	3992

20,250 components needed in a month. The capacity in components per month is given by

$$\frac{\text{Hours\_per\_mo (availability)}}{\text{Test\_time\_per\_unit}}$$

Assume the test time (including loading and unloading the DUT) is 2 min. A standard 3-shift 40-h work week is also assumed with an availability of 0.8 due to operator breaks and equipment maintenance. At 4.3 wk/mo the tester can be counted on for 12,384 components/mo. This is considerably less than the 20,250 needed, so a second tester will have to be purchased unless the test time can be shortened to less than 1.25 min. A second tester adds cost under all columns except the test programming cost since it assumes that the second ATE will be the same model as the first.

Clearly, even if all the costs were doubled in this case, the choice would be to test components since a savings of \$3.4 million/yr would be realized. A number of more subtle factors such as interest, inflation, and taxes have been neglected in these quick comparisons. A more thorough examination of the situation would be in order if the choice were not so obvious.

Subassembly Test. The basic decision-making process used in the component testing example can be applied to nearly any testing scenario. In the subassembly test area the questions to be answered are more complex than that of whether or not to test. Major alternatives exist in the methods of test, and the economics of applying them separately or in combination must be evaluated.

The three major types of subassembly or pc-board-level ATE commonly available are functional testers, in-circuit testers, and memory testers. A wide range of equipment is available in each category. Memory testers range in cost from \$100,000–\$500,000, in-circuit testers from \$300,000–\$800,000, and functional testers can be found which cost as little as \$50,000 and as much as \$2 million. Tester speed, flexibility, and interface size (number of pins which can be controlled or monitored) vary with the price. The choice is not so clear as it would seem since several models of testers exist that are hybrid functional–in-circuit or functional–memory testers. Product characteristics and volume mix of circuit types must be used to try to achieve an optimum combination.

Printed Circuit Board Alternative Selection. If the test strategy has not already been specified by the customer requirements, several choices must be made among the alternatives which can be formed by combinations of the memory, in-circuit, and functional testers. The vague decision among complex alternatives can be reduced to a set of sharply defined decisions among a few possibilities by applying some engineering judgment. If a full investigation is desired, any one of the several available economic simulation software packages should be used.

An initial simplification is based on the reasoning that a tester which does only one type of test is probably more efficient at that test than a more general-purpose tester. ATE dedicated to memory testing may be matched in performance by a high-performance functional tester which includes algorithmic control capability in the pin cards, but the functional tester manufacturer is unlikely to give away the general logic test functions which are absent from the memory tester. The cost of memory test capability will be lower, therefore, on the dedicated tester, provided that both machines are in full use.

If the functional tester is to be purchased on its own accord but will not be utilized fully, the excess capacity could be used to test memories for only the added cost of the difference between a functional tester without the algorithmic pins and a functional tester with the algorithmic pins. Memory test sequences are generally quite long (several minutes to tens of minutes), which makes it difficult to fit them into spare time on a general-purpose tester. The required capacities for both testers should be calculated before drawing a conclusion.

Modeling the most probable test scenarios is a tractable approach for manual solution of the cost comparisons. Having set aside the issue of memory testing for the moment, the remaining alternatives include testing the pc cards in one of the following test flows (as diagrammed in Figure 12.4) [5].

1. **System Test Only.** This alternative is the base calculation for the tradeoff corresponding to the "cost of not testing" determined for component test earlier. The assumption is the absence of pc-board-level testing, with all boards installed directly into the final product where they are tested as a system and troubleshot by a technician using board extenders, a logic analyzer, and the system diagnostic software.

2. **In-Circuit and System Tests.** The bed-of-nails-style in-circuit tester is used to check the boards for correct component installation, component functionality, and circuit board shorts before the boards are submitted to system test. Faults found at the in-circuit test are repaired, and the boards are recycled through the in-circuit test which they must pass before moving on to the system test.

3. **Functional Board Test and System Test.** Newly assembled pc boards are exercised and checked using deterministically generated test vectors applied through the edge connectors by a functional tester. The test usually detects a high degree of stuck-at faults and operates the board as a subassembly. Failed boards are troubleshot using the board tester and sent to repair, from whence they are returned to be retested. Passing boards are submitted to the system test.

4. **In-Circuit, Then Functional, Then System Tests.** Both of the board tests are applied to all boards, with the failures from each being troubleshot on the tester that detected the failure and returned after repair to that tester.

5. **Functional and System tests with In-Circuit Used for Troubleshooting Functional Failures Only.** This flow takes advantage of the inherent ease

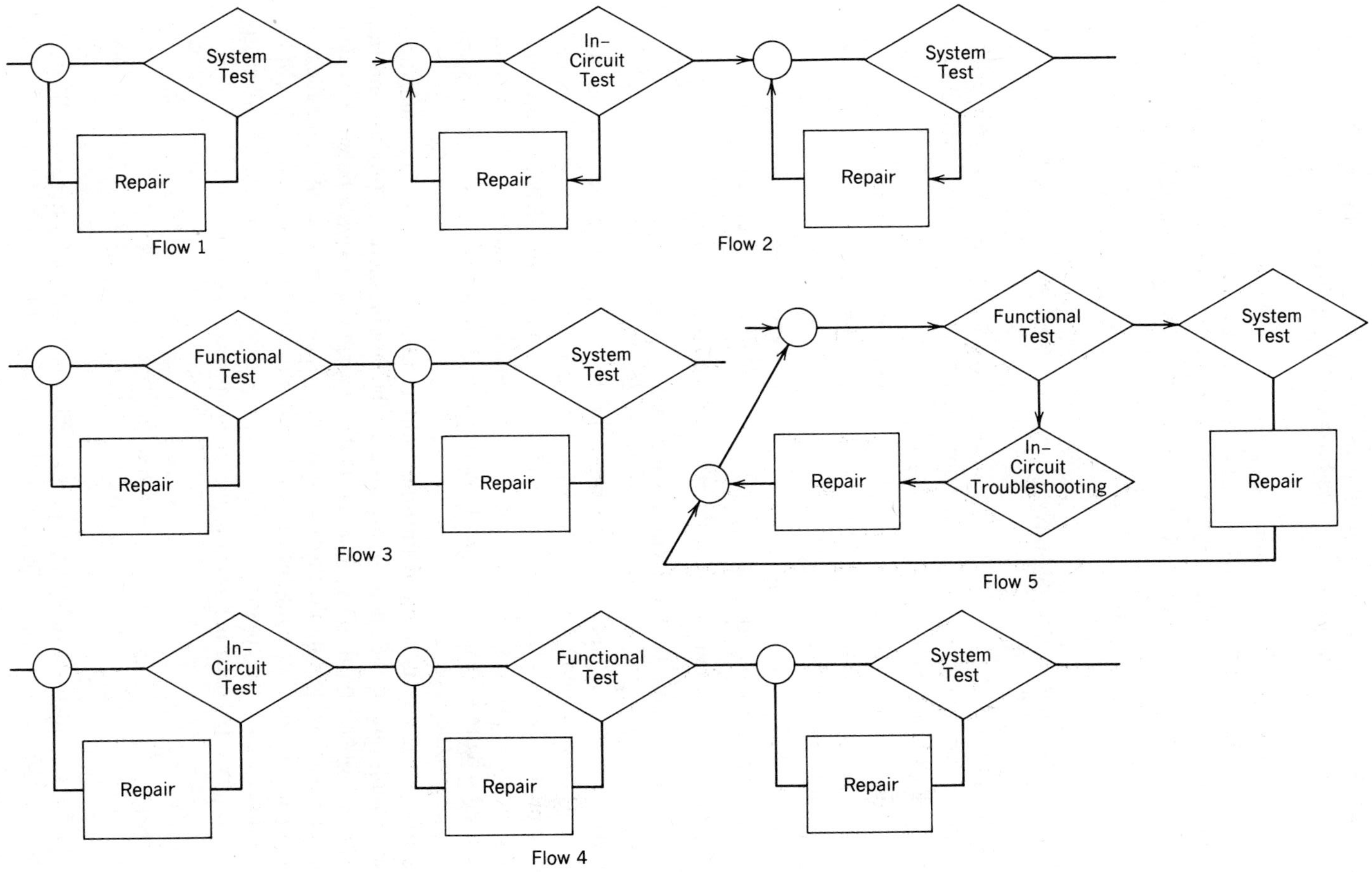

Figure 12.4. Flowcharts for alternative test approaches.

of troubleshooting for most failures on the in-circuit tester while reducing the number of steps in the good board production process.

The method of analysis is a simplified calculation of the test and troubleshooting costs of each flow. Each test step within each flow is evaluated for cost separately based on the input yields from the previous step. Then the costs are totaled for each flow and compared. For each test method the costs are represented by

$$\text{\$Step} = L\left\{\left[(2 - Y)t_T\right] + \left[(1 - Y)(t_D + t_R)\right]\right\}$$

where the following definitions apply:

L   Is the prevailing "burdened" labor rate for the technician who is to perform the tests or troubleshooting. The burdened rate is the labor base cost plus the overhead charges prorated per employee hour.

Y   Is the yield of the test or assembly step immediately preceding this one. Yield is defined as the number of good boards divided by the total number of boards output from that previous step.

t_T Represents the time required to test an average board using only a go–nogo test. It includes the loading and unloading time for the ATE involved.

t_D Is the time required to troubleshoot the fault to a component or site.

t_R Is the repair time for the average faulty board.

Notice the differing factors involving Y in the two parts of the formula. The $(1 - Y)$ factor is merely the number of defective boards which must be repaired. The number of boards which must be tested is the sum of all the boards produced by the previous process step and the $(1 - Y)$ boards that failed the first test and have been repaired. Thus, $(2 - Y)$ boards will be tested and $(1 - Y)$ boards will be diagnosed and repaired.

An inaccuracy is introduced into the formula in the case of system and functional testing due to the sequential nature of the troubleshooting process and the statistical probability of more than one fault occurring on a single board. For system and functional testing the procedure commonly used involves stopping the test and diagnosing the first fault encountered. When a board has two faults, it will be tested three times and troubleshot twice. The number of faults that are likely to occur on any one board is related to the yield approximately as

$$\#F/B = \ln\frac{1}{Y}$$

which suggests that for yields of over 50% the effect will be rather small. For the purpose at hand the simple formula will suffice.

The average yield of the board mix described in Table 12.1 is 76.6%, which results in a $(1 - Y)$ of 0.234 and a $(2 - Y)$ of 1.234. Assume the repair time for an average fault is 0.2 h (remove and replace an IC). For the example calculations a table of test and diagnosis times (Table 12.5) will be used.

The test times include removing the board from its processing routeing container, attaching it to the ATE, and reversing the procedure when the test is completed. Troubleshooting time includes the time needed to print out a diagnosis or write one down for repair guidance in addition to the hookup time of any auxiliary test equipment. Labor charges including overhead might be $40 for a skilled technician.

For the simplest flow scenario the system test cost alone makes up the total. The yield in the formula is the yield of the item being tested, which in this case is the product. In the example, the product contains a basic configuration of boards and a number of options which may be ordered and integrated before test. The basic unit has a microcontroller, two memory arrays, one serial I/O board, a graphics controller, and a clock–keyboard combination board. On the average the configuration which is purchased is

1.0 Microcontroller

0.5 Arithmetic unit

4.0 Memory arrays

1.9 Serial I/O

0.8 Parallel I/O

1.0 Graphics controller

1.0 Clock–keyboard

The yield of the overall product (ignoring failures due to the backpanel wiring) is the weighted product of the board yields. Weighting is accomplished by raising each board yield to the power of its frequency of occurrence in the system. A board with a yield of 0.8 which occurs twice in the average system will result in a contribution of 0.64 to the overall yield.

$$Y_P = 0.6^1(0.65^{0.5})\,0.8^4(0.75^{1.9})\,0.9^{0.75}(0.6^1)\,0.85^1 = 0.054$$

TABLE 12.5. Example Test and Troubleshooting Times

Method	Go–Nogo Test Time (h)	Time to Diagnose (h)
In-circuit	0.02	0.03
Functional	0.10	0.50
System	1.50	3.50

This rather surprising result means that there is virtual certainty of failing the system test on the first pass and that it will take many tries to get a system through the test. Since the yield at the input to system test is so low, a formula which accounts for multiple fault troubleshooting in a sequential fashion should be used to estimate the time.

$$\$\text{sys} = L\left[(1 + F)\,t_T + F(t_D + t_R)\right]$$

where F is the number of faults expected for each new system into test. F is obtainable from Y using the natural logarithm relation and is 2.9 in this case. Filling in the example numbers gives the following result.

$$\$\text{sys} = \$40\left[(3.9)1.5 + 2.9(3.7)\right] = \$663 \text{ per system}$$

The second scenario uses an in-circuit tester to test all boards, with the good ones going on to system test. The yield at the input to the in-circuit test is the manufacturing yield used above (76.6%), but the yield at the input to the system test will be the comprehensiveness of the in-circuit test. Assume that the in-circuit test has a 90% fault coverage for the types of faults represented in the manufacturing yield. The yield at the input to system test will then be 0.90.

If the results of the in-circuit test are considered to be uniform across the board types, the system yield formulas reduces to

$$Y_P = 0.90^{(1 + 0.5 + 4 + 1.9 + 0.75 + 1 + 1)} = 0.90^{10.15} = 0.34 \quad \text{or} \quad 34\%$$

Since the yield is still below 50%, the cost will be calculated using F as before. In this case $F = 1.07$, which implies that either formula would work reasonably well. The cost for the system will be

$$\$\text{sys} = \$40\left[(1 + 1.07)1.5 + 1.07(3.7)\right] = \$282 \text{ per system}$$

The $381 savings per system adds up to $9,144,000 for the 24,000 systems to be produced each year, but the cost of in-circuit testing must be subtracted before the actual savings realized can be counted.

Applying the yield driven cost formula for the in-circuit test step gives

$$\$\text{ict} = \$40\left[(2 - 0.766)0.02 + (1 - 0.766)0.23\right] = \$3.14 \text{ per board}$$

Since this number is a cost per board, it must be multiplied by the number of boards in a system, which in the example is an average of 10.125. This gives a cost of $32 per system. The net savings over scenario 1 is therefore $349 per system or $8,376,000 annually.

Investigating scenario 3, which substitutes a functional tester with a typical coverage of 97% for the in-circuit tester, gives the following results.

$$Y_P = 0.97^{10.15} = 0.73 \quad \text{or} \quad 73\%$$

$sys $= \$40[(2 - 0.73)1.5 + (1 - 0.73)3.7] = \$116$ per system
$fcn $= \$40[(2 - 0.766)0.1 + (1 - 0.766) 0.7(10.125)] = \$116$ per system

The total cost of scenario 3 is \$232 per system.

Continuing the same procedure for scenario 4, in which the in-circuit and functional tests are applied in tandem before system test, the yield used for system test is the same as used in the last case (73%) since the functional test feeds the system test in this scenario also. The functional test calculations are made using the in-circuit fault coverage (90%) rather than the manufacturing yield (76.6%).

$sys $= \$116$ per system (from above)

$fcn $= \$40[(2 - 0.9) 0.1 + (1 - 0.9) 0.7(10.125)] = \$73$ per system

$ict $= \$3.14$ per board (from scenario 2)   or   \$32 per system

The total for this scenario is \$221 per system or \$5,304,000 annually.

The final scenario requires different formulas since only the defective boards are being tested by the in-circuit tester. The input yield to system test is still equal to the coverage of the functional tester. The number of boards that will be submitted to troubleshooting is still $1 - Y$ since all boards which fail functional need to be fixed. The functional test and in-circuit test team can be described by the usual formula if the in-circuit troubleshooting time is used in the $1 - y$ term and the functional test time is used for the first term.

$sys $= \$116$ per system
$hybrid $= \$40[(2 - 0.766)0.1 + (1 - 0.766) 0.23(10.125)] = \$72$ per system

The total for scenario 5 is \$188 per system.

Remember that for each of the scenarios the incoming pc boards have a yield of 76.6% and that the output after system test is essentially the same. The difference in test strategies is a matter of cost as long as there are no extenuating circumstances such as contract deadlines or specific contractual test requirements. A summary of the methods is shown in Table 12.6.

A glance at the chart suggests an easy decision, but the cost of implementing the test strategy has not been included at this point. Only the cost of applying the test strategy in terms of labor has been calculated. Before the tests can be applied, equipment must be purchased and installed, interface adapters must be designed and built, and test programs must be written. Costs for these capital and service investments vary widely, and the test and troubleshooting times can be expected to depend somewhat on the amount spent in development, but the relationship is not a strong one. The capabilities of a particular type of ATE usually can be judged on a technical basis to narrow the choice to a few. Then the price–performance

TABLE 12.6. Test Scenario Cost Comparison

Scenario	Cost per System ($)	Annual Cost ($ in Millions)
1. System only	663	15.9
2. In-circuit and system	314	7.5
3. Functional and system	232	5.6
4. In-circuit, functional, system	221	5.3
5. Functional (in-circuit) system	188	4.5

criterion becomes one of a few factors in the final selection. Details of development planning will be covered in the next section of this chapter.

In order to maintain the continuity of the comparison exercise, assume that the following costs apply. In-circuit test capability requires the purchase of a tester at a cost of $700,000 and the procurement of seven bed-of-nails fixtures at $5000 each, or $35,000. Programming the in-circuit tester is 6 w of work for an engineer per board type for a total of 1680 h, which at $50/h (burdened rate) will cost $84,000.

Functional ATE will cost $1,100,000 in this case, but the interface adapter cost will only be $5000 since one well made adapter will test any board type. Test programming requires fault simulation for which a computer must be used and considerable engineering labor. The average costs per board will be $20,000 in labor (about 10 w) and $15,000 in computer resources. The total for the 7 board types comes to $245,000 in programming costs.

A capacity study must be performed next to determine the ability of the suggested equipment to the production levels projected. In scenarios 2 and 4 the entire output of pc boards must be tested by the in-circuit tester. As seen in the earlier formulas, each board spends an average of 0.0785 h on the in-circuit tester. During a 3-shift day with 6 h per day effective operation, the tester can handle 229 boards. Since this adds up to only 4580 boards a month it is obvious that 5 in-circuit testers will be needed to test the 20,250 cards expected each month. The actual cost of establishing scenario 2 is therefore equal to the programming and fixture cost of $119,000 plus the cost of 5 testers (the adapters will be shared) for a total of $3,619,000.

Scenarios 3, 4, and 5 require the functional tester to test the entire output of production. At a flow rate of 3.5 boards per hour only 63 boards can be tested and troubleshot in a day. Using the same 3 shift, 6 hr. assumptions as before, 16 testers will be needed for the functional test portions of those scenarios. Since scenario 3 needs only a functional tester, its cost is equal to the programming cost of $245,000 plus the tester and adapter cost of 16 times $1,105,000 or a total of $17,925,000.

Due to the sequential use of both testers, scenario 4 will cost $21,544,000 to get underway. The load on the in-circuit test in scenario 5 will be only those boards which fail functional test. The 23% of production boards which will be defective

TABLE 12.7. Test Scenario ROI Comparison

Scenario	Cost per System ($)	Annual Cost ($ in Millions)	Annual Saving ($ in Millions)	Estimated Cost ($ in Millions)	Average ROI (%)
1. System only	663	15.9	—	0	—
2. In-circuit and system	314	7.5	8.4	3.6	233
3. Functional and system	232	5.6	10.3	17.9	60
4. In-circuit, functional, system	221	5.3	10.6	21.5	49
5. Functional (in-circuit) system	188	4.5	11.4	18.7	61

amounts to 4658 per year. A single in-circuit tester can keep up. The cost for this scenario is $18,744,000.

Table 12.7 can now be improved by the addition of an "establishment cost" column and a return on investment column which uses a simple formula of income over investment. It should be viewed as a rough comparison tool since it does not account for interest or depreciation. Since the system test is a base for comparison, its establishment cost is considered to be zero.

For the example, the best strategy is a simple in-circuit test for all boards before sending them to be installed in the system. The payback on investment of any of the alternatives is respectable, and other factors such as the use of the equipment by other product lines may tip the scales to a more complex alternative.

The memory test decision which was deferred earlier can safely be made on its own accord now since functional testers are not in the favorite scenario. A capacity study of the memory test will need to be made, of course. Several other alternatives should be considered for inclusion in a high-volume production facility such as the one described. Continuity testing of the bare circuit boards is rapid and easily programmed by a learning procedure from a KGB. Its place in the flow would parallel the incoming component testing. A more exotic but rising technology for possible inclusion is the automated optical inspection system [6]. Optical inspection can be applied as a screen in the same flow line as an in-circuit tester to detect missing or misaligned components, unsoldered or missing leads (by inspecting the bottom of the boards), and solder shorts between paths. As a troubleshooting aid a robotic optical inspection system might closely inspect an area of the board in which a functional test has identified a fault for the physical cause of the fault.

Financial considerations are traditionally the drivers of business decisions, but the level of decision may be broader than that just presented. If the field support effort, either contractually or economically driven, calls for functional tests for all pc boards, then the programming costs will have to be borne anyway, and the factory alternative may be greatly improved. The example has been treated as a stand-alone operation, but in many cases test planning pertains to the addition of a product line to an existing testing operation. Similar decisions are made, but an overriding of compatibility in both equipment and test generation methods must be dealt with simultaneously. If the existing test method supports functional testing, the continuance of that method may be less traumatic to schedule and budget than the establishment of a new test method.

ESTABLISHING A TEST FACILITY

Implementing a test strategy is a much larger task than deciding on one. Although service companies sell testing and test generation on a contract basis and the option of buying an entire test capability is available, logistics usually dictate partial or full installation of testing on site. Shipping all of the product to a separable laboratory for testing is infeasible in all but the smallest manufacturing operations.

Establishing a test facility is a product which requires coordination of a variety of dissimilar tasks. Hardware, software, personnel, buildings, and services must be arranged. Among the many details to be scheduled, there are a few major subjects of concern.

ATE. Major ATE purchases usually include options and configuration decisions such as instruments to be included on the IEEE-488 bus, the number of pins with which the ATE can interface, and the capability each block of pins will have. Software development and information management computer suites which match the computer in the ATE are often available to off-load nontesting tasks.

Interface Adapters. Although the configuration of the ATE used determines the adapter configuration partially, so does the DUT configuration. Adapter kits may be offered by the ATE vendor or a third-party house, but overall design and construction must be managed by the user.

Facilities. Adequate space, power, and cooling must be provided for ATE. Cooling requirements are particularly critical since most ATE is manufactured to commercial specifications only, and temperatures over 80° may cause erratic operation or functional failure of the equipment. In some cases, additional cooling must be provided directly to the DUT as refrigerated air or liquid. Floor space must include room for in-process storage of DUTs and adequate space for auxiliary test equipment such as oscilloscopes in the area of the ATE test head. Requirements for material handling and electrostatic discharge control must be observed for sensitive products.

Personnel. Testing the product will require a technician of only moderate skill, but troubleshooting a complex pc board requires current, in-depth technical skills found in experienced technicians. Even with skilled technicians on the floor, engineering backup will be needed for the extreme cases involving complex VLSI chips or unusual mixes of electronics. Test generation is a task requiring engineering analysis of electronic circuits and familiarity with computers as tools and as controllers. Although the job is often called "test programming," many of the skills needed are weak or absent in the average computer scientist. Additional specific training beyond the BSEE degree will be needed for electrical engineers, but the basic background is best suited for this task.

Test Generation Tools. Computing resources and software packages such as ATG, fault simulation, and translator–compilers will be required to support the test programming effort. In some cases the test generation tools provided with an ATE computer may be adequate, but in any case they are usually sold as separate options or items.

Maintenance Support. Complex electronic systems such as ATE need periodic calibration and maintenance. If the service is not performed within the framework of the manufacturing operation, an outside service can be contracted to perform the work.

A more thorough discussion of these topics follows as two sections, divided into hardware and software parts for convenience.

ATE Acquisition

Once the strategy for testing has been established, the method and source of acquiring test capability must be addressed. Actually, since there is an entry in the strategy planning process for the approximate cost of ATE, some of the preliminary work of defining the requirements and locating suitable sources of ATE must already have been performed. The assumption here is that only the strategy planning information as to the requirements of the ATE is available and that no sources have been pinpointed.

In order to deal with a multifaceted selection among multiple candidates a decision matrix can be constructed. The decision matrix is a table which displays candidates on one axis and comparison criteria on the other. A score is placed in each intersection to represent how well the candidate meets the criterion. A priority

TABLE 12.8.   Decision Matrix for ATE Selection

Criteria	Weight	Tester A	Tester B	Tester C	Tester D
Capabilities					
<300 pins I/O	1.0	10/10	7/7	3/3	9/9
Voltage programmable	0.6	10/6	3/1.8	6/3.6	5/3
Drive current	0.8	4/3.2	5/4	10/8	6/4.8
Vector rate	0.7	6/4.2	8/5.6	4/2.8	7/4.9
Time resolution	0.5	6/3.0	10/5	1/0.5	8/4.0
Guided probe	0.9	5/4.5	9/8.1	0/0	6/5.4
General Hardware					
Size (smallest)	0.3	2/0.6	4/1.2	9/2.7	6/1.8
Ergonomic	0.5	8/4	7/3.5	3/1.5	7/3.5
Reliability	0.9	9/8.1	10/9	8/7.2	4/3.6
Repairability	0.9	9/8.1	9/8.1	6/5.4	7/6.3
Expandability	0.7	4/2.8	10/7	2/1.4	6/4.2
Support					
Adapters available	0.6	4/2.4	8/4.8	6/3.6	9/5.4
Software tools	0.8	9/7.2	10/8	2/1.6	5/4.0
Training classes	0.5	7/3.5	8/4	4/2	10/5.0
Documentation	0.5	7/3.5	10/5	6/3	7/3.5
Total weighted score		71.1	82.1	46.3	68.4

weight is also assigned to each criterion, and the net score for each intersection is its raw score times the weight assigned to that criterion. After completing the ratings, the candidate's net scores are added and compared to one another as a basis for the decision. Comparison criteria are not necessarily all objective and are seldom rigorous enough to provide an ironclad selection. The matrix is merely a method for organizing the observations for first-order analysis.

In the example of Table 12.8 the four candidate testers are compared on subjects which represent common concerns in ATE selection. A typical matrix might have many more criteria and would include specific requirements such as pulse-width measurement capability, portability, or ATLAS compatibility which are derived from the product and the environment.

Common concerns can be divided into the three areas shown, although the criteria within the areas change from case to case. Electronic specifications include speed, flexibility, and strength of the pin electronics. Speed is actually a simplistic word for a complex set of timing specifications. The maximum rate at which any one pin can be changed from a 1 to a 0 and back is called the clocking speed. A given clocking speed does not mean necessarily that new test vectors can be applied at that rate. Broadside pattern or vector speed, which is the maximum test step application rate, depends on the structure of pin electronics control as well as on its timing capability. Other timing criteria include the accuracy with which a change or edge can be defined for a signal and the resolution with which it can be programmed.

Flexibility is related to the ease of control of the parameters and timing of the pin electronics. The granularity with which the pin electronics can be controlled (only in groups of eight pins or one at a time) and the availability of built-in programmable loads are flexibility features. Strength is usually measured in dc current drive capability for each pin, but the ac characteristics such as output impedance can severely degrade a test at high speeds. If adequate strength is not available for driving large fan-outs on complex cards, a more complex interface adapter with buffers installed will be needed and some of the system flexibility will be lost.

Overall hardware concerns might appear to be insignificant in view of technical criteria, but once the tests have been developed and the system is in place, the performance against reliability and maintainability specifications becomes translated to an economic driver. Down time of a multimillion-dollar ATE suite can waste thousands of dollars a day in lost production and cost similar amounts in repair charges. Most ATE systems have extensive diagnostic programs which aid in repair. Less prevalent are built-in automatic calibration systems that compensate for drift in the pin electronics over the operating life of the tester. Regular recalibration can reduce the availability of the tester by several hours per week. Features such as an instrument bus control interface and architectural room for pin electronics expansion are important for future products and the lengthening of the ATE's useful life, even if they are unneeded in the present situation.

Support features pertain to the test development or test maintenance tasks.

Changes in product configuration, updates in the product, or later additions to the test sequences for improved test performance are less expensive if database management tools such as editors and file managers are available. Security features such as restricted access to test files and file backup utilities are vital to preventing accidental loss or corruption of test files after they have been verified and placed in use. The high cost of test development places emphasis on protecting the results.

Both hardware and software support hinge upon understanding the intricacies of complex ATE systems. The training and documentation provided with or available as an option for a major piece of ATE can have a large impact on the cost of test development and the up time of the machine.

Test Development

Creating a test development capability is a matter of bringing together competent test engineers and test generation tools in a controlled and directed fashion. The first step, that of acquiring the services of competent test engineers, is the hardest. Hiring a potential test engineer is a matter of looking for the right mix of electronics knowledge, software familiarity, and problem-solving skills. Very few universities teach test engineering and those that do have only a shallow curriculum in the subject. It is therefore the ability to become a test engineer (and the interest, since test engineering is often looked down upon by designers) that is sought. An eclectic engineer with an interest in broad hands-on responsibilities has the characteristics needed.

Specific training in the use of the test generation tools and the ATE with which work must be performed will be needed as soon as the neophyte has been oriented to the general procedures and the requirements of the job. Classes are commonly available from ATE or tool manufacturers, but they inevitably must be followed up by close supervision and possibly by one-on-one training from a senior test engineer. Crosstraining of personnel to familiarize each engineer with several ATE systems (if several different systems are to be used in the test operations) will prevent gaps in production if a test engineer changes jobs or becomes incapacitated for a time. Test engineering cannot be learned overnight!

Establish a Method. There are many ways to generate tests, even when the target ATE and the test tools are the same. In order to reduce the inefficiencies of experimentation and the maintenance headaches of "Tower-of-Babel" test programs which are readable only by the author, a standard or guideline procedure for the generation and verification of a test program should be established. A method for test generation should include the following items.

1. General Flow Chart. This is a chart of the typical operations encountered in generating a test with the resources involved and the important interdepartmental contacts noted. It includes statements like "Get the latest (check DUT part number update history) circuit description file from the computer automated design department and transfer it to your workspace."

2. Description of Inputs. List the information inputs that can be expected for each test generation effort and give examples of how they are to be used.

3. Specifications for Results. Each output from the test generation process should be listed and defined in terms of completeness, form, documentation, and to whom it must be presented or where it must be installed to signify completion of the task. They provide examples to be used as shell programs or guidelines when possible.

4. Quality Control. Define the verification and validation techniques and criteria to be used against test programs. Outline the configuration control method to be used to keep test programs in compliance with hardware changes.

Since test generation contains many aspects which do not appear in software design and since the test program is a very restricted type of program, ordinary software planning and software quality procedures fit only loosely. Test programs for digital electronics usually consist of a simple shell for controlling the power supplies and setting up timing in the ATE, followed by a block of test vectors which were translated from the output of a fault simulation program or an ATG program. The syntax and structure of the test program are defined almost entirely by the ATE on which it is to be executed.

The most applicable standard for the test generation process output is the military "test program set" or TPS definition [7]. A TPS consists of the following components:

Test Program. Is a listing of the sequence of instructions in the appropriate ATE language with a table of the digital input and output vectors. A flowchart may be required for analog tests, but is seldom needed for digital versions.

Test Program Instructions. Are instructions written in prose for the test technician to follow in setting up and executing the test, including the way to install the interface adapter and a definition of the program options if applicable.

DUT Documentation. (Often called a Test Requirements Document). Is schematics, theory of operation, and pertinent numerical data describing the normal operation of the DUT.

Troubleshooting Information. Is notes pertaining to circuit idiosyncrasies or test subroutines which might aid in interpreting the ATE test results.

Interface Device Description. Is a set of mechanical and electrical drawings and any notes needed to allow the correct interface adapter to be identified (and in some cases enough to reconstruct the adapter) and to aid the tracking of signals from the DUT to the ATE and back.

Adopting a similar set of guidelines and insisting on their use will produce a uniform test program result which can be examined for correctness by other test engineers and which will be easily applied by a technician.

Before the test engineer can produce test programs for complex circuits, the proper set of tools must be available. A test engineer's tools are a mix of instruments and software, depending on the nature of the circuits to be tested. The software tools usually include a fault simulation or fault grading program and may include an ATG program. Translators, preprocessors, and postprocessors abound since information must be moved and manipulated from diverse inputs such as the circuit description in the CAD files, the vendor's description of a component operation, and the microcode program which resides in the ROM on the DUT.

If the tools are purchased, an update and maintenance agreement is a must since technology will obsolete the average test generation system in a few years. Many of the translators, pre- and postprocessors can be written and maintained within the test engineering organization or at least within the company by competent computer scientists. Internal software support is valuable in situations where a number of different projects must be supported over a few years' time and the need for special translators, and so on, must be met in a few months each.

SUMMARY

Much of the test planning process is identical which good management practice. The items which require a technical judgment are the initial requirements analysis, and ATE evaluation, and the hiring and training of personnel to generate test programs. Economic analysis techniques are valuable as guidance in selecting a test strategy, but many of the factors involved in test planning decisions are technology driven, requiring an understanding of the nature and direction of circuit technology and the capabilities of sophisticated test equipment. The technical influences are often subtle compared to the multimillion-dollar costs which must be justified. In many corporate environments the test plan must fit within management policies or contractual requirements which override other factors.

Once the test requirements are ascertained from management policy, customer requirements, and the product's characteristics, the alternative test flows which meet or exceed the test requirements are delineated for comparison. The consequences of each test flow are evaluated in terms of the total cost of testing necessary to achieve the quality goals. If a stage in the test flow is deleted, the fault level per board must be assumed to carry to the next level of test.

Before accepting the cost analysis as the final decision, extenuating circumstances such as schedule pressures upon the test generation time, design stability, and the relationship of the product line being planned to other lines or manufacturing operations must be evaluated. A head start in test generation may be afforded by utilization of design verification tests or self tests intended for system maintenance. Careful weighing of the entire situation will result in effective testing and long term efficiency. Rapid development of a test facility is difficult in any case due to the long personnel training and equipment procurement times. A good plan looks several years ahead.

REFERENCES

1. P. Crosby, *Quality Is Free*, McGraw-Hill, New York, 1979, p. 15.
2. H. F. McAleer, "A Look at Automatic Testing," in F. Ligouri (Ed.), *Automatic Test Equipment: Hardware, Software, and Management*, IEEE, New York, pp. 31–46.
3. B. Davis, *The Economics of Automatic Testing*, McGraw-Hill, U.K., 1982, p. 12.
4. J. L. Riggs, *Economic Decision Models for Engineers and Managers*, McGraw-Hill, New York, 1968, p. 201.
5. *Circuit Test Systems: Financial Justification*, Hewlett-Packard, Palo Alto, CA, Mar. 1980, p. 4.
6. S. T. Jones, "Flexible Inspection Systems (FIS) for Printed Circuit Board Production . . . ," in *Proc. 1985 Int. Test Conf.*, IEEE Comput. Soc., pp. 403–412.
7. S. M. Schlosser and M. N. Graieri, "Managing a Test Programming Activity," in *ATE Reference Handbook, vol. 1*, Morgan-Grampian, Boston, MA, 1983, pp. I-70–I-74.

GLOSSARY

Adapter   An apparatus which serves to physically and electrically connect the device under test to automatic test equipment.

AHPL   A hardware programming language, an early variant of the general-purpose computer language APL modified to allow succinct description of the operation of digital computer circuits.

ASICs   Application-specific integrated circuits.

Backtrace   The process of determining which sets of inputs to a circuit can cause an observed or desired output.

Bed of Nails   An adapter consisting of a box with spring-loaded contacts protruding from its top, reminiscent of the hindu bed of nails.

Bidirectional   Able to send or receive logic signals.

Breadth-First Search   An investigative technique in which all the possibilities at the present rank or level are explored before moving to the next rank.

Broadside   Synchronized application of a number of signals (usually device-under-test stimuli) at once.

Combinational   A circuit whose outputs are only dependent on its present inputs.

Concurrent Fault Simulation   A software method of computing the correct response of a circuit to a series of stimuli while evaluating the effects of faults by carrying along those effects which are active as each rank of logic is solved. At each node the solution is calculated for the good circuit and all faults active on the nodal inputs, and only those fault effects which produce differing answers are propagated.

Cone   All circuit nodes which influence the state of a node (particularly a primary output) are said to be in the cone of the node (output).

Connectivity   A listing or data structure which identifies which circuit elements are wired together. A listing of the fan-outs from a logic driver constitute its connectivity.

Controllability   The ease with which the logic value of a node can be manipulated from the primary inputs.

Crosspoint   The internal connection in a programmable logic array which determines whether an input or term is included in a particular output equation.

Crowbar   A protection circuit which limits the output voltage of a power supply by shorting the output.

CUT   Circuit under test, the subject of a digital test.

D–A Converter   A circuit which controls an analog voltage or current in linear accordance with the logical values presented to its digital input.

Daughter Board   A small printed circuit board attached to a larger one.

Deductive Fault Simulation   A software method of determining the correct outputs of a circuit for a set of stimuli while determining at each node which faults or fault effects may change the value of the node.

Defect   Any deviation in physical structure of an electronic component or circuit which causes it not to conform to applicable requirements.

Depth-First Search   An investigative technique in which each path of relationships to an item is checked to its end before the next is examined.

Deterministic   Intentionally calculated or structured toward a goal.

Distinguishing Sequence   A series of input stimuli which, when applied to a circuit, produce outputs that allow an observer to deduce the state of the circuit.

DUT   Device under test, a general term for an electronic assembly being tested.

Edge Connector (or Edge Board) Testing   A test approach which exercises a circuit and attempts to determine its correctness by observation using only the circuit connections used for normal circuit function (functional testing).

Failure   An instance of incorrect output or result from a circuit.

Fault   A defect in a circuit which has the capability to cause a failure.

Fault Activation   The process of achieving a logic state opposite the fault condition at a faulted node.

Fault Collapsing   The process of eliminating from further consideration all faults which are equivalent to or covered by a prime fault.

Four-Wire Connection   The use of remote voltage-sensing wires in parallel to the power wires from a power supply to a load to compensate for ohmic voltage drop in the load-bearing wires.

Functional Test   A procedure to determine the operability of a circuit using only those connections intended to be used in the final installation.

Glitch   A short-lived change in a signal value such as a short pulse, normally not intentionally created.

GPIB   General-purpose instrument bus, a standard interface for computer-controllable instruments.

Hazard   Any change in a circuit's inputs which, while traversing from a beginning state to a final state, may pass though other states due to input skew.

Hierarchical Simulation   A software process which deduces the outputs of a circuit in response to given input sequences using representations of circuit portions at several levels of abstraction.

Homing Sequence   A series of input stimuli which will at some time in the sequence bring the circuit to a known state.

Identification Sequence   A series of input stimuli which when applied to a circuit will result in outputs that are unique to that circuit.

IEEE 488   A standard defining a hardware interface and software protocol for a bus linking computer-controllable instruments and controllers. Based on the GPIB.

In-Circuit Test   A method of printed circuit board testing utilizing a bed-of-nails adapter to contact all circuit nodes and allow each integrated circuit to be tested separately in situ.

Initialization   The process of forcing a circuit into a known state.

Intermittent   Appearing and disappearing at intervals in time. A circuit condition which changes back and forth as the environment changes.

KGB   Known good board, an example of a printed circuit board which functions correctly and is thus used as a standard for comparison.

LFSR   Linear feedback shift register, a series of flip-flops with exclusive-OR coupling which creates a polynomial division circuit.

LISTENER   The programmable device on an IEEE-488 bus which is receiving data at the moment.

LSSD   Level sensitive scan design, a methodology for circuit design which restricts the use of sequential elements to the latches in scan registers which can be accessed serially for test purposes.

Manufacturability   The degree of ease of manufacture of a given circuit or device.

Microprobes   Wires sharpened to a point of a few micrometers used to contact integrated circuit die directly.

MIMD   Multiple-instruction–multiple-data, a computer architecture which uses parallel processors which run different programs on differing data to accomplish an overall task.

Model   A representation of a circuit in software which supports calculation of the circuit's response to stimuli.

Modem   Modulator–demodulator, a device for encoding data onto a carrier such as an audible tone and decoding data from a carrier.

Neighborhood Fault   Loss of data in a memory cell due to interaction with surrounding cells.

Observability   The ease with which an internal circuit state can be propagated to a primary output.

OEM   Original equipment manufacturer.

Overcurrent   A condition of loading that requires more than the expected maximum current to maintain a specified voltage.

Overstressed   Having been subjected to voltage or current which exceeds the manufacturer's specified maximum levels.

Overvoltage   A condition of excursion of supply voltage above the expected maximum.

Parallel Fault Simulation   A software method of determining the outputs of a circuit to a given series of inputs while predicting all possible faulty outputs by maintaining parallel models of the circuit and solving each node for all parallel models simultaneously.

Partition   A subcircuit which can be isolated mechanically or electronically from the remaining circuit.

Path Sensitization   The process of determining input stimuli which will allow the effects of a fault to be propagated to a primary output of the circuit.

Personality Board   Part of an adapter which provides interconnection or loading particular to a given device under test.

Pin Electronics   The circuitry of automatic test equipment which interfaces electronically with a single connector pin of the device under test. Usually consists of a driver and a receiver.

PODEM   Path-oriented decision making, a method of determining which primary inputs must be controlled to activate a fault and propagate its effects.

PPM   Parts per million.

Primitive   A representation of a circuit element which has no internal structure, but is described by functional rules.

Pseudorandom   A repeatable sequence in which there is no apparent relationship between one number and the next.

Reconvergent Fan-Out   Circuitry in which a signal follows more than one path, but at least two of the paths are inputs to a given node.

Rehosting   The process of modifying software to run the program on a different computer.

RS-232   A standard of serial communication primarily specifying the hardware and low-level protocol.

RTL   Register transfer language, a method of describing a circuit's behavior in terms of data transfers between storage elements.

s-a-0   Notation for a fault condition which causes a node to appear to be stuck at the logic 0 value.

s-a-1   Notation for a fault condition which causes a node to appear to be stuck at the logic 1 value.

s-a-Off, s-a-Z   Notation for a fault condition which causes a node to appear to be stuck in the 3-state nondriving or high-impedance state.

s-a-Open   Notation for a fault condition which causes a switch element to appear to be stuck in the nonconducting state.

Schmoo   An oddly shaped graph which relates the test results to two independent variables.

Sequential   Circuitry containing storage elements, whose state may depend on past inputs as well as the present input stimuli.

Serial Fault Simulation   A software method of calculating the correct outputs of a circuit in response to a series of stimuli which also calculates the outputs of possible faulty circuits by repeating the procedure using modified versions of the original circuit.

Signature Analysis   A data compression technique which uses polynomial division to produce a coded value representing a long serial string of logic values.

SIMD   Single-instruction–multiple-data, a computer architecture which performs the same sequence of operations on two or more separate data streams simultaneously.

SISD   Single-instruction–single-data, a computer architecture which performs a sequence of operations on a stream of data without parallelism.

Soft Failures   An error condition which results from an intermittent or temporary cause and once corrected does not immediately reoccur.

Stacking Fault   A defect in the crystalline structure of an integrated circuit which may result in short circuits.

Statistical Fault Grading   A non-fault-simulation method for determining the approximate coverage of a test by examining controllability and observability values and the good circuit activity.

Strength   The amount of current a circuit element can exert to maintain its logic value.

Synchronizing Sequence   A series of stimuli which forces a circuit into a pre-determined state regardless of the initial state.

TALKER   An instrument or controller on an IEEE-488 bus which is transmitting at the moment.

Testable   Arranged in such a fashion so as to allow manipulation and observation of circuit functions in a test environment.

Threshold   Voltage level separating two logic states.

Transition Counting   A method of data compaction which consists of counting the number of times during a test sequence that a logic node changes state.

Transparency   The degree to which circuitry between primary ports and the nodes to be tested can be characterized by a simple transfer function.

UART   Universal asynchronous receiver–transmitter, a device which handles the serial protocol of a communication line.

Unit Delay   The time required to propagate a logic value through a single rank of logic.

UUT   Unit under test, the circuit to be tested by an automatic test system.

Vector   A set of logic values to be applied to a circuit or observed from a circuit in a single step.

Verification   A broad application of tests and test environments aimed at assuring that a circuit was properly designed to specification.

VFET   Vertical channel field effect transistor, a high-current member of the field effect transistor family.

ZIF   Zero insertion force.

INDEX

Boldface denotes primary reference.